After Eden

Books by Michael Tobias

Tsa

Deva

The Mountain Spirit
(ed. with Harold Drasdo)

Deep Ecology, ed.

Mountain People—Profiles of
Twentieth-Century Adaptation
(ed. with Tessa Taylor Tobias)

After Eden

History, Ecology and Conscience

Michael Tobias

Published in 1985 by Avant Books
3719 Sixth Avenue
San Diego, California 92103
(Write for free catalog)

ISBN 0-932238-28-9
Library of Congress Catalog Card Number 83-73257

Library of Congress Cataloging in Publication Data

Tobias, Michael.
After Eden.

Includes bibliographies and index.
1. Man. 2. Human ecology. 3. Civilization – History
I. Title.
BD450.T595 1985 909 83-73257
ISBN 0-932238-28-9

Produced by The Word Shop, San Diego
Cover and book design by Ed Roxburgh
Cover photo by Michael Tobias

Printed in the United States of America
First Printing

Contents

Acknowledgments

I owe an extreme debt of thanks to all of those authors referred to in the selected bibliography, of whose works I am an unabashed fan. And I'd like especially to mention those people whose prolific and inspired thinking has helped me to draw connections — albeit diminutive ones — of my own. These folks include: Michael Bostick, Lester Brown, Tom Brown, Jr., Eric Davis, Irenaus Eibl Eibesfeldt, Paul and Anne Ehrlich, William Everson, Kimon Friar, Clarence Glacken, Eric Alfred Hoffman, Carol and Jimmy Katz, Helen Kazantzakis, the late Loraine Kuck, Donald Lach, Jeffrey Long, J.E. Lovelock, Adrian Malone, Sue Mansfield, Lynn Margulis, Ashley Montagu, Sophia Morgan, John Nance, Rod Nash, Noel Perrin, Frank Poirier, Jonathan Schell, Robert Sheahan, Paul Shepard, George Steiner, William Irwin Thompson, the late Arnold Toynbee, Yi-Fu Tuan, Carol Wonsavage and Gerardo Zampaglione. Thanks goes to my all-enduring editor, Larry Platt, and to my friend Robert Radin, whose own enthusiasm and vision proved instrumental to the production of this volume. Finally, I wish to thank Michael Gosney, the spirited brainchild behind Avant Books.

This book is dedicated to my parents, Betty and Bill Tobias.

"SAKYA COMING OUT OF HIS MOUNTAIN RETREAT"

By Kato Moriuke, School of Tsununubu,
Muromachi, 1683, Japan
Private Collection

Preface

Between its celebrations of privilege, and the angst of its reckonings, human life gathers unto itself a chaos of contradictions. The region of this duality is protean, here a battlefield, there a quiet churchyard. These are the landscapes after eden, upon which human consciousness has focused.

The ecology of good and evil obeys a facile and infuriating set of predictions, mathematical inanities as monstrous as they are compelling. We share things, and this fact alone recommends our species. Yet for all of our wisdom, we may murder; in the heart of feeling, we can be cruel. Where our arrogance is liable to denude and degrade, our capacity for faith, our abandonment to beauty can conversely reinstate all existence, lend renewed dignity to original things. If we are ceaseless tamperers, we are also from time to time unobtrusive. Though we shout, so may we whisper. Our tears tell us everything about the ocean's salinity, out of which we were fashioned. And somewhere in the bacterial gut, we share every sentiment with the beasts of the field.

After Eden sets out to trace these two patterns of thought and conflict according to that body of fact and feeling, the myriad data which human beings – in their perception of Nature – have professed. I mean to examine these pathways in a collection of essays all tied to the argument that human culture is conditioned to thinking and behaving in terms of the *aftermath*. This may at first seem peculiar. As a foundation for inquiry, however, the aftermath motif is indeed a charismatic one. In psychological terms, it supports a narcissism capable of fixing almost non-chalantly on apocalypse. Eccelsiastically, such aftermath is endowed with powerful emblems of heaven, as well as the grueling and damnable crusades to get there. Biologically, the aftermath endorses the supposition of resolved population dynamics, of homeostasis. In quantum mechanics, the dissipative structure, balance through entropy, emerges as Nature's paradigm. In literature and music, such stability is a catch-all for the achievement of characterization, voice, harmony. Technology and mysticism are oddly merged in the blueprints of tomorrow, while the anthropological record resists them on the premise of extinction. What this aftermath really refers to is ecology, an ecology of conscience. An ecology that encompasses every twinge and mood known to life; cobalts and greys, flattery and desolation, the stinting and audacious. All that I have feared and longed for. This ecology is neither orderly nor intelligible. Whatever laws of form it may intimate, there is at least one species likely to contradict them. The amalgam of these contradictions has a storyline – suited to drifters like myself – that is lodged worrisomely between an exponential future and allegorical past.

These essays are rather ramshackle, soft-spoken, unacademic. Recklessly stemming the sheer walls of antiquity and science fiction, I have set out to forward some queries long-lingering from my own boyhood. And green-eyed, by a hominid sort of photosynthesis, sluggish, plodding, I have sought – not answers – but companionship-in mystery. Bushwhacking to the heart of it.

Michael Tobias, Namolokama, Kauai, Summer 1984

PART 1

Flower, Protein, Woman

Michael Tobias

Chapter 1

The Paradise Factor

The Earth is alive, given to experiment. Two provocative biological systems—the prokaryote and eukaryote cells—have provided a sufficient hospice out in the wild for the eccentric inventions of time, mortality, and eventual conscience. Homo sapiens long ago defied Nature. More recently, science has begun to decipher a mysterious code in the biosphere, a purpose to life, the paradise factor.

Etosha, Africa: mud flats, oxygenless. Flocks of pink flamingo—tens-of-thousands, cavorting, graceful. They gather where the anaerobic muck has merged into a great mirage, clustered, the burning sun, the blue, the heart of Africa. The flamingos are elusive birds, mating in the depths of a continent, migrating at night when they are impossible to track. They build their nests in the rich mud, feeding on crustacea. Such stuff of earth—primal, cloying, dark, and moist—is the beginning of us all. The flamingos need nearly 90 kilograms of mud to fashion a nest. These are elegant exemplars of life on the planet. They prefer privacy when they mate, and share sitting on the nest. The crustaceans they devour are ribboned with carotene, lending the characteristic pink to the flamingo's body. These animals need rain and so does the blue-green algae, a varicolored sheen of nutrient for the crustaceans. This bubbling, fermentative soup is close to our origin. It is the critical basis for life.

Water, algae, and mud – a Namibian salt pan. A configuration of inexhaustible magic. As for the flamingos, they've had their problems. The Romans killed them for their tongues.

As the sun hits the mud, light beams are polarized. They twist counterclockwise, evidencing life at work: proteins, amino acids, optically vital substances endowed with organic properties. When sun hits certain types of clay, minute flashes of light are emitted. Such light fuels the conversion of ammonia (NH_3), methane (CH_4), nitrogen (N_2), and hydrogen (H_2) into more complex creations, engendering all of the first organic precursors.

Life seems to be conspiring everywhere in the universe. In 1972 a meteorite smacked into Australia. It contained the twenty amino acids found in living organisms on Earth. Two of the four DNA–RNA bases, guanine and adenine, were also detected in the alien stone. Titan, Saturn's mysterious moon, has a surface temperature of $-195°C$ – mellow by the standards of the Milky Way. Its atmosphere contains methane. Titan shares much of the Earth's original ambience. Organic molecules are perpetually coming into being across its stark, reddish deserts. Whether life as we know it will ever grow there, or on Jupiter's moon Europa is unknown. Earth's two nearest siblings, Mars and Venus, are composed of deadly acids. But Earth – whom the Greeks called *gaia*, earth goddess – maintains the most delightful habitat for life that any of us are ever likely to know.

Spun off in the fiery catastrophe, infant Earth landed deep center in a star system with no obvious penchants for life. But it was given precarious shelter in its cocoon of gases and lay awake like a great moth with folded wings, waiting to move. While the Sun's heat escalated (nearly 30 percent in the last 3.5 billion years) the Earth's munificent surface stayed the same temperature, gently given to organic experiment and regulated in much the same manner as our bodies. How?

Until recently it was believed that the Earth's original atmosphere worked the miracle of life – that it formulated organic molecules, amino acid chains, DNA; that it produced the balanced pH and maintained a temperature range conducive to the watery state throughout those critical, life-engendering latitudes. It is more likely that life re-engineered the planet for its own imperatives. Aboriginally, the days were 22 hours long, only five of them sun-illuminated. Oxygen, liberated in the 2400–2600-angstrom wave lengths of ultraviolet radiation, composed but 1 percent of the atmosphere. Yet it was enough for an ozone layer to develop. Ozone protected emergent life from the power of the Sun – life that shrank from the oxygen, stayed in the mud, fermented, and produced oxygen and methane. The methane rose into the atmosphere, counterbalancing poisonous gases en route. Up high, the methane oxidized into carbon dioxide and water vapor, then into oxygen and hydrogen. The oxygen returned to Earth. But methane production — which consumed oxygen — kept it from over-accumulating, a process that would have destroyed the Earth by fire.

The percentages of increase and decrease of oxygen over millennia followed nascent laws — perfect equations that baffle modern science. Methane was apparently at the controls. It regulated the Earth's oxygen, delicately urging the proliferation of single-celled forms of life that retreated from, and contributed to, the oxygen sypply. Oxygen would remain an enigma to scientists until well into the eighteenth century. And why it should have come into the scheme of things on Earth is unknown. But what is certain is that life itself took responsibility for its own engineering feats. It produced oxygen and could easily have been destroyed by it. Instead, there was balance. The precise quantities of methane and oxygen required to maintain an organic balance were bizarre, improbable, for a persistant disequilibrium occurs among atmospheric gases — .03 percent carbon dioxide, 79 percent nitrogen, 21 percent oxygen, and 1 percent argon. If ever there were a deity, it is oxygen, whose presence is precise, understated, immensely rewarding. In the beginning, it was wrathful: Life had to grow to like it.

At first, the Sun put out 30 percent less heat than it does now. How did the infant biosphere, feeding on the atmospheric blanket above, cope with the cold? The oceans provide much of the answer, illustrating inherent homeostatic, self-regulating mechanisms at work then and today. I am speaking of life's own imperatives, of a theory suggesting that life itself was at the helm of creation. Scientist James Lovelock revolutionized our understanding of this alternate scenario.[1] His "gaia" hypothesis merges biology and determinism. A. Graham Cains-Smith furthered the effect of the hypothesis by introducing the idea of genetic crystals — the mineral origin of life. Within 200 million years of the Earth's outgassing (4 billion years ago), the planet's surface temperature had cooled sufficiently to condense the atmospheric water vapor. The result was rain, millions of years of monsoons. Igneous mountains were thus eroded, and hydrochloric pools of rainwater provided the nearly infinite possibilities for mineral crystals to evolve significant relationships. Kaolinite clay templates offered perhaps the best prospects for mineral replication.

It was not the vast number of possibilities for life that is impressive, but the fact that an actual *choice* was forged here on Earth. Theories abound on the details, but the formation of life seems to have been lodged in a primordial clay of long protein chains. Other scenarios for the origin of life have been suggested. Mike McElroy traces the life spark to the recently discovered hot sea-vents at the ocean bottom. Francis Crick, in the tradition of astronomer Fred Hoyle, looks to exobiological sources of life — pockets of moisture inside comets that struck the Earth 4 billion years ago. But of these various hypotheses, the fundamental mineral structure — composed of carbon and silicate — is the most compelling.

The ice ages, resulting from small changes in the Earth's orbit around the Sun, actually affected only those areas outside latitudes 45° north and south — or one-third of the planet. The continental shelves — the wetlands, the delicate zones — were largely spared the inclement weather. And those regions were gaia's vital organs.

The oxidation of nitrogen and sulpher produces immense amounts of acid. Acid is balanced by ammonia, of which the Earth produces more than 1000 megatons a year to control acidity. Why that particular amount? It so happens that 1000 megatons is just sufficient to sustain a rainfall pH of 8, which is the optimum for life. Without ammonia, rain would taste of vinegar. That same ledger sheet can be drawn and balanced for every other major chemical.

How have oceans maintained their exact salt content, Lovelock asks, when a mere 80 million years of time is sufficient to replenish these saline seas with the salt run-off from land? Why is there no massive surplus? The answer may be in the huge beds of chalk and limestone, natural sinks of negatively charged silicon and aluminum ions that balance the excessive sodium and magnesium ions of salt.

Such checks and balances indicate the imperative of life. Without that balance, the atmosphere would have long ago become mostly carbon dioxide, as on Venus. And it is through the burial of carbon in the anaerobic muds of the seabed that the net increment of oxygen is perpetuated. Such burial leaves one perennial oxygen molecule adrift for every atom of carbon extracted from circulation by phytosynthesis. We can still discover carbon remains of microorganisms in the form of minute carbonaceous aggregates, the microfossils from lower Precambrian sediment, including those oldest of rocks at Isua, Greenland 3.8 billion years; or from the North Pole in western Australia, stromatolitic accretions of fine grain. Elsewhere, a colorful array presents the earliest evidence of multicellular life, in chert, paler barite, arched nodules, gypsum crystals of silt and mud, breccia pebbles, a plethora of wrinkled laminates, all fossilized colonies of once aspiring life. There is a tension between the production of too much carbon and too little; between heat and cold, ammonia and acidity, glaciers and oceans. The balance is manifest in every square inch of Earth wherever life has prevailed. Only of late, since the dawn of human technology, has the balance shifted. Mankind has altered every natural cycle. Dramatically, technology has increased the atmospheric carbon dioxide by nearly 10 percent.

We started with mud, replete with its methane-manufacturing creatures. The organisms are now called *anaerobes* because they share a pronounced distaste of oxygen. When the oxygen got too dense, these organisms retreated. Like biochemical renegades, they hid deep in the mire of prehistoric river bottoms and salt marshes. Now they are back again, only this time in the bellies of all beasts, bugs and rhinocerous alike. Lovelock's colleague Lynn Margulis suggests that life's very destiny is somehow anaerobic. They can thrive in radiation and are maximally secure, cozy one might venture, in the niche of man's intestinal tract. Eighty percent of all past life is intimately wound around the anaerobe's single-celled proceedings.

With the advent of oxygen and a global atmospheric pressure equal to that of today's, emerging life experienced its first trauma: Oxygen was as lethal as chlorine. So how did life cope? By cleverly converting the free oxygen into energy, which could, in turn, be used to manufacture the evolving organism's foodstuffs. The Earth became a massive energy brokerage house, an open system of reproducibility. Three and one-half billion years ago, the first self-sustaining cells began competing. These microorganisms no longer relied on minerals. Life had come out of its chemical regime into the biological. A great gobble of photosynthesis was at hand, spelling a vegetative rage for order. But it was chaos! Ultraviolet light stimulated mutations, which in turn embroidered those structures most able to survive an atmosphere of oxygen. So stable were the original life permutations that today they remain unchanged. I am referring to the bacteria and algae.

The mutation rate slowed down as the atmosphere stabilized. Such random diversification needed some replacement. It was sex, the great moment of opportunism. The libidinous impulse to consume – to come together, wriggle for its instant, then divide – wrought the miracle of natural selection we call evolution. Using Oxygen in energy conversions, the rate of energy liberation (of entropy), was increased by eighteen times what previously had been obtained from the same nutrients under fermentation. Between 1.3 and 2 billion years ago, the first distinctly modern sexual, multicellular organisms burst into being and copulated with a frenzy – flatworms, ciliates, camptostroma. These oldest known metazoa were thin and fragile, soft bodied: eukaryotic cells containing true nuclei. They represent the zenith of cellular complexity to date. The eukaryotes evolved out of a most peculiar association we today term *mutualism*. The earliest unicellular (prokaryotic) cells became increasingly specialized, ready to reap the benefits of photosynthesis but incapable of marshalling the necessary talents. By incorporating a chlorophyll-laden neighbor, such cells could have achieved the startling capability. This integration is called *endosymbiosis* and may mark the most important cellular transition in the history of life.

Nature is energy. With the coming of the chloroplast – eukaryotic *organelles* that embed membranes, pigments, and the enzymes of photosynthesis – emerged the energy transport systems, cytochrome electron carriers, and adenosine triphosphate (ATP). These sophisticated strategies were quicksilvered 2 billion years ago (the Earth came into being 4.6 billion years ago). The hereditary mechanism at work in these light shuttles was mitosis, a square dance, all the first nucleic acids do-si-doing, making for the diploid number of chromosomes in each daughter nucleus, with a range of possible pairs multiplied by a logarithm of 10^{80}. This orgasmic anarchy resulted in the multiplication of tissue and dazzling new organisms – slimes, clams, snails, lamp shells, moss, worms, starfish. A host of invertebrate fauna insinuated into mangrove and seagrass clusters. In sedimentary rocks, certain sea snails took to land. Such creatures possessed eventual lung-like organs. The South African lung-fish, *Polypterus,* with its double system for exchange of respiratory gases might have characterized these first land creatures, the vertebrates. In tropical environs, lobe-finned fish moved landward from pool to pool, feeding on insects that had proliferated by the late Devonian period, 350 million years ago. By that time much had happened on land.

Fossilized plant spores from oil-drilling cores in Libya indicate land flora as early as 470 million years ago. Within another 100 million years, ferns were predominant in all of the forests. Club mosses carpeted damp, cool-shadowed glens. Vascular plants with internal conducting tissue for fluids and nutrients, first among them *Cooksonia caldeonica,* had taken root. As for the soil, land masses were submerged and drained at least fifty times. Little creatures inhabited the mudflats, while an organic tangle underfoot began compacting into future coal deposits. Topside, 1 square meter of the soil yielded – as it yields today – 45,000 small earthworm relatives, 10 million roundworms, 48,000 mites and other tiny insects. A gram of fertile soil contained 30,000 unicellular animals, 50,000 algae, 400,000 fungi, and 2.5 billion bacteria.

Amid this fantasia-like burgeoning of color and form, an extraordinary experiment was going on among the vertebrates, the first fishes. This trial – by ingenuity – was the brain. A holographic image of the fish brain will reveal frontal olfactory receptors and an optic nerve carrying information to a rear corner. Behind the forvision section is one for balance, to prove critical when the miniature stumulus–response organ found itself propped up on powerful fins in the guise of a migrating *crossopterygian,* the first fish out of water. An early period of droughts had likely prompted the expedition into the full light of day. The brain of this creature could not have been overworked at the time. Its nervous reactions were those of the instant, lacking any flexibility. But after 25 million generations, it began to learn. It had become an amphibian.

Another 25 million generations and the amphibian had become king, a full-fledged reptile, holding undisputed morphological title to the land, at the turbulent crossoads of dinosaurs, birds, and mammals. The amphibians covered their delicate, nutrient-filled eggs with shell, which protected them from the continued period of drought that we know persisted throughout the Mesozoic era. The amniotic egg was essentially the same sack from which we all were pushed and pulled out, wet with a fluid reminiscent of the sea from which we originally came, a fluid of 1 percent salt.

The planet was ripe for nudity: the temperature, diffuse food sources, nurturing sun. One hundred and thirty million years ago the first flowering plants blossomed, angiosperms, outgrowth of the pollinators, and of the earlier naked-seed-bearing gymnosperms. By the early Cenozoic era, selection for pollen dispersal had engendered the grand spectacle of flowering life cycles from female gametophyte to mature sporophyte: *Filicinophtya,* horsetails, blooms of saguaro cactus brush, orchids, Scotch broom, glorybower, cattails, Arctic lupine, Stephanotis, and thousands of other familiar wonders. With them came the bumblebees, Gila woodpeckers, and hummingbirds. The choroplast arrays were situated to track for a maximum of sunlight, and the seeds were enclosed in ovaries. When the ovary ripened, it was fruit. A miracle!

By the late Cretaceous period, from Antarctica to central Mongolia, Earth was indeed a paradise, at least as the name implies for human beings. All of the major groups now known among plants were in existence. The last 100 million years of the preceding Triassic and Jurassic epochs had witnessed the spectacular ascendancy of dinosaurs, and of peculiar new adaptations among them: Their teeth had grown into molars and premolars; their scales started changing into hair. Some of them, the very small ones in transition, began giving milk to their young. The fossil record does not permit a precise interpolation, but by the end of the Mesozoic era, at the apogee of herbivorous evolution, a catastrophe took place. Not only had the dinosaurs abruptly vanished, but the smelling brain of the small emergent mammals had intensified. A thin coating of grey matter had covered that portion of the mammal's nostril-devoted mind, just as ozone had covered the early Earth. Why smell? Because the mammals were scared of the big guys in the field, who hunted for the most part during the day, relying on vision. So the first mammals took to night, evolving these olfactory capabilities over infrared ones. Perhaps the sudden profusion of odoriferous flowers urged this natural selection preference. The new grey matter, the neopallium, absorbed the overflow of incoming data, information pertaining to smell, catalysts that spawned heightened activity. The couriers of these new data banks were endothermic, warmblooded, generating four times the heat internally as did dinosaurs, who had been tragically stuck in too narrow a range of adaptations. The simple brain circuitry of a frog illustrates the closed-cycle dilemma of the dinosaurs; the frog will starve to death — though lanced insect meat be suspended before it — unless the prey moves. For all of their loftiness, the dinosaurs had no chance of adapting. Their sperm may have been rendered useless by a temperature change; the food pyramid crashed. They were unable to explore alternate diets, nor could they burrow anywhere. Whatever the cause of environmental change, the dinosaurs could not endure. Instead, with zealous craft, a host of daring mammalian species occupied all of the paramount ecological niches, stole the behemoths' eggs, gregariously food-grabbed night after night. Lost amid a welter of Cenozoic evolution is the scant record of small animals with short muzzles able to see in three dimension. They had opposable thumbs.[2]

The meek had inherited the Earth's narrow food zone, a minute crust of activity between huge inorganic stores of gas, solid, and liquid. Shrew-like insectivores, wonderful noshers, obscure, skinny and quick-footed, hung on throughout the upheavals of the Cretaceous period, came of age, went into the forests, climbed the trees, and there began the inconspicuous transition toward primates, over a period of 30–50 million years. Their hand could grasp things, they could see color, and in the round, in as much as a miscalculation of depth-perception might be fatal to a brachiator.*

The muzzle, in the mean time, was reduced, the face flattened. Now the cranium stood sentinel over its grey matter. The brain had become lord and master over all the forests in Asia and Africa by 30 million years ago. Man's lineage was working its way up through the medley of vines, through screeches and shadow play, tropic and rain, canopied, sequestered, on high. Because of morphological and anatomical resemblance, we assumed that our ancestors were African apes. But recent DNA sequence evidence suggests that we developed in Asia. The genetic material records evolutionary history. The viral gene sequences of Asian baboons, seen under the electron microscope, yield stronger similarities to *Homo sapiens* than do those of African apes. In the jungles of China, across the thick, temperate, deciduous forest of the Himalayas, in the deep upper Burmese hinterland, our Pliocene forebears may have grown up. They were, no doubt, a jolly, noisy, sparring bunch, with a compelling future awaiting them.

If we did in fact evolve in those eastern highlands, the Bible was not far off. Paradise was said to lie to the East, and most Biblical commentary down through the Renaissance solemnly speculated on its precise location. The general consensus had it well hidden deep in the Asian mountains. The picture of Genesis is that of a paradise in which peace persists between man and animal, man presiding. No flesh is eaten by

*Not infrequently, gibbons fall from trees. The eye had to be accurate at gauging distance and detail; to accomplish such adaptations, the eyes came forward and took on the intense look monkeys have today.

any living being nor is there any marked aspect of exploitation or dominion. In the twelfth book of Isaiah this is the paradise of the future, as it was in the beginning.

We are unsure what the last pre-primate looked like. Possibly it resembled the elephant shrew, small, lemur-like predators inhabiting the lower strata of tropical rain forests. To make a monkey from a lemur you need only enlarge the brain, shorten the face, craft more agile hands, and bring in the eyes. In the Pondaung Hills of Northern Burma, four mandibular fragments have been found that belonged to a form of monkey very similar to today's Burmese gibbon. This region may very well be the point of origin of the Anthropoids, of two genera in particular, the *Pondaungia* and *Amphipithecus*, whose bones have been dated at some 40 million years. This region, also known as the Golden Triangle, is an attractive site for such speculation. Shrouded in mist, its northernmost edge is curtailed in spectacular uplift; it is remote, jungly, bound by glaciers, and still largely unexplored. But there is heated competition for the claim of origins. On Purgatory Hill in eastern Montana, for example, fifty isolated teeth from a single Paleocene quarry site have been recovered, putting back the primate trajectory another 15 million years. *Purgatorious* left dental evidence suggesting that he was an eater of fruits and vegetables. The paradise factor?

The Earth, as we have been describing, *is* a paradise, constantly changing, tenaciously remaining. It is true that 90 percent of all species that ever graced a habitat are now gone, leaving between 5- and 10-million species of plant and animal today. The extinctions were *natural*, and I introduce that word here for the first time in conjunction with no special pleading, but rather with the simple logic of etymology, a logic devoid of bias.

"Physique . . . of man, of beast, of herb and stone." This was the meaning of nature to the fourteenth-century English imagination, as coined by the Kentish landowner John Gower. The word had come from the Latin, through the French and into the twelfth century. The word is feminine, from the Latin *nasci*, to be born.[3] *To die* is also the meaning of nature – which takes us back to the question of extinct species. We saw how the Earth – after 3.8 billion years of organic activity – had given birth to its largest creatures, the dinosaurs; and how – within 100 million or perhaps as few as 20 million years – had extinguished them, filling the void with demure successors. There was no extrication from their tarpit in time. But their domains did not go to waste. They were reoccupied, brought back to an optimum level of activity. Life selected for evolution, for individuals with consciousness. *Survival* was a priority only in the beginning. Thereafter, the Earth had something else in mind.

Extinction curves, size, and tropic elegance work together to maintain the fruitful balance between life and death. Mammalian extinction rates are high, much higher than the rate for diatoms and invertebrates. The bias is in the size. More than 80 percent of all creatures are shorter than 10 millimeters.[4] Paleontologists know of dozens of major periods of attrition. Extinction, until the present time has been noted within family parameters, two families per million years disappearing, carrying with them a host of genera and species. As Stephen Jay Gould has pointed out, a rather odd gaia mechanism has been at work since Cambrian times, when 4.6 families were lost each million years.[5] That extinction rate has eased off 2.3 times. Had it been maintained, 710 more extinctions would have occurred. We do not know which families would have been marked. The migrations, continental shifts, volcanic eruptions and fires, the asteroid, food scarcities, and the Earth's magnetic quirks have skewed and masked details. But we do know that 680 new marine families have emerged since the Cambrian period. Neither Gould nor anyone else can adequately assess the approximate coincidence. But there is a tremendous long-term balance speaking in its near equivalency.

Where is today's balance? Our species poses the first challenge to the Earth's homostasis in the planet's entire history. In the year 2000, one out of four presently existing species on Earth will have been expunged by *Homo sapiens*. At the same time, our species is consuming about 5 percent of all the photosynthesis that has ever occurred, and is occurring, throughout the biosphere.

Aeschylus called upon the "premeval prophetess . . . who brings all things to life, rears and takes again into her womb." In our death, we are headed backward, not into any future; always backward, where life eternally begins and ends. This is the true law of gravity, where organic and inorganic agendas combine. In the heart of symbiosis has come a differential, degrees of upset to which life has persistently responded in creative ways. By *creative* we can only mean life-fostering. By such definition, the atmosphere, the Pangaean rift, collisions between continents, planetary outgassing, the disappearance of microscopic *Globotruncata* from the oceans, the emergence of the Rocky Mountains, of stripes on the tiger's back, had in each instance the paragon of balance for a model. This is gaia's clairvoyant legacy.

Human beings have always recognized a fundamental problem in their special situation of consciousness. We ask *why*? In the Revelation of St. John, the triumph of God over evil, visions of heaven, trumpet blasts, angels, judgments, martyrdoms, are all incitements to eventual reunion and balance that preoccupied the writer of apocalypse. Rather childish clues, in fact, as weighed against the primordial giant of the oceans, the aurora borealis, all of Africa and Asia, the trillions of governed molecules working in split-second harmonies for billions of years. What is this weak vision of reconciliation, this soul-rending challenge that all of our past, our heart, our belief in the future has enforced? We are awkwardly aware of ourselves, struggling to fix the priority, the very basis for our continuing evolution in some graspable mode. But in 6000 years of devoted scribbling, of documentary dark nights of the rationale, of pinioned thinkers with troubled aspect, no new insight has changed our relation to the first day of our consciousness. Of course, certain box springs have come into being by our brilliance – two plus twos, silicon chips – but for all our agonizing and self-congratulation, we lie dazed in the same bed of Earth, capable of good, capable of evil.

From the earliest calculations, coaxing of fire, inscription of petroglyphs, to gene splicing, the categorical point of vantage afforded *Homo sapiens*, its rough and tumble doubts and deliberations, has perpetually probed a paradise factor in ourselves. We have pondered every abstract and sensory phenomenon to come our way, begging the issue of transcendence, but with little success. Basic parameters of idea and action refuse to change – take the size of a bed, the shape and desire of a kiss.

Our species has lived in various camps on anywhere from 800 to 7000 calories a day, but the fundaments of survival, the verbs and vexations, never change. And chief among them is that maxim in all mentation, the sense of separateness. Frail, without the surest footing, reimmersed in the likeness that nature has conceded to us but more often kept monstrously aware of itself, is this rebel of evolution, the hero, the outsider, fire-starter, myth-maker, king, and executioner.

By our elevation we impose the rights of special hierarchy and nostalgia. And with these two we have invented paradise. Caught in the full of malignant yearning, we would go back to the beginning bargains only to make ourselves over again in the image of an ongoing future – paradise regained. Mankind's two most ancient stories – "Gilgamesh," and the *Egyptian Book of the Dead*, had no other concern than the epic consideration of truth and immortality. The drama is old but has missed the point: Paradise is here and now, its covenant – the rainbow – first declared in Genesis. For the poet John Keats, Isaac Newton's discovery of the prism destroyed the rainbow. Clumsily, science searches for nature.

NOTES

[1] J.F. Lovelock, *Gaia – A New Look at Life on Earth* (New York: Oxford University Press, 1979).

[2] Robert Jastrow, *The Enchanted Loom – Mind in the Universe* (New York: Simon & Schuster, 1981).

[3] Lewis Thomas, "Are We Fit To Fit In?" *The Amicus Journal*, Summer, (1981): 29–33.

[4]Roger Lewin, "Extinction Leaves Its Mark on Ecology," *Science* October, (1982): 42–43.
[5]Stephen J. Gould, "Free To Be Extinct," *Natural History Magazine* (1982): 12–16.

Chapter 2

Rhapsody of Embers

Through the discovery and control of fire, Homo erectus completely altered his biological clock. This resulted in a spectacular individualism which has guided us in our irruptive journey ever since. Language, art, osteology, and the size of our neocortex have been transformed as a result of these earliest fires in the night.

The living pygmy chimp may most closely resemble our earliest primate progenitors. But despite any similarities, an 8-million-year gap exists in human lineage. It occured 4–12 million years ago, during which time all evolutionary lineage is mere surmise. But paleontologists are fast filling in these blanks on the map of human origins. In 1982, Kiptalam Chepboi discovered two pieces of jawbone in Kenya's Samburu Hills. The bones are 8 million years old and may represent the missing vector between man and gorilla. But they are not gorilla bones per se. Nor are they chimp or human, but of some creature straddling the three.

Other research on early hominids is being conducted at Sahabi in the Libyan Sahara. It is the best-dated site in North Africa, with rich evidence of Pliocene swamps, estuaries, and subtropical savanna — just the right vacation spot for our ancestors. Mastodons lived at Sahabi 6 million years ago, as did monkeys of the Macaca and *Libypithecus* genera; and dolphins, gerbils, sabertooth tigers, even the first bear in Africa. Fossilized acacias show evidence of fire scarring. In 1979 and 1981, a clavicle and fibula were found, both hominid indicating a brain size smaller than the pygmy chimp's. More fossils are needed to make a judgment, and doubtless they will be uncovered. But the Sahabi and Samburu primates are pieces of an increasingly loaded question: Who are we?[1]

We wrestle with the past, pick over the runes and dolmens in the hopes of greater self-awareness. We sift through the *abracadabra* of ourselves to discover what makes us different, why we have drifted apart from that which we perceive around us — all of life. Lodged in the limbo of memory and dream, we reinvent the ghosts whose genes we carry, scratching at the dust that has devoured them. We are like dogs tormented by a scent we cannot reach, deep in the earth; or gnats pinioned in a drop of amber from a tree that no longer exists. Hominids are restless.

Pan troglodytes, the chimpanzee, is capable of murdering its own kind, perhaps from a similar moment of terrible doubt and rage and limitation — as human beings are wont to experience. Several such incidents were documented some years ago by Jan Van Lawick-Goodall at the Gombé Reserve in Tanzania.[2] But that capability is rarely indulged. The chimps have cranial capacities averaging 400 cubic centimeters. Clever and curious, they secure enduring attachments, a special kind of love. The chimp has demonstrated an incipient willingness to come over to our side, in absence of any conceivable remuneration, to learn sign language and join the learned symbols of *Homo sapiens* to the vocabulary of behavior. Its range of vocalizations and gestures is provocative: it can feign, intimate, deceive, wield weaponry, hunt and kill cooperatively. Its greetings and expressions of deference are remarkably similar to our own. Goodall has reported incidents of pronounced, articulated conscience among chimps. They perform what has been likened to a rain dance and Harold Bauer has observed a solitary chimp trek to a waterfall, apparently for the sheer pleasure of communing with it. Melvin Konner has compared this to the origin of human awe.[3]

A comparison of human blood proteins (albumins), cytochrome *c*, hemoglobins, and DNA with that of the chimp's reveals a 99% sequence similarity in evolution. This makes us as related to the chimpanzee as horses are to zebras. Why then have the chimps not evolved in 5 million years? One explanation may lie in the fragility of our framework for assessment. Why must we assume that our own evolution is the logical thrust of nature? We have seen that bacteria may have higher standing on Earth. The dolphins beat us to the vantage of consciousness by 50 million years. But beyond such appeals, there are certain judgments to be levied. An *apparent* gap exists between the chimp's mental facility to undertake basic tasks (i.e., termite fishing) and its conceptual ability to think about the future.

Our early forebears, *Australopithecus*, were also limited in this way. How limited we cannot say. Yet *Australopithecus* grew up, and within the space of 2 million years his brain size doubled. Could it be that the chimps are indeed willful in their adaptations, confined to a range of culture that they markedly prefer, a narrow yet satisfactory space of subsistence precluding the later complexities of *Homo*? Inventions of the hand, of complicated handiwork in stone, alteration of diet, of tissue, stance, gait, osteology; language, art, religion: Could it be that chimps were somehow dissuaded from all the work, the trouble of such advents? Glynn Isaac has argued that what profoundly singles out our species is food sharing. The chimps rarely share.[4]

The most primitive so-called, precomplicated society yet observed was a now-extinct group of aboriginal Tasmanians. Their tools were stunningly in tune, adaptive, frugal – wooden clubs, spears, digging sticks, cutting tools of finely chipped stone for fashioning containers. They had controlled fires; they shared everything – possessions, child rearing, sexual access, and food. Such reciprocity demands more than mere advanced mental work, the ability to calculate into the coming days. It demands quality, a lyric gentleness in the catch-all of socialized life – patting one another on the shoulder, caring for your neighbor, the sheer poetry of Being.

The prejudice of modern man toward his ancestors and surviving aboriginal peers is dismaying. The Tasmanians eventually became a victim of ninteenth-century genocide. But more than mere ignorance is involved. The modern mind has rejected the supposed ennui of our forefathers, the unpredictable night, unsheltered sun, the open-game judicial process by which all creatures partook more equally in the measure of abundance. We fear our past and yet we are intensely curious children lodged in the zoological halfway-house of civilization. In the surviving tribal clusters of today is the stirring reminder of our earlier dignity and genius. Such genius – the original language of culture, the close-knit tribe – is *still* with us, which is the important fact. Over the course of biological evolution, the tribe, by reinforcing behavior and imagination, conferred new weight and neural complexity to the grey matter in the brain's neocortex. Among other things, the pleasurable calculations of handcrafted objects, and of carrying devices, resulted in one distinctive acquisition: the establishing of a stable place where food could be shared.

The hominoid home base was indeed critical for extending the period of maternal care during which time the infant had more opportunity to socialize. Under such conditions learning became mandatory. Only among humans is there such a base camp. Baboon troops, for example, have no policy of nursing the injured back to health beside the hearth. The sick and infirm are left behind when they can no longer move with the troop, abandoned and easy prey. By selecting only the sick to kill, the predators, in turn, purge the troop of its blemishes. This has traditionally been proferred as nature's bittersweet masterplan. But there is more to it: Invariably, those ill-equipped to survive will not mate and pass on disease or genetic liability. They will live and die on the quiet fringe of obscurity. Because predators are not essential to the genetical protection of the troop, the trophic pyramid might thus have generalized those herbivorous impulses preferred by the Great Apes for its prototype. But it did not, and no complaint will much matter.

And so we have inherited a distinct bias for civilization, with its bold and stubborn rationalization of certain supposed fruits. "No arts; no letters; no society," observed Thomas Hobbes in the seventeenth century. "And which is worst of all, continual fear and danger of violent death; and the life of man, solitary, poor, nasty, brutish, and short."[5] But Hobbes' description of the alternative to civilization bears absolutely no resemblance to the human being in nature. This was the "nature" that Europeans romanticized upon discovering that their civilization guaranteed no joy, no success. The unfettered and newly discovered wild people in America and Asia were just possibly happier and closer to paradise. Rousseau admonished that our sole joy would come from perpetually rediscovering the animal in us.

But there is a middle ground. It occurs among the !Kung san bushmen of Africa's Kalahari Desert. The !Kung are fully egalitarian in their sexual division of labor. They possess great wit and wisdom, as well as natural sciences encompassing familiarity with hundreds of plants and animals and seasonal details. Each !Kung owns about 12 kilograms of private property, travels in a small band, and expertly follows the tracks of Kudu, gemsbok, eland, and roan antelope. But more than 60 percent of their food is vegetable. Earlier hunter–gatherers probably relied on 80 percent vegetal matter. And this system of depending on plant *and* animal foodstuffs enhanced the hunter–gatherers' coping strategy. Not only do the !Kung eat well, but they appear to be a happy lot.[7]

In the *Origin of the Species* Darwin referred to the "struggle for existence" but he did not like Herbert Spencer's later phrase, "survival of the fittest." For Darwin, the "struggle" was a metaphor pertaining to "dependence of one being on another, and including (which is more important — sic) not only the life of the individual, but success in leaving progeny." Natural selection favors cooperation as opposed to struggle; to survival of the fit, as opposed to the fittest. This crucial insight is easily validated throughout the animal kingdom, whether it be with *Hyponomeuta* caterpillars that show strong preference for being with each other, or Van Tieghem's *myxamoebae Dictyostelium,* cells that maintain individuality but form – like coral – a single aggregate; or herring gulls, sea urchins and protozoans, which all survive with greater chances in large groups than in small. Similarly, goldfish placed together in suspension survive longer that when so placed alone. Even planarian worms know to stick together: When exposed to ultraviolet light, they disintegrate more rapidly when by themselves. Lack of mother's love will kill any child. Pathologists called it *marasmus* when a motherless child slowly dies from not digesting his food. A monkey will go insane without at least some surrogate mother; even a pitiful hunk of metal will do. The point is that animals need each other. The hunter-gatherers have known this for at least 3.5 million years. No other primate seems to have paid quite as much tribute to the confessional needs of the ego as has *Homo sapiens*.

But let us go back to the beginning of *Homo*, to truly picture his origins, his earliest environment, and his motives.

On a raft down the Omo River, Ethiopia, we pass waterfalls, slurching hippos; alligators repose on sandy banks beneath the climbing canopy of forest in which Australopithecines once camped. The flora of tropical Africa has changed very little since the Miocene epoch. It was here on the Omo that Camille Arambourg's team uncovered 125 hominids in 1933. *Australopithecus* was a ground dweller who lived scattered across the bushlands, favoring the banks of streams and lakes. Comparative bone anatomy proposes the picture of a slightly stooped but upright creature. He appeared in Africa about 4 million years ago, weighed 70 pounds. His 1 pound brain was the size of a fist. Because he had a small body, this size brain was actually substantial. He had excess grey matter available in the temporal and parietal lobes, and could plan, perhaps abstract.

Early deposits of *Austrulopithecus* include the bones of antelope, horse, hippo, giraffe, rhino, warthog, and baboon. He had other, stranger creatures to contend with and possibly eat: elephant-like Deinotherium; three-toed Hipparions moving in immense herds; Chalicotheres – huge horse beasts endowed with massive claws; antlered giraffe; gigantic pigs; Dirk-tooth leopards, larger than today's; and lion-sized Dioneflis cats. All animals will naturally aspire after nutritionally balanced intakes of food. Under the competitive conditions to which early hominids had to accede, meat helped greatly to enlarge vegetable diets, increasing at once the rationale behind tool-making. The taste, the added vigor, and the dramaturgy of hunting – its collective energies – all contributed to meat mongering.

Australopithecus afarensus most likely came upon large animals already dead or dying, and proceeded to cut away at them. The objects from KBS* site, 15 miles east of Lake Turkana, include stone cores,

---

**The site is named for its discoverer, Kay Behrensmeyer.*

flakes, and chipped stone. These hominids were small-time butchers wielding highly effective obsidian cleavers and hand axes — tools with razor-sharp edges. The earliest of meat-eaters might have obtained animals weighing as much as 30 kilograms. Whereas the great apes (of which there are no more than 1000 left in the world) fed on the spot, Australopithecines could — with a carrying device — transport a sufficient amount of food to make sharing at a home base worthwhile. Such a carrying device made all the difference; the first shoulder satchel and the hands to hold it. To walk while transporting dinner was the first energy conservation, a veritable industry of highest caliber.

We pick up a stone. Our right hand controls the size and shape of the emergent tool that we will awkwardly create, striking off flakes; while the feeling, left hand, actually picks up the rock and holds it down for crafting. We have nothing else to do in this place that can possibly feed us as efficiently, surrounded by the semi-arid turf of Afar, Northern Ethiopia, at the site of Lucy.

A 454-gram rock would produce a cutting edge of about 5 centimeters, an edge sharper than teeth for cutting skins. The left hand has its feeling function, the right hand its harsher, analytic duty. Both brain hemispheres will contribute to the decisions and action in producing the final implement. In this way the primates came to see a world in which the constant encounter with objects, prepossessed tools, surpassed in importance the earlier space–time continuum of life. This sounds vague. Yet the evidence is clear: at Olduvai, Laetolil, Koobi Fora, throughout the Omo Valley, and in the Afar Triangle, a place known as Hadar where Don Johanson and his colleagues made a spectacular find of at least thirteen individuals, in November 1975. The knee, pelvis, and foot of these people were modern, suited to bipedalism. In other words, they were able to carry food some distance.[8] Mary Leakey's discovery of a 23-meter trail made by two individuals 3.5 million years ago furthers our conviction that there is indeed kin altruism, with a well-established natural history, and the bread-winner's trail to prove it.

The *flexor pollicis longus* is the most important muscle in the thumb, enabling it to oppose the other four fingers by rotating at the wrist joint. So equipped, the hominid hand acquires the broader base for grasping and a precision grip. This, combined with a complex nervous sytstem throughout the hand (lacking among the earliest tree dwellers), gave the hominid tremendous tactile sensitivity. The thumb is represented in the cortex by an area almost as large as that which controls the hip and leg. As the hand took over tasks previously reserved for the jaw, the cranial morphology changed; and the hand developed an increasing power grip in its role as chief exploratory agent of the body. The sweat glands in the thumb and fingers do not respond to heat, unlike the body's other sweat emitters. Instead, the thumb ducts react to emotional stimuli by secreting a fine film, urea, which must have ensured a strong grip in the trees later on to counteract the dangerous fall to the ground, where predators roamed. At least 200 imprints of hands with missing thumbs, interspersed with pictures of flora and fauna, were found in the Tiberan Cave of Northwestern France. The theory has been put forth that the thumbs were ritually amputated, sacrifices to some primeval god of anatomical evolution.

Larry Kelley's microware analyses of wood, bone, hide, meat, antler, and nonwoody plant, at high magnification, enables us to determine what tools were used, and on what materials. The Australopithecines were great chiselers. Eleven major tool types have been discovered. Humans are the only mammals who depend absolutely on tools for survival. Stripped bare in the Kalihari, left alone, without contrivance, I might die within a few weeks. I am young, strong, accustomed to the outdoors. I have been trained to find water and edible tubers, to fashion and use a bow-drill device for starting fire and to erect a solar sill in the desert sand for collecting condensate. With enough frantic foraging, I might survive a month. But with no help on the way, it is doubtful I would have the store of energy to experiment and save myself. This dependence on tools sustained all the expertise of a desperate science between generations. With bipedalism, erect posture, and the free, left–right brain use of hands, tool manipulation became a crucial part of the physiology. In truth, the tool is an extension of our body. Without it, I am stranded, right here, right now, strangely naked, unfitted to the Earth: I grapple with a crude piece of stone, unwilling to admit my feebleness. I cry out into the stinging night and emptied dawn, where I lay fetally in the sand, inept, fearful for my life.

At Rwanda National Park and in the remote outback of West Gabon, we can peer through a mountain forest at foraging gorillas. The apes have only one baby in 5 years. With increasing competition from other hominids, this was not a terribly successful strategy for maintaining a stable population. The problem then was how to speed up the birthrate. The apes failed to address this bit of mathematics, but some of their descendents did. By moving around less, the mother would be able to devote more time to multiple young. But then the food requirement was intensified just when less mobility made it harder to come by. So, believes Owen Lovejoy, the males got into the act, fighting over less accessible females. If each male won a female, food sharing would have to come into play. And as the parents became less mobile – freed from distance by the tools in hand – they also had the increasing advantage of uprightness, of food in the hand and the means of getting back to the family.

Ironically, Kenya today has the world's highest annual increase of birthrate, 3.9 percent, with 8.3 kids per mother. In addition, women produce 80 percent of Kenya's food. Family planners admit that the women plead for contraceptives, but the men are indifferent. This would explain Birute Galdikas' observation, after studying orangutans in northern Sumatra, that the female dominates and is worshipped by less-revered males. Why not? Females do all the work. Facing congestion with the great apes in the primeval forests, our hominid ancestors may have moved camp out into the savanna. By that time they were fully erect, voracious, able to hunt, and carry weight. Sometime between 4 and 1.5 million years ago, tool usage went from being wholly secondary to being crucial. At the same time, our population increased.

The backbone has its own critical transition, marked by erect posture. Once so formed the lower backbone faced major reorganizational challenges. The result was all manner of grievance, from weakened abdominal wall to varicose veins. Major changes occurred in brain, postcrania, face, skin, sex, vocal tract, and the birth process; a rearrangement of pelvic structure, of gluteal muscles, foot skeleton, and back. But these extrasomatic innovations borne of competing *Hominidae* genera paid off. We suddenly had our hands, and the pure opportunism apparent in the shapes of the earliest stone tools gave way to more orderly arrays for form that would begin appearing throughout the Old Stone Age. Form, in turn, which came to exert unimaginable power over thought; form from which a new world order emerged in every facet of art, ritual, and mime. Form indisputably followed function.

What we know about the look of our ancestors has been largely reconstructed from a labyrinth of worn bones. Peering back into ashes, employing microwear technology to date and assess the variety of tool uses, the flora consumed, and animals chased down, applying potassium argon and carbon-fourteen dating criteria, and pollen studies, we begin to surmise substantive, cultural designs. But it is all of a twilight, faceless.

Our immediate predecessor is an enigma. From the Solo River in Java came forth remains unearthed by a young Dutch anatomist, Eugene Dubois, in 1891. Seven hundred thousand years old, marked by a small, flat cranium of 815 cubic centimeters this first *Homo erectus* find suggested culture. Solo Man weighed about 198 kilograms, putting his brain size in a proportion with his body size that fits our own.

Fossils not unlike those from Java came to light at the Choukoutien Caves, 25 miles south of Peking, in the 1920s and 1930s, following the work of Swedish scientists assigned as advisers to the Chinese government. Digging for 6 months, Birgir Bohlin was rewarded royally with a single tooth, which he had set in a large gold watch chain so as to allow for easy show-and-tell. Bohlin first showed the tooth to Dr. Davidson Black at the Peking Union Medical College. Black compared it to one other tooth earlier recovered from the site. On the evidence of two teeth, Black established a new genus, *Sinanthropus*, with species name of *pekinesis*.

Returning to the cave, the Swedes recovered more teeth and parts of a skull. Dr. Franz Weidenreich then took over the dig, but China's unstable politics prompted him to remove the fossil materials from Peking. Arrangements were made to transport the bones. But they never reached the United States. Peking Man disappeared.

Since that time, other remains have emerged. In all, 6 complete skull caps, 9 skull fragments, 6 pieces of facial bone, 152 teeth, mandibles, and limb bone fragments have been discovered — the most complete of all *Homo erectus* finds. The date? From 300,000 to 700,000 years ago. His brain size averaged 1040 cubic centimeters. What accounts for the broader cranium? The size itself offers few clues. Richard Leakey's discovery of a 3-million-year-old fossil (ER 1470) with an essentially human neurological brain organization tells us that human behavior patterns, including the ability to learn language symbols and to utilize them were long ago implicit in our makeup. The skulls of *Homo erectus* and ER 1470 not only share the shape of *Homo sapien's* brain, but possess the first distinct impression of Broca's region — that area for coordinating the muscles of the mouth, tongue, and throat during speech.

Let us now turn toward the plains beneath radiant Mt. Kenya, where someone so endowed with tools and thoughts sat at sunset once, roasting a shank of meat in an open fire pit. Prometheus stole it from Hephaestus. In nearly all cultures the stealing of fire constitutes a creative, revolutionary act — from the Dogon people of Mali, to New Zealand's aboriginal Maori. In East Africa hunters supposedly obtained fire from the subterranean world of the dead, though in some cases that cavernous abode was actually aloft in the mountains, where lightning storms threw up their magic, and glowing embers could be gathered like wild strawberries. In many archaic civilizations the origins of fire imply sexuality, as we will later examine, the shaft furnace being female, the bellows male. The sacred fire would be tended by virgins. And among the Parsiis of India, the modern Zoroastrians, temple fires are perpetually kept going. It would be sacreligious to let them ever die.

For our ancestors it would have been suicide to cease the maintenance of fire. But then fire had its own method for exacting vigilance from its human proprietors. That method was the total alteration of man's 24-hour biological clock. Its implications are far reaching. This chronometer in all animals works with the precision of a machine. It is meant to put the animal to sleep — not when it is necessarily tired, but when the body deems rest important — and to wake it up before its predators can catch it off guard. The clock works with an on–off inflexibility built into the animal's natural adaptive strategy over millions of years. It regulates nearly every involuntary vital force, from hormonal secretions to pulse rate. A rat lacking optic nerves, congenitally blind, nonetheless responds to the on—off switch of its clock each twelve hours.[9]

Not far from Mt. Kenya, at Chesowanja, scientists uncovered the remains of controlled fire usage dating to 1,400,000 years ago. The remains were those of *Homo erectus*, the first human-like primate: a big-time meat eater graced with fully erect posture; a world traveler, an artisan, a thinker. Fire had a mind-altering effect on him. Fire was not only a supreme hallucinogenic (the earliest myths and supersititions are borne of flames), but it actually imparted additional grey matter to his brain — nearly 454 grams of it in less than 1 million years. This correlation takes the cranial evolution of *Homo erectus* as a function of his unique and unprecedented fire culture, a culture absolutely pervading his life. Fire was no mere catalyst: it provided him the protein, the source of self-reflection and of comfort. With additional neurons crowding into his skull, his forehead bulged upward and out. For 3 million years his brain had come about slow and ploddingly, with great care. But 700 cubic centimeters were acquired hastily by the neocortex during *Homo erectus'* reign. The cranial Rubicon dividing Hominidae from ape was somewhere around 800 cubic centimeters. Modern man's brain extends in varying sizes from 1000 to 2000 cubic centimeters. By 750,000 years ago, the brain of *Homo erectus* was packed with 40 million neurons per cubic centimeter, for a total of 10 billion neurons and 100 trillion circuit connections.

The *Homo erectus* fetal life showed extraordinary growth curves never to be enjoyed by the more soft-witted primates. During gestation the brain grew at the rate of 20,000 neurons a minute. During the last 3 months of gestation it accumulated 2.2 milligrams a minute, weighing 382 grams at birth. By the end of the third year the *Homo erectus* child had achieved a mental bulk of more than 1 kilogram, 4/5th the size of an adult. The juvenile skulls of *Homo erectus*, as well as those of Neanderthal man, exhibit features much closer to those of modern man than to their own mature forms. The point is that natural selection chose in *Homo erectus* for grey matter, for the contemplation and energy of youth looking ahead. In so doing, natural selection extended his day into his night, altering the encephlographic record of his dreams, shortening his alpha state by context of consolidated hours of sleep, rendering him independent of his biological clock to a greater extent.

To fully grasp the significance of fire — not only for the human diet, but for the psyche — we need only to consider the shock-phase hypothesis of Curt Richter, a Johns Hopkins Medical School investigator who has studied the effect of fire on biological rhythms. Head trauma can suddenly bring about a return to the on–off phases. The emergence of the primeval biological clock greatly modified behavior and might offer clues to the diagnosis of certain brain pathologies like schizophrenia. Since the advent of fire, all subsequent human physiology has been out of phase — some active, some inactive. And these discrepancies are smoothed over in a performance we perceive to be normal. A blow on the head quite literally knocks sense into us, puts us back in phase. The farmer's regimen offers something of that reconstitution, and later we will consider its psychological rhythms.In darkness we return, reattune. The cave is its symbology.

We have seen the stonework, the tight tapestry of cause and effect singling out those characteristics most apt to further a purpose: bipedalism, a social network of defined sexual roles, the caring of young, a home base, and a territory. From Piliocene to early Pleistocene times, our forebears — Lucy's descendants — walked with an essentially modern stride, their foot, eye, hand, and hope for tomorrow basically our own hope.

Homo erectus was a big game hunter. Today his affairs are managed from safari Range Rovers, with high-powered rifles and infrared scopes, or by the last remaining pockets of hunter-gatherers in the Northwest Territory, Greenland, Surinam, Zaire, Angola, central India, Thailand, Australia, Brazil, and West Irian. Such hunters dissipate cholesterol in ways not readily available to those of us today better suited to pushing shopping carts. Great bursts of adrenaline are vented in stalking, killing, sharing and digesting prey. This fact has been documented among the !Kung, the Mbuti Pygmies, the Eskimo, and the aborigines of Arnheim Land and of the Andamans.

From finds at Koro Toro, Chad, and Ternifine, we know that he was feeling his wings, dispersed, possessed of protein surplus and running with it. With a thick, supraorbital receding forehead, he was the original lowbrow; wide-nosed, prominent jawed. As the first agriculturalists storehoused their grain — acquiring in 3 weeks what took the hunters a year to collect — so *Homo erectus* and his band enjoyed the luxury of big meals, of metaphor. And when they cooked the meat, they had culture.

We know from the record of the ash that the Choukoutien Caves were inhabited after the successful use of fire at their entrance. Could it be that the very cave was tamed in the eerie light; that such magic was the prerequisite to cave residency? Other forms of magic prevailed here as well. Amid chopping tools, stone flakes, and bone artifacts are thousands of hominid and animal bones, mixed. Every cranium shows damage to the region surrounding the foramen magnum. This cannibalistic mutilation persists into the twentieth century. A person weighing 50 kilograms produced 30 kilograms of edible muscle. A well-cooked *Homo erectus* could serve sixty participants at a feast, all feeding methodically from skull bowls. His diet was supplemented by at least ninety species of other mammals that he preyed upon. Mice, elephants, even charred ostrich eggshells were found in the rubble and 18 feet of ash. His favorite food appears to have been cooked venison. His palate was sophisticated and certainly more diversified than that of most twentieth-century dilettantes. Some of his tools show continuous refinement and many were ornamental. Forty percent of the dated bones were of youths under 14 years old; 2.6 percent of the bones represented the 50-60 age bracket; 7 percent between 15 and 30; 8 percent between 40 and 50. Ten percent, in other words, survived into old age to transmit culture.

The fire at first erupted from compressed organic matter, guavo, middens, cast-off wastes; or was captured in the previously mentioned embers. In the record of the flames – burnt clay evidencing controlled temperatures of 752F – is the far-off attribution of deep, pensive fascination, perhaps synonymous with the earliest religion. The fossils at Choukoutien are not the oldest of *Homo erectus* finds by any means. We have referred to earlier discoveries in East Africa. And even within China there are other *erectus* artifacts that predate those at Choukoutien by nearly 1 million years (at Chenchiaow and Kungwangling, near the town of Lantrian). But Choukoutien is a special haven. When the full moon illuminates the hills, one knows in the heart that the whole world is on fire, breathless; that every living being throughout the mountains of the Mongolian watershed to the north is attuned, vital, poised in the elegant discovery of life, life that is wonderful and freely granted to perception. This was a revelation unique to those endowed with fire. They were able to stay awake at night and ponder such a moon, such silhouettes, optical nuance, late evening umbers and mauves, all of the colors and breezes of tranquil philosophy.

Fire gave us both speed and pause in our hasty evolution. Wood was consumed, presumably stored, and such surplus predated the impetus for agricultural surplus. The 3000 visible stars, seen for the first time from the hearth's security, gave shapes, the pleasures of infinity, to the mind. Ordinarily, primates are not star gazers. Their glance is limited to more pressing matters at home. But *Homo erectus* was offered the first late-night purview of the firmament. On the cave wall the eye picks out a host of astrological signs; seizes upon symbol with night-after-night consistency until, slowly emergent in the morning, are all the first numbers, save for zero. If *Homo erectus* entered the cave only after successfully containing fire at its entrance, we can understand his trepidation: The cave was itself an animal – dark and forbidding, with no end to its gape or digestive track. But during storms, early man must have taken emergency shelter in the beast's gut. In so doing, he saw through the illusion, domesticated his fear, and fused dreaming and vaginal mythology with folk science. In those caves given to great depth, a veritable participation mystique evolved, *rites de passage* analogous to alimentary fixation by dint of intestinally circuitous routes reaching back into the cave's genitalia. Through these wending chambers, early man aspired, torch–phallus in hand, to light up, inseminate. Fire, art, and daring sexual advance engendered together all progress, economics, propitiation, and eventual agricultural organization. All of early man's fertility — like a star — shone brightly in his darkest moments, the night gatherings of cave domination.

We imagine fierce warriors at the brink of the last great interglacial, 500,000 years ago; beetle-browed ancestors on the verge of becoming you and me, settled beside a roaring fire, entranced by shadow play on the damp cave walls. Such a moment would have constituted the first instance of self-contentment, of individualism in its literal, survivalist sense: the sense of being warm and cozy unto oneself. Until arrays of form replete with mineral fell into the eyes; the iris glanced off the blown night firmament, took in the full stellar circus, then fabricated a dipper from a stag's soft braincase; searching, from the depths of his right hemisphere, for a horizon line to the universe, where astral game might be running. In the gurgles of intuition comes the first sweep of foresight.

Imagine if you will this man and his long-haired woman, dressed in fancy skins, cutting into giant panda with their quartzite manufacture, throwing out the liver, keeping the brain and muscle, sucking out bone marrow, their cave covered in luxurious fur. Poking their fingers through accumulated ash and making the faint, tandem outlines of some future they desire for themselves. They are neither hungry nor cold nor wanting for anything. Only curiosity moves them now. They have their cave, their heat, and one another. In narrow row after row rest the crania of others, all of whom have given meat of themselves and whose opened brains were blessings upon the heavens.

Man and woman have their appointed tasks the coming day. Chase the rhino down, collect ginger root. Beware of cats, of course; stay to the vulture's orbit; go no further than the ridge, that blue escarpment. Beyond it is danger. But this night, there is no such peril. The hunter has an idea; his mate has an idea, into the wee hours of the night until emerges a spine-tingling notion, cascading with possibilities. Up and Adam, great volition all at once, the fire ablaze, rock shadows speaking oracles of congratulation and challenge, provocation to uproot. Imagination is finally set free from its constraining genotype of habit. The similes and routes into chorus, the trigger of inspiration are all the same as they will ever be. Contradictions of heart, of the lazy and the fired, the comfortable and anarchistic; preferences of taste, task, temperature, in the narrow margins of human personality, fall to the side before so mammoth a design. In the early morning, keen to every sight and sound and smell, they leave 500 generations behind them; both in their teens, heading for the mountains.

NOTES

[1] Noel Boaz and Douglas Cramer, "Fossils of the Libyan Sahara," *Natural History Magazine* (August 1982): 35–40.

[2] Jane van Lawick-Goodall, *In the Shadow of Man* (New York: Dell, 1971).

[3] Melvin Konner, *The Tangled Wing: Biological Constraints in the Human Spirit* (New York: Holt, Rinehart & Winston, 1982).

[4] Glynn Isaac, "Food Sharing and Human Evolution: Archaeological Evidence from the Plio-Pleistocene of East Africa," *Journal of Anthropological Research* 34 (1978): 311-325.

[5] Thomas Hobbes, *Leviathan* (London, 1651).

[6] Frank E. Poirier, *Fossil Evidence – The Human Evolutionary Journey*, 3rd ed., C.V. Mosby Company, St. Louis, 1981.

[7] Richard Leakey, *The Making of Mankind* (New York: E.P. Dutton, 1981), and Konner, *Tangled Wing*.

[8] Donald Johanson and Maitland Edey, *Lucy: The Beginnings of Mankind* (New York: Simon & Schuster, 1981).

[9] Curt P. Richter, "Discovery of Fire by Man – Its Effects on his 24-Hour Clock and Intellectual and Cultural Evolution," *Johns Hopkins Medical Journal* 141 (1977): 229–232.

Chapter 3

A Lost Metaphor

Paleolithic culture was determined by the intimate connection between the female and the forces of nature. Throughout the world women had their periods in unison, controlled the diet, the fire, the very essence of Ice Age spirituality. But the male-dominated agricultural revolution changed all that. The new model of human reality was engendered by the militant farmer, first embodied in the Biblical murderer, Cain. The Mind could not escape the confusion of these two prototypes. The result was conscience.

Christianity denied the very wilderness into which the Jews had fled; denied it in the hope of reaching the *true* wilderness, namely Biblical paradise. But that paradise had long before been flooded out. Let us go back and see what actually happened, how changing glacial environments and additional grey matter in the neocortex supported the Neanderthal's acceleration of animal extinction rates, with subsequent protein scarcities and the rise of gardening and domestication.

The paradigmatic relation is very much with us today. The cave paintings and bone ornaments depict Cro-Magnon man's valiant if fated effort to exert shamanic control over animal migrations, to enter into the stream of sentience, returning to the womb of the matriarch whose society controlled male egocentrism. The woman, in every way—sexually, economically, politically—was a goddess, mated to the moon, bleeding in unison throughout the tribe during collective menstruation.

Her consort's reverence for her fertility, her gathered grain, anticipated the tactical obsession with agriculture soon to engulf the Mediterranean basin lands, endowing the Bible with its similes and virgins. In the traumatic transition from totemic survival – hunting camps armed to the hilt, preying over free distance – to the mind of the consolidator, human beings had crossed the river of separation, had begun to reckon their disturbing free fall away from the wellspring. In China the separation would cast its shadow over the very language. *Hsin* versus *Hsing* – mind against nature. There was compelling proximity alright; irrevocable schism.

From the remains of Terra Amata (present-day Nice) and Torralba in Spain, the hunter–gatherer's life style can be reconstructed. At the French Riviera he slept in cozy quarters. His diet included birds, turtles, stag, elephant, wild boar, ibex, Merck's rhino, oysters, mussels, limpets, and a staggering variety of fishlets. All that cooking may have promoted certain sedentary tendencies, at least among the females. Controlling one's spontaneity – inhibition, in other words – is a mark of evolutionary sophistication. They seem to have been happy. Their domiciles were equipped with indoor bathrooms, and they slept near a central hearth. Their huts at Terra Amata were built on a hill that today hosts the Chateau de Rosemont, expensive real estate overlooking the sea. The King of Yugoslavia once stayed there. Not too shabby for 300,000 years ago.

Continental drift, the rending of Pangaea – the earth's supercontinent – volcanic intrusions along mid-ocean ridges: Such plate tectonics, sea drainages, and bipolar episodes had forged stunning migration corridors for all species. *Homo erectus* radiated out, turned up in the Middle East and in Europe where environmental pressures selected for one child prodigy after another, until Neanderthal was walking between Cape Prince of Wales and Cape Dezhev in Siberia. During the Wisconsin glacier, over a period of 9000 years, he walked south to Tierra Del Fuego, stopping in La Jolla 45,000 years ago (can you blame him?). At the Jayamachay Cave, near Ayachuco, Peru, excavations reveal a succession of unbroken remains spanning the life of the hunter and that of the farmer. Elsewhere, Neanderthal took up boating. At least he went the distances between Java, Timor, Borneo, the Celebes, and Moluccas. At Mixnitz, Austria the Neanderthal were busy at the Dragon's Cave, killing bears beneath a waterfall. And at Shanidar, Iraq (Cave IV) a burial took place 60,000 years ago on a bed of Zagros grape hyacinth, Batchelor's button, hollyhock, and yellow flowering groundsel – all woven into the branches of a pine shrub. At another Shanidar cave, a Neanderthal was killed by rock fall, apparently while being nursed back to health from previous wounds: paleolithic conscience at work.

Neanderthal mortuary evidence has yielded curious facts. Normally adults were buried in the front of caves, children in the rear.[1] We obtain glimpses of the Aurignacian culture's reverence in the wonderful burial adornments. Two Negroid skeletons of the Grimaldi des Enfants cavern were layed to rest with a crown of flowers, shelfish bracelets, and pendants of bone and ivory held by red-painted clay — consonant artistry, a terrific sense of the afterlife, "immanence of the soul."[2] At the same time, fear of death and of the dead themselves often bade these anonymous gravediggers to bind the bodies in fetal positions where they would be prevented from coming back to haunt the living. We cannot gauge the relative motives, for such symbolic positions might also indicate a desire for the dead to be properly reborn into infancy. Already the yearning for beginnings? As Arnold Toynbee put it, the cities of the dead antedated the cities of the living.

There were other burials at La Chapelle and La Ferrassie. At Teshik Tash a ring of goat horns surrounded a child's grave. In one instance parents were buried head to head, their children interred nearby. And there is no question that the culture of early burial produced intense intellects, by any standards; men, women, and children of powerful curiosity, passion, imbued with the desire to *know*. Before their extinction by ranchers and Carmelite holy men in the eighteenth century, the Esselen Indians of Big Sur, California buried their dead in semicircular arrays, all of the bodies elegantly stretched out flat on their backs, palms kindly extended in a posture of prayer, facing the Pacific.

Eighteen thousand years ago in Upper Egypt and at Lascaux, in southwestern Europe, the impulse to remain sedentary had pervaded Paleolithic consciousness like a drug. If property is theft, as Pierre Proudhon insisted, then language, logic, efficiency, and love conspired to heist the whole garden of the world and remake it in the pattern of control. Hogans, calendars, formalized ritual, a priestly hierarchy of painters — spiritual shamans — places of precise and charismatic interest: This cultural hegemony had engendered a symbiotic coexistence of man, woman, and nature. The earliest paradise myths probably arose at this matrix of artistic genius, cried out to be heard before the chaotic onslaught of congested kinship, congested numbers, waning sources of protein, changing weather, and all the usual inclemencies. During the last major advance of the final Pleistocene glaciers, large herds of reindeer roamed western European tundra, offering themselves, enriching the human insight, bolstering its muscle and its population impact like never before. Man had entered into the life of the reindeer. Pleistocene populations were shaped by the creatures that were hunted. To be a more effective spear thrower, man became a herd animal. He felt mass culture, enjoined himself to it. Cultural evolution now resided in the hands of the hunter as opposed to the less equipped competitors of his species. We have no evidence for mass murder during this time; the reindeer were indeed slaughtered, but the tendency was not extended to other human beings, or not that we know of, not yet. All that we can see suggests colorful hubris. At Lascaux, during the Gravettian and Magdalenian renaissance, some fifty paintings were set down: strange human figures clad in animal skins and adorned in ocres, tallow yellows, burnt sienna, lambent blue; purposeful, powerful, and animated. But the horizon – thick with ungulates – looked different than before.

Pygmies are so conditioned to a certain size and shadow within their forests that when asked to measure an outside entity – such as grazing buffalo in a distant field – they have likened them to ants. A ninteenth-century explorer took photographs of Australian bush people and then showed each his individual rendering. Apparently not one person could recognize himself. The Aivilik hunters drew their island outlines well from acquaintance. But the women drew the settlements with greater detail. All evidence points to the fact that women know their immediate surroundings better than men; but men are more comfortable in their acquaintance with far away places. What unites people in their respective landscapes is *apperception*, thinking about thinking; having a purpose in mind, the enzyme of deliberation, and the space of nature to focus the concern. The animal reacts. But man thinks and rethinks his response. It gives him pause. It allows him to reflect hierarchically, to make conscious his every action, to say, "This is beautiful! This is mine!" Man's unique consciousness, in sum, provides him his environmental angst and delectation.

Apperception originally characterized our species; it acted out intentionality that further estranged the man from his friends, the animal spirits once kindred to him, but which he now had the foreknowledge to exploit. We can see a late Paleolithic intelligence beset with primeval doubts – for all of its ingenuity and modern materialism. These anxieties stemmed from the rather sudden separation of the human heart from the world: This separation produced a severe talent for calculation, a self mastery that had otherwise inhered, sustained itself on impulse, and traversed the forest floor free of worry and introspection. The veneration of bison on cave walls echoes this sensation of important loss, the fateful achievement of delicate conscience and the fearful symmetry of newfound will to power.

Nature slid away from him and became the other. No longer were we so conveniently a part of the grander scheme. *Homo sapiens* now had to manipulate the elements to retain even a precarious foothold. Of course we had waged those same struggles, stalked the same game previously. Now we had an inner mandate to do so, but one entangled in debate. At the Magdalenian El Juyo Cave, a ceremonial stone head bears a countenance half-human, half-lion. The face is melancholic. With apperception our ancestors seem to have managed quite well, to have matured beyond the first crises, only to be stripped in the dark, confronted with their own emerging reason — reason which cried out, grovelled to know more.

Paleolithic culture was the original affluent society, in steady-state. These people knew their territory, and we can draw contemporary examples to suggest the range of their acuities. The Yakuts in Siberia can distinguish stars in the Pleiades that would ordinarily require a telescope. The African Dogon celebrate a similar pole star prescience once every 60 years. The Salteaux Indians of Manitoba possess a spectacular nomenclature for their winter and summer hunting grounds. The Tasaday of Mindinao live 40 miles from the sea but have never seen it, have no word for it. Why? Because, they say, it would be wrong to leave the forest, where there is good sleeping, good quiet, and good eating. The living quarters of the Kalahari bushmen allow for 188 square feet per occupant (just half the American Public Health Office's minimum recommendation). But the bushmen show no apparent stress to their immune system; nor does their production of cortisone and lactic acids, their adrenaline and calcium, belie any maleffects. Quite the contrary. In such dense social conurbations, women are protected. Only recently, since the !Kung began adopting South African agricultural practices and enlisting in the military, has there been a resulting separation of housing. !Kung women have learned what it is to be beaten by their man, without instant recourse to compatriots nearby. When the Chukchi sketch their hunting territory with reindeer blood on a piece of broken wood, every detail is intact. The Bedouin possess 800 words for camel; and Inupiat hunters of the Arctic have a similar vocal nomenclature for snow. Brazilian Indians in the Matto Grosso can wield more verb tenses than could Plato.

We can speculate unfruitfully and forever on the lore of lingo in the backwoods, what Cro-Magnon men concerned themselves with, who cracked the first joke, and who shed tears for the invisible spectres around him. All the first inquiries launched the legacy of metaphor, between the inner eye and the environment — a beeline to conscience, a terrible reconciliation in the here and now. Among migratory Polynesians, there appears to be a 2000-league directional accuracy. And among Hopi, Klingit, and Hoonah Indians, space-and-time delineations come oddly into fashion according to the calculations of sub-atomic quantum theory. The agility and diversity of myth, medicine, and moodiness contrasts with the universality of muscle. The healer uses the language of the reindeer and the hereafter to make his charges receptive to the earlier magic. But he is thinking beyond the nuts and bolts of immediate clarity. He is abstracting with obscure and awkward elocutions that have descended from the gothic nave of inspiration; come down in the floating light — an inkling, a poetic.

The *look* of fossils begins to meet our own. Eighty thousand years ago in a skeleton from the Kobariah Valley near Aswan. And then the entire face at last, haunting, exposed, sullen. She is a small stone carving called the Lady of Brassempouy, at Lascaux. Beneath her veil is the trembling look of power modified by humility; of beauty aware of itself. It is quite likely that women of the late Paleolithic era were in striking synchrony, uniform menarche, collective pregnancy, rhythmic with the Moon and the tides. This social phasing was recorded notatonally, as on the engraved *baton de commandement*. Nineteen-thousand years ago in the Dordogne, lunar astrology was intimated in the incised bison horn in the hand of the Venus of Laussel. Thirteen or fourteen notations invite ovulatory, menstrual, and lunar symbolization. Leroi-Gourhan's key-frame computer study of Lascaux imagery has yielded sexual paleozoics.[3] They are best perceived in the painting "Shaft of the Dead Man," with its erection and spear.[4] A bird is positioned beneath the man, apparently merging the animal and human soul in flight – a shaman's transport. A "coat of arms" further intimates a color-code geometry that appeals to speculation on the origins of language. The structure transcends its details.

Philip Lieberman's supralaryngeal anatomical studies of Neanderthal endocasts indicates that one particular "supervowel" emerged as the key to language. It was the *i*, and children everywhere imitate this sound in infancy with a higher regularity than that of any other sound. Along with *a* and *u*, these vowels apparently have selective advantage. There is data exhange without speech (as among deaf mutes). But *with* speech, there is power. At Lascaux, unquestionably an advanced dialogue was taking place, its vocabulary depicted across the walls of this subterranean Sistine. It was the last time prehistory would declare its awe so unabashedly. Lascaux was the zenith of nature religion. But such passion was a prolific prelude to the lethal evolution of dichotomy: the pastoral versus the coming police states of the Near East. Lascaux was a common prayer posed in the cathedral of free air, down under, where man was born.

Language structure indeed evolved over millions of years. Of course we do not know that for certain. Some have argued that it came about suddenly, perhaps 100,000 years ago, maybe 20,000. But these figures are insults. The house of diction, with all of its increments in place, came together, provided shelter, early in its genesis, when the chopper had its haft; when the child's eager imitation had its bare vowels and modifiers and secondary consonants. But spoken language may have exerted rather recent pressures on the brain's morphology. Marjorie Le May has examined crania and deciphered bulges on the inner surface of the skull at the right front and left rear. She believes that the right–left brain assymetry emerged approximately 30,000 years ago. Nothing we have fashioned technologically has ever matched the profound cybernetics of language advancing us into the future. If the past 10 trillion generations of bacterial life have also been driven; if the very neuron reacts towards the future, then language set out to manufacture it, to control it absolutely. The Paleolithic *language of paradise* was the hinge of tenses in the throat.

You notice in the mirror a smudge on your cheek and act by reflex to remove it. Similarly, language seizes chaos and orders it. With language, the mind set itself out to control. Could man remember a nonlinguistic past? Doubtful. *Homo erectus* probably manifested some useful speech. What the late Paleolithic breakthrough imparted – we see it all around us – was a container for the hundreds of thousands of years worth of symbols, neural reckonings that had accumulated in the mute, rear caves of impression. Suddenly, the colors exploded with double-entendre. Language facilitated numerology and thus prepared our species for coming tonnages of wheat, which would have to be stored and accounted for. Language protected our acquisitions.

Even during the Magdalenian, we have a record of what might be the earliest agriculture in history, at Kostenki-Borshevo in European Russia's Don Valley. And it came right out of the ground in the form of vegetable tubers, coaxed with digging sticks that might have been used to herd reindeer. Cord-marked ceramic shards indicate the consumption of beans, peas, and water chestnuts at the Spirit Cave in Northern Thailand as early as 14,000 years ago. Enkor and emmer wheat strains were crossed and cropped with Natufian sickles of flint at Shanidar 5000 years later. And at the same time, common beans, chili peppers, and foxtail millet were cultivated throughout Central America; barley at Ali Kosh. Pigs were domesticated, sheep and cattle tamed. Dogs entered a powerful communion with the exploiters. In Upper Egypt at Al-Badari, there was grain cultivation, animal domestication, stone walls, wheeled vehicles, bronze, extensive irrigation – all keeping apace of the emerging Mesopotamian city states. But the entire mechanism of produce and subordination was still linked to the twilight of Ice Age spirituality. Animals were buried on the edge of town, wrapped in linen.

The practices of paternity, forced insemination, and war grew out of these urban assemblages of mankind, where all of the Earth goddesses were shouldered off their pulpits, converted into sexual objects, degraded, and eventually traded like chattel. William Thompson's trenchart analysis of the destruction of femininity of the world that had, for a million years been built around the compelling mother goddess, is of especial importance here. Thompson dissected the inner meanings of Catal Hüyük's shrine VI.A.10, the central chamber, where bison pour forth from the cornucopia of a goddess' womb. The shrine has none of Lascaux's subtlety. The goddess was larger than life, and all of the statuettes excavated here on the dry Anatolian Konya Plain support such status. Catal Hüyük's period of dominance as a Near Eastern trading center ended abruptly around 5400 B.C. The peaceful trade and worship vanished in the ashen skies of war and fire. The new mythology would be expressed at Ur, in the tense struggle between male and female, Enki and Ninhursag — water and stony earth. The myth of "Enki and the World Order," quoted by Thompson, suggests a canticle to the penis, the "rampant bull" with its "sweet-tasting wine."

Catal Hüyük was the last center of the old religion, where women relished the providence they alone provided — the grain, the child, the erotic clasp. Women were buried with children and the best jewelry — obsidian mirrors quarried from the nearby volcano Hasan Dag — right under the sleeping quarters. But men were interred in smaller, out-of-the-way corners. Couples made love directly over their deceased women. In this participation mystique was meant to be sustained the hierarchy of fertility that women had come, as gardeners, to invent. Men had been thrown out of the inner clique, from sensuality to nomadic campsite, where herds had dwindled. There was little killing to be done, and times were indisputably hard on the ego. The male must have felt terrible resentment, impotency for no fault of his own, a diminished power base. Did he ever really have it to begin with? Hunting and gathering was the principle means of livelihood of 90 percent of all people who have ever lived. A terribly successful strategy, the gathering, given to great contemplation and modesty; while hunting vented the political inequality of sexes.

The mural of a reindeer hunt — any of hundreds like it painted throughout the Stone Age — looks nostalgically back to the freedom once enjoyed by the hunter. But his hegemony, now threatened by the trickster in climate, geophysics, plant succession, population dynamics, and emotional entanglement, was no more. The conservative hierarchy of women, with their clay storage bins of grain needed protection: Wealth required proof, entrenchment, a sanguine second generation to play overlord. The unemployed male, with his hunting agenda, became the martial warden, dressed in epaulets, a bounty hunter. As in a wrestling match, the power grip suddenly reverted to the policeman. It was a quick change. The women, for all of their Earth-touch and relatedness, had lost the language of politics closing in around them. Races mingled. This losing of intimacy may be the crucial metaphor separating the psychology of all prehistory from the modern annals; a record first inscribed on the hot and dusty plains of modern-day Turkey.

Throughout the third mellenium, the Chaldean masters of Ur (in southeastern Iraq) supported an arsenal of produce derivative of slaves, thousands of them. Three thousand acres of prime agricultural lands were cultivated, and 3000 left fallow every year. Most of the food was allocated to temple personnel, who were ruled by the ziggurate zealots — priests and priestesses whose sexual forays atop the seven-tiered mountains brought heaven and Earth together. Three thousand head of cattle were slaughtered annually. The horse, first broken by pastoral nomads in the Ukraine, was introduced. The old numerology, now set down in the form of economic tabulation, was wrought in the Sumerian and Proto-Elamite pictographs, as were written records of stock, inventories of slaves, other machinery, and horse-grooming techniques. Non-Semitic Asian nomads, the Sumerians, with their considerable artistry, gold and copper, irrigation canals and cuneiform, also invented the lever and devoted much of their energy to constructing clay maps. With such information, man could effectively wage war on the unknown by maintaining inventories both of his force and his food, as well as charting the course of his rampage.

The great warriors of Mesopotamia were Sargon (ca. 2600 B.C.) and Naram-Sin from northern Akkadia (2159–2133 B.C.). Naram-Sin was celebrated on a military victory stele. There is the bristling sun, the tree of life, and the mountain. Women praise him from beneath as he ascends along the shoulders of his army. He has conquered, set flame to nature. Only the mountain stands taller, by a few heads, but not for long. Naram-Sin is still climbing toward heaven. With his horned helmet, bow and arrow, sandalled feet, fierce muscle, he seems utterly determined. Nothing can stop him. And as we look back, we realize that nothing has.

Still, mere conquest was not enough. The Sumerians were nostalgic for their own better past, and with Gilgamesh invented the first documented paradise mythology – of Dilmun, the land of eternal life, and Ut-napishtim, the one human ancestor who achieved it. Gilgamesh's closest friend, Enkidu, has a foot in each realm, is a wild man, half beast. Twelve tablets, 3000 lines (about half of the epic) were found in the library of Assurbani-pal at Nineveh. In addition, a remarkable statue of Gilgamesh himself was discovered at Khorsabad, Iraq, and dating from the eighth century B.C. With his whip, pet lion, and Odyssean bearded locks, he is the sovereign melancholic: King of Uruk, but unable to attain his true desire, which is to reach paradise, where there "was no widow." Alexander the Great inherited this hubris.

Alexander fought his most hideous battle in Anatolia, at Arbela, with 47,000 men in phalanx, wearing steel-ribbed skirts, greaves, and breast-plates. They marched in one morning, prepared to do battle with nearly 500,000 Persians. Alexander's troops crunched through the lines. The Persian commander Darius despaired and retreated, leaving a blood bath along the way. From Arbela, there was no stopping the young student of Aristotle. He went all the way to Taxila, in India, where he encountered Hindu ascetics, gymnosophists who apparently softened him and sent him home. By that time Alexander's men were tired of war, on the verge of mutiny.

Thales, founder of the Ionian school of philosophy, was born in Miletus some 300 years before Alexander. Thales believed that the Earth was made entirely of water and that its foremost property was its living soul. Consider the Babylonian Enki, also water, but not very friendly (male, aggressive, inflexible). Thales first conceived of the gaia principle, of an entire planet functioning as a living organism with needs and frailties, even its own dreams for the future. The Earth was not only alive, but it was intelligent, a vast animal with a soul of its own. The Greek belief in the physical, intellectual, and religious unity of all life bespeaks a brilliance out of kilter with its military record. In Aristotle's biology, throughout the passionate corpus of philosophical nature poetics that includes the writing of Panaetius and Posidonius, and their successor in the Roman Lucretius, there is ample evidence that the Greeks, at heart, felt good about life, believed in the pastoral ethos, in farming, and in the lofty interconnectedness of all things, animate and inanimate.

But there was a problem that one of Aristotle's students, Dicaearchus unconsciously enunciated. In the *Life of Greece,* Dicaearchus referred to Greece's Golden Age. Herodotus and Hesiod, as well as Plato, had similarly — after Homer — looked with yearning to the Hesperdes. European civilization would in turn peer back towards Arcadia (a central Greek mountainous region), its epitomized image of paradise. Dicaearchus did not find much interest in the Arcadia of his own time. Quite the contrary, we get the name of Megalopolis, with its disturbing, overpopulated cacophany, from Arcadia's very own center, its largest seat of political power; a seat which never set well with the indigenous montagnards. In Dicaearchus' glance backward, he recognized a time when men were much happier, because they did not desire all the fruits of *excess* labor. (Note the comparison to Wallace's feeling of evolution having conferred *excess* language on the mind). Dicaearchus felt fondly toward that prehistoric dawn of calm, when aspirations were few, there were no wars, the wolf did not rend flesh, nor the lion devour its prey. But from that time on (again, squarely, the vision is not actually fixed in *time*, but in space), according to Dicaearchus, the pastoral mode came into being. Animals were captured against their will, the women domesticated along with the Mesopotamian onager (wild asas), and all leading inexorably to agriculture and the stripping of Earth.

This critique went against the grain of Apollonian ethics, of the graceful sheepherder with his windwood instruments. But agricultural prose writers would question practices that so fervently set into operation a host of tilling, plant breeding, plowing and drainage, manuring, even insect control techniques. The destruction of the foothills and littoral of the Mediterranean, of the Syrian plateau, of vast regions throughout the Near East — in Iran as well as in Sinai — resulted from the Neolithic shift to agriculture. Juniper forests in Persia were destroyed at the same time that Hacilar II in Anatolia was burned to the ground, 7000 years ago. Erosion produced an untenable silt load in the

Tigris and Euphrates, which to this day persists. Vegetational destruction, dry farming, charcoal burning, massive land clearance and cultivation, and the general uncontrolled abuse of the now bald mountains and foothills throughout so much of Greece attests to the nervousness of Plato in the *Critias* (110C-111) and later of Cicero, who spoke out against topsoil erosion.

But the tension and discrepancies in the Greek record – from the god-tanned Aphrodite of Praxiteles to Xenophon's fascinating, if painful discussions of the early Peloponnesian War – have a deeper meaning for us. Yes, the Greeks were human after all, with both masks, and we should be gratified that they managed their particular achievements, which are so ingrained in all of the world's philosophical heritage and self-consciousness; in the very nature of religious metaphor and animism, of perceiving cranky, human, felicitous dieties. And we know that the Greeks were modest, capable of much greater ruination.

In 75 A.D. the Greek inventor Hero made a toy steam engine that he called an aeolipile. It consisted of a hollow metal ball filled with steam from a fire underneath. When the steam was forcibly expelled through simple vents, bearings were made to spin. Anticipating the third law of Newton, that for every action there is an equal and opposite reaction, Hero took his minor breakthrough absolutely nowhere – which we should find provocative.

In seeking a truer answer to the Greek dialectical nature, we come to the most revered wellspring of ripostes in all the ancient world, to Phocis at the foot of the south slope of Mt. Parnassus at Delphi. The Delphic Oracle gabbed generously, told all. Ascribed to Apollo – god of light, poetry, music, prophecy, shepherds, flocks, beekeepers, and wild animals – he was a beautiful young man, played the lyre, carried a crook, and strolled the mountains down. The god of healing, father to Asclepius, he was eventually identified with the Sun, was androgynous, and inherited the oracle from – who else? Gaia (gaea). Apollo speaks for the Greek personality: wildness tempered. In the sixth century B.C., the oracle was delivered by a priestess seated on a golden tripod. Called Pythia, she went into trance and uttered her frenzied sibilants. In the political fragmentation of Greece, all paid serious attention to Delphi. Dionysus was buried there, wildness in his person having united with the nonchalant rationality of sun-drenched Apollo. But in the seemingly flawless mating there is confusion, discord, deforestation. And it stemmed from farming, the over-domestication of Apollo, the castration of wild animals, the Earth sacked. It is important to isolate the culture, with its lovely olive shepherdesses, gleaming white Acropoli, and windswept porticos; its Sounions and Panathenaic frieze and Apollo Belvedere at the Vatican, from the slaves kept, the incessant massacres, and long-time recognition from its writers and scientists that something important was missing; that each increment of harmony was shadowed by an earlier, more enriching sunlight. Later on, we will see how Aristophanes tore his hair out trying to keep a sense of humor against the backdrop of numerous Greek internecine warfare.

Paul Shepard has acutely psychoanalyzed the farmer.[6] His conclusions bear powerfully on that integration of opposite persona and behavior in Greek mythology. Shepard believes that the farmer's world view is dualistic – loving nature while hating nature. It is dangerous, self-demeaning, and has led historically to repression, sexual anxiety, even schizophrenia. The farmer applauded these same denials in the hunter, whose lifestyle Shepard would have us reawaken in the modern world.[7]

In early agricultural societies, a woman commonly bore a dozen offspring. She thus had less time for her many progeny. The child's all-important connection with his or her mother suffered as a result. Over the years, the child felt this important loss and would turn to a surrogate – nature. But the information from nature had already been shackled, penned, and herded. Even the slaves, feudal laborers, had lost their humanity in the child's eye, had become mere objects for use in conquest. The young person began to suspect himself of being made, Shepard goes on to say, as his world and village was made. Earlier hunter-gatherers had examined nature and extended the metaphor of totemic taxonomy to human interrelationships – mythopoetic marriages, ecological food chains, sensitive connections. But the farmer, equally intense, put all of his energies into the future, for he was endlessly accountable to tomorrow. He hoarded his seeds, traded in chattel, derived power from the enforced bondage of plants, animals, and fellow humans. By breaking the metaphor with the good Earth, he also broke down the age-old covenant between his body and everything feminine in the world. This left him out of touch, distrustful of his sexuality and manhood, paranoid with power, and unable to gain succor from his mother. Schizoid, driven, his obsession with power palliated all other grievances. Power characteristically is polarized between good and evil, strange and familiar, exploiter and exploited. God is forced into the position of outside creator, as opposed to indwelling spirit. Vesalius' model of the human being as a machine – Descartes' smug manner of torturing animals to the delight of his friends on the shared assumption that animals could feel no pain – sums up the dilemma that the farmer encouraged. Man had become master, workhorse, motherless, barren, despite his annual harvests. Where hunter-gatherers had perceived species as interdependent, the farmer knew himself to be uncongenial, wringing every radish, every possible bit of green chard from his turf. For thousands of years he has missed out on the truer connection. His dualistic thinking is further complicated by the fact that he *is* in continual relation with the Earth, as a farmer, and therefore appears to have dirt on his hands. Of course he has dirt on his hands. But what Shepard is suggesting is the deeper psychological dysfunction dating back to man's earliest fantasies and fears, his abilities to subordinate plant and animal life, to make them work for him, to bolster his ego until he is ready and eager to destroy the world if need be. This is basic immaturity. It carries the gravest of consequences.

We sense that a sedentary farmer will be less dangerous than the nomadic hunter who can ride his horse, descend steppes, pillage; but recall that Cain is the farmer, Abel the mountain pastoralist. The gates of the palace of Ashurnasi-apal, the ruler of Assyria from 885 to 865 B.C. tell the story: He was a conqueror who took territory by greed and madness all the way to the Mediterranean. But he also established centralized government. The gates show carved warriors with eagle's heads. But they also carry spathes for fertilizing date palms by hand. The Assyrians were among the first to recognize plant sexuality and the techniques of hand pollination.[8]

Hellenistic writers such as Lucretius and Vergil expressed the alienation of man from nature, as cities grew immense at the expense of slaves. They shared deep-seated reservations but ultimately applauded the human capacity to better nature through its modification. The tools of this changeover included a variety of spades, anvils, shears, the beam press, and a three-horsepowered hydromula, the Vitruvian mill, enlisted for grinding grain all across the Roman Empire. While the Romans terraced, irrigated, cut forests to build ships to fight the Carthaginians, overgrazed, they were forced farther and farther out from their center, as had the hunters been at Catal Hüyük. Four hundred thousand slaves were kept during the period of the Emperor Augustus, when Rome had a total population of nearly 1,000,000. Without those slaves, the aqueducts could not have been built; the road to Ostia, over the Alps, the great ships, the extensive agricultural complexes — none of it would have arisen. The Roman disquietude between technology and countryside went well beyond that of the Greek. Flamboyant interest in landscape architecture, decorative planting, tree-lined promenades, even natural enclaves within cities, promoted the first gardens with sculptural edifaces — little boys pissing into pools — at a time when the surrounding wilderness was sternly oppressed. We should not be surprised by the ratio. The dualism neatly defines mind as it has grown up in nature, grown up unchecked.

The purpose of creation — whatever it may be — has been fraught with discord, more virulently since the advent of human machines than in all preceding eons. Basic questions have persisted. They are queries that cannot go away. Clarence Glacken has brilliantly recorded Western civilization's experience with that dualism, from Mesopotamia to Boston.[9] Let us follow its important highlights, encapsulating the course of Western slave–master ethics, technofix, transitory politics, and nature mysticism as these elements have come into conflict.

In seeking to understand the human preoccupation with mind over nature, Glacken examines the hierarchy of hubris and its excuse in human technology: the farmer is keener than his cows, who, in turn, have it all over the daisies. According to this hierarchy – applauded by Socrates – man is the chosen one, the measure of all things. The belief in man-as-paragon reached its apogee during the eighteenth century in Georges Buffon's massive *Histoire naturelle (Natural History)*.[10] Buffon wanted to know what mankind had really done to the Earth. He divided the planet's history into seven epochs. The seventh epoch was reserved for man's *seconding* of nature. So convinced was Buffon of the human catalytic agency in the Old World, that he decided that America must be a grim place, inhabited as it was by Indians who lacked the refinements of European civilization. If Indians were uncouth, their environment would be so as well, impugned the natural scientist. When Thomas Jefferson visited Buffon in France to protest the scholar's treatment of American wilderness, Buffon said "Prove it!"[11] Jefferson shipped off the whole lot of a giant moose to Europe, thus forcing Buffon to take note of the American wildlife in all of its magnitude. The naturalist had assumed that the French reindeer was king. Jefferson's evidence (homely though it must have been) startled and won over Buffon to American landscape. Alas, he died before he could rewrite his forty-four volumes.

Socrates had explained to his friend Euthydemus that what the gods had created in all of their glory, they did explicitly for man and man's comfort alone. Another of Aristotle's students, Theophrastus, was not so sure. Invoking an agnostic scientific method, he set out to prove that there was order (God) in the seeming chaos of nature and that man was merely an inconspicuous fragment of that coherence. The Sumerian Enki, the Egyptian gods Amon-Re and Aten, and Plato's Demiurge had each engendered in the world a very special biological diversity that grew up (in spite of mankind), enjoyed myriad alterations on its own accord, then perished. God and nature were one, a cyclical unison, organic and lyrical, the very stuff of all human culture. By Hellenistic times, this visionary solution could be obtained only in rural life. The city had betrayed it; the farmer was on his way to doing so. The earliest Greeks had been hopelessly ahead of their time with optimism. Epicurus seized on thermodynamic laws, Aristotle legitimized the science of biology, remarking upon over 500 species. Plato emphasized the beauty in nature, referring to yet higher callings that the everyday blossom reflected. The corpus of Hippocratic medical philosophy had, by the fifth century B.C., made marvelous connections between the health of the land and of human beings.[12] The difference between cultures was perceived on the basis of environmental determinism, making such logic one of the oldest inherently scientific doctrines in Western civilization.

Extending Plato's critique of deforestation ("Tearing the soil of her, year by year") Lucretius Carus, a Roman poet born in 95 B.C., recognized in his fantastic poem *De rerum natura* (The Nature of the Universe) that the Earth, like man, was mortal.[13] His near contemporary Junius Columella wrote a twelve-volume agricultural treatise [*De re rustica (On Agriculture)*] linking mother earth to a human mother, setting this comparison down in an "equable law of fertility." The emphasis on arable land in Vergil, its dangerous conversion into grazing lands cited in Cicero, all contributed to the Roman mission: horticulture, irrigation, the design argument. This emphasis was fully opposed to Dicaearchus' earlier schema, in which domestication was perceived as the first reversion from the golden age, the zenith of intelligence turned inward. But was not the taming of wilderness the very same task of God's first caretakers in Eden? And by extension, had not God done the same thing, creating light and beauty and order out of cold chaos? Cicero, for one, wondered about it. Was there purpose or was there none? Was mankind the microcosm, a feeble, secondary spin-off of the creation, or did he embody the whole pastrami sandwich, its gravest principles? "In fine," wrote Cicero, "by means of our hands we essay to create as it were a second world within the world of nature."[14] *Psalms* 1:24, 104 and *Romans* 1:20 made it clear that not only was God behind everything, but that – true to form – he had implanted purpose in it, with man as his gardener. There was wisdom, the gift of domination, and everywhere in nature His invisible love was manifest. The *Hexaemeron* of St. Basil (ca. 331–379) and the succeeding songs of St. Ambrose furthered this acknowledgement – of little consequence – that nature blushed with balance. St. Basil was convinced of the "fitness" of life and intimated principles of ecology in his songs. The sea kisses its beholder; the smallest plant shouts out the Creator's name.[15] Basil was the Greek Walt Whitman. Augustine carried the commentary on Genesis 1:31 ("and behold, it was very good") by inciting that "their loveliness is their confession." Henri Bergson or George Santayana could not have stated it with more transcendentalist conviction. Lucretius had believed in the inalienable rights and self-consciousness of animals. And in his *Canticle of Brother Sun*, St. Francis of Assisi (1182–1226) poured out his empathy, humanizing birds.[16] He made a covenant with a wolf at Gubbio, and later, at Cannara, all the swallows shut up until he finished his sermon. Glacken's predominant concerns in tracing this history comprise three main theses: the argument of the designed Earth, the notion of environmental impact on culture, and man's role as modifier. These are symptoms of a mixed motive that tell a long story of mind reshaping its original intuitions, while always clinging, if subliminally, to symbiotic insights voiced at the beginning of time.

The Roman sense of nature is witnessed in the writings of Catullus and Livius Andronicus, in Rome's emulation of the Persian garden with its *chahar-bagh* ("four gardens"), narcissus blooms, sheared hedges, terraces, and fountains. The Muslim Arabs who conquered Persia imported these formal effects, canals and water jets from oases in the Orient that would turn up in Frascati, at villas on Lake Como, and in many of the decked-out estates surrounding Rome. Chinese influence was felt in other ways. There had been trade, exploration, and the general influx of aesthetic sensibility from Asia. Hippalus in 79 A.D. learned from the Arabs of seasonal monsoon changes that in turn allowed him to exploit the wind to fix shipping routes between the Red Sea and Indian Ocean. All previous transcontinental forays had hugged the shore. And under the Byzantine emperor Justinian, two Nestorian monks traveled to China and back, the first documented Westerners to do so. Many others were in transit – soldiers left over from Alexander's swath across inner Asia.

The earth goddess Gaia was ravaged by the madman Nero, whose forces destroyed Delphi. This was ironic, because in Rome Gaia's own incarnation held magnetic power over the entire body of pagan beliefs. She was Cybele, the Anatolian great mother of the gods, her holiest of cities – Pessinus – not discovered until 1834. But her cult had been introduced into Rome in 204 B.C. The Phyrgians who originally worshipped Cybele in the ninth century B.C. were fixed in the geographic and racial vortex of those forces that had first ordained, in Paleolithic times, the supremacy of the female. Cybele, born of cliffs in Anatolia not far from Catal Hüyük, was both Asian and Etruscan. The Phrygians – inventors of cymbals, flutes, triangles and pan-pipes – passed on to Greece and Rome, the myth of Arcadian prehistoric origins, of hunters who had dedicated themselves to Artemis and her companion Pan. But the original dyad was Cybele and Attis, a maternal wild couple with the male self-sacrificing element of castration and renewal. Dionysius and Elysium, the dithyrambic pageantry of bulls, self-mutilation, half-demonic corybantes, and the re-creation of fertility chains: Such was Cybele's inheritance, celebrated in Rome during the Ides of March. The Romans marched along the Tiber River with a silver statue of the goddess and the sacred stone of her consort. Lions proceeded them, drummers went in the rear. Dancers upstaged the solemnity by ripping into their own flesh with sharpened nails. Cybele was the last remnant of the Catal Hüyük goddess, perverted by five millenia of male-dominated politics, her urgings skewed, neurotic, no longer translatable. All-night orgies were as close her celebrants could come to reliving her power. The rest of the year, Cybele had become the Virgin Mary.

"Wheresoever was Eve, *there* was Eden."[17]

NOTES

[1] Karen Watson, "Neanderthal and Upper Paleolithic Burial Patterns: A Re-Examination," *Mankind* (1970): 302-306.

[2] Sally Binford, "A Structural Comparison of Disposal of the Head in the Mousterian and Upper Paleolithic," *Southwest Journal of Anthropology* 24: 139-*154.*

[3] Leroi-Gourham, *Treasures of Prehistoric Art* (New York: Harry N. Abrams Publishers, 1967); and Alexander Marshack, *Roots of Civilization: Cognitive Beginnings of Man's First Art Symbol and Notation* (New York: McGraw Hill, 1967).

[4] William Irwin Thompson, *The Time Falling Bodies Take To Light: Mythology, Sexuality, and the Origins of Culture* (New York: St. Martins, 1981).

[5] ibid.

[6] Paul Shepard, *Nature and Madness* (San Francisco: Sierra Club, 1983).

[7] Paul Shepard, *The Tender Carnivore and the Sacred Game* (New York, 1973).

[8] Daniel Botkin, "Of Moose and Men," mimeographed (University of California – Santa Barbara, Environmental Studies Program, 1981).

[9] Clarence J. Glacken, *Traces on the Rhodian Shore: Nature and Culture Century* (Berkeley: University of California Press, 1967).

[10] Buffon, Compte de Georges-Louis Leclerc, *Histoire naturelle, génèrale et particulière* (Paris, 1749 – 1804).

[11] Cited in Glacken, *Traces on the Rhodian Shore.*

[12] In fact, one of the most famous Hippocratic treatises is called "Airs, Waters, Places." Hippocrates may or may not have written it. [*Works of Hippocrates*, trans. W.H.S. Stone (Cambridge: Harvard University Press, 1948)].

[13] Lucretius, *The Nature of the Universe*, trans. R.E. Latham (Harmondsworth, England: Penguin, 1951).

[14] Cicero, *On the Nature of the Gods*, trans. H.M. Poteat (Chicago: University of Chicago Press, 1950).

[15] Basil the Great, "On the Hexaemeron," in *Patrologiae cursus completus, Series Graeca*, ed. J.P. Migne (Paris, 1857-1866).

[16] St. Francis of Assisi, "Cantique de moine soleil" (Canticle of Brother Sun), in *Les opuscules de S. Francis d'Assise*, ed. U. d'Alencon (Paris, 1905).

[17] Mark Twain, *The Diaries of Adam and Eve.*

PART 2

Coping with Abundance

UNITED NATIONS

Chapter 4

The Journey to Cambaluc

The world's most intriguing and perilous detective story began throughout southern Sumeria and continued in the T'ien-t'ai mountains of China, and the Echizen peaks of Japan. It encompassed the melancholics of St. Catherine's Monastery in Sinai, and a galvanized array of Renaissance explorers. In the end, the world was left with a compelling image – of Heaven on Earth – painted by a Persian mystic in the early 16th century. But having captured the ecology of immortality, Sultan-Muhammad, like Lazarus before him, took a vow of silence and disappeared.

We enter the dense woods behind Kweilin in southern China. A waterfall plunges from clustered karst limestone fissures high in the mountains. I carry red silk, peach wood, a collection of herbs, and wear an old mirror strapped to my back, the reflective portion facing out to scare would-be assailants, demons. There before me is a Chinese cromlech, an ancient stone with a vertical array of carvings. The words read *Ching Shan Hui Ching Tzu* meaning "river mountain meet view on this place." The sacred configuration endows the spot with religious power. It is here that the weary traveler is urged to stop, rest, and take in the subtle essences of nature. The mountains, the river, the trees – all are considered inviolate in Chinese tradition. The human body is its microcosm, through which the five elements circulate. There are human birds in the trees, *hsien jen,* immortals who consume gold and cinnabar, perform esoteric alchemy that captures the energy of the Yang particles soaring up to heaven and the female Yin particles plunging to Earth. Everything we see, if properly prescribed and penetrated, will bring deathlessness. Ancient Taoists wore such a mirror when entering the mountains.

> *Men have always spoken and will always speak of the beauty of mountains and streams. High peaks that go soaring into the clouds: translucent torrents, clear to their very bottoms, flanked on either side by cliffs of stone, whose five-fold colours glitter in the sun; green forests and bamboos of kingfisher-blue, verdant through every season of the year. As the mists of dawn roll aside, the birds and monkeys cry discordantly. As the evening sun sinks to rest, the fishes vie at leaping from their deep pools. Here is the true Paradise of the Region of Earthly Desires. Yet, since the time of Hsieh, no one has been able to feel at one with these wonders, as he did.*[1]

This was written in a letter to one of Hsieh Ling-yün's relatives by the venerable Taoist hermit T'ao Hung-ching in the early sixth century A.D. Hsieh, born in 385 to one of the most powerful families in all of the Six Dynasties period, was to become a legendary figure. He was poet, farmer, scholar, mountain climber, and founder of the *shan-shui* aesthetic, meaning mountain/water, the most characteristic combination throughout all oriental art, poetry, religion, and philosophy. Shan-shui, from aboriginal times, was an ingrained way of perceiving the world, a concerted effort to morally *become* the resplendent view, the periphery of domestic life where wilderness set the perfect example for times past, the everyday, and the hereafter. In tropical Asia, the very basic root crops were more in harmony with the inward, passive mysteries of the feminine, shan-shui principle than were Western cereals. Look at the five most critical staples in the world today: corn in Mexico, the potato in the Andes, wheat and barley in the Near East, and rice. It is this latter grain whose paddy soaks and fosters the rich muck of original creation. And it was through this interaction with rice that the Chinese developed their close relationship with the Earth and the rain clouds, incarnations of Buddha.

Hsieh Ling-yün's family moved to southern China when the young Hsieh entered into diplomatic affairs. But he was greatly dissatisfied with government, preferring the out-of-doors. By 422 he had gone into exile, written a major piece on satori (instant enlightenment), and visited his grandfather's rolling estate, taking an active interest in gardening. His was not the normal removed interest of the prevailing Confucist bureaucrats, the wealthier of whom kept mountain retreats with trains of servants. Rather, the Taoist and Buddhist curiosity bade him take off his shirt and toil in the mud. In the spring of 422 Hsieh invented the first hob-nailed boots (crampons), which he used to climb the most slippery of mountains. Apparently, he gave a pair to the ascetic T'an-lung, whom he encountered on one of many hikes. The ascetic disappeared atop Incense-Burner Peak at Stone Gate and vanished for 6 years. Hsieh was fascinated by the notion of such disappearance and sought to emulate it. But his quest was cut short by the emperor who pursued and persuaded him to take an office. A few years later he was gone again, carousing, writing poetry, and farming with some seriousness. He was convinced that the landscape was the most perfect embodiment of Buddha. Landscape contemplation thus constituted for him a religious technique, bringing him into the body of the Dharma or law itself. Toward the end of his life he fell victim to unsavory political maneuverings and a jealous governor and was sentenced to death. On the eve of his beheading he wrote, "Thick and green the cypress, heedless of the frost, Soaked with dew the mushroom, suffering in the wind. What does a happy life amount to after all? I am not troubled by its brevity." On the way to execution he cut off his goatee and presented it to a monastery. His body was buried among the Kuei-chi mountains he so loved.[2]

In the eighth century, the T'ang poet–hermit Han-shan (meaning Cold Mountain) lived with his minstrel friend Feng-kan for 30 years in the T'ien-t'ai mountians of Chekiang. Han-shan was married, but one day – the Gauguin factor – he just packed up and walked into the hills.

> *At my ease, idle among white clouds – do you have to buy a mountain to enjoy it? Valley streams quiet, limpid and clear – joys and delights that never end! Bright moon shining, white clouds all around, sitting alone, one old man, no life-and-death for him!"*[3]

Han-shan's character would be adopted by the Zen Buddhists of Japan who saw in his lifestyle the Pure Land of Buddha.

One such Zen master was Ju-ching. He made a 40-year pilgrimage to all the sacred mountains of China, flirted with enlightenment, and returned home. His student Dōgen sailed to China, studied at a mountain monastery, and one day asked the cook, "What is the practice of Tao?" To which the cook replied, "Nothing is concealed throughout the entire universe." Dōgen returned to Japan and eventually settled in the Echizen Mountains. He lived in the Eiheiji Temple (Temple of Eternal Peace) and died in the posture of zazen, or tranquil meditation, August 28, 1253. For Dōgen, as for Han-shan, man and nature shared an identical destiny, a "radical love" as Dōgen called it. Writing from Eiheiji, Dogen composed the *Shobogenzo*, the most remarkable piece of religious speculation on nature in the history of Japan. The 29th section of the work, Sansuikyo, or Mountains and Rivers Sutra, is the clearest expression from a Zen master of what he sees when he scrutinizes his environment. We become lost in a blue physics, of mountains moving over the face of the waters, all metaphor igniting with true perception. Dōgen, founder of Soto Zen, asks us to enter into nature and be transformed.

The poets, painters, and court savants of Dōgen's time rejected technology for a more passive, religious kinetics. The Taoist term *feng-shui* relates to the art of adapting the residences of the living and dead so as to foster harmony. One of *feng-shui*'s attributes is its distaste of geometry and over-complexity. When China's first railway was completed in 1876, a local was killed in an accident with the locomotive. The old *feng-shui* sentiments ran high. Within a year the government gave in and abolished the train.[4]

But the Chinese always had technology at their beck and call and well in advance of the West. The piping of natural gas, collar harnesses, magnetic compasses, stern rudders, seismographs, silk reelers, paper manufacture, torque — these were but some of the medieval advents in China. Yet such devices had nothing to do with the Six Noble Arts of a gentleman, nothing whatsoever to do with popular culture and the abundant reverence of nature that prevailed in every small detail of common life.

Soot was required for the black ink that fueled the noble art of calligraphy. Soot came from burnt pine. We are told that even before the T'ang Dynasty in the early Middle Ages, many mountains had been reduced to carbon. If the Mayan, Khmer, and Teotihuacan cultures suddenly vanished as a result of lime burning and the attendant deforestation, the Chinese gentlemen, through their very art, contributed to the destruction of watersheds. By the third century B.C. the *Chou-li* manuscript described two types of officials: one was charged with conservation in mountains, and the other with conservation in forests. Lao-tzu, Chuang-tzu, and Mencius — China's premier philosophers — all cautioned against tampering with nature. Officials themselves, throughout China's history, have been aware of the dangerous consequences of deforestation. Yi-Fu Tuan records the words of a Ming Dynasty scholar who reported on a flood in north China.

> *At the beginning of the reign of Chia-ching (1522-66) people vied with each other to build houses, and wood from the southern mountains was cut without a year's rest. The natives took advantage of the barren mountain surface and converted it into farms. . . . If heaven sends down a torrent, there is nothing to obstruct the flow of water. In the morning it falls on the southern mountains; in the evening, when it reaches the plains, its angry waves swell in volume and break embankments causing frequent changes in the course of the river.*[5]

The Yang is male, cosmos. His are the regular patterns of the stars and solar bodies. His designs have forever been translated into Chinese architecture, following the prototype of Yu, legendary founder of the Hsia Dynasty. Yu "opened up the rivers of the Nine Provinces and fixed the outlets of the nine marshes." All subsequent rulers had a rather hard act to follow, in the absence of their own army corps of engineers and especially because of the resistance put up by rural Chinese farmers to massive constructions. But Ch'ang-an, capital of the T'ang period, and Hang-chou, capital of the southern Sung, managed to employ a monstrous labor force. Both cities eventually swelled to over a million. Their expansions resulted in serious forest depletions and resource draw-downs in the surrounding hills. Many farmers actually gave up rice cultivation for forestry. Losing touch with rice proved nearly disastrous in 1012 when drought struck China. Millions of lives were saved only when an imported drought-resistant strain was imported from Champa in Indochina. The rice matured quickly; new strains were explored with zeal. Several hundred years later, a 30-day maturation seed was worked out. Rice planting once again dominated many of China's upper river basins. Yet another cycle of soil depletion ensued, resulting in the floods of 1981. Millions of families were left homeless.

The construction of Ch'ang-an during the Sui and T'ang dynasties reveals the partial triumph of the Yang over the Yin, a victory that Mao Tse-tung effectively reaffirmed in his day. With unprecedented scale, this new capital – near the modern Zi'an – was selected on the basis of historical and astrological merit. The North Star and the Sun's shadow were measured on successive days, walls were aligned to receive 22 miles of moonlight. Villages and trees were uprooted, rectilinear avenues laid out, and the trees then replanted in strict line. Man's will had been imposed, with the sanction of heaven. The Chinese always did things *properly*. A single old locust tree, in whose shadows the chief architect of Ch'ang-an reclined, overseeing demolition and construction, was spared on special order of the emperor to honor the architect. Such tokens are ingrained in that age-old disparity that keeps the artists and the politicians at odds. Fortunately, bones speak. Ch'ang-an was built to be a paradise on Earth. And medieval Europeans wanted to believe it as much as the Chinese.

St. Catherine's Monastery in the Sinai Peninsula was constructed early in the sixth century. Behind it is the legendary mountain Jebel Musa. Though no one knows for sure, thirteen other mountains throughout the Peninsula have been identified as possible candidates for Mt. Sinai. It was here, down through the Renaissance, that ascetics and pilgrims came to pay homage to an idea.

In the Monastery's central courtyard, a single water source issues, driven from an underground spring that originates within the holy mountain. Adjoining it is the Basilica of the Transfiguration of Christ. In its gleaming arched tessera, Christ's arms extend through the cool, sloping cupola. Cypress and date palms sway in the garden – an island in the desert. Until recently, a basket and pulley hoisted visitors up the thick granitic battlements into the dank abbatial dens. Astronauts trained nearby for moon landings in the early 1970s.

It has been said that the Greek saint contemplates, and the Western saint acts. A comparable equation, more mythological, says that the man acts, but the woman *is*. Both insights will crack open whatever nut you happen to have in hand. The Greek Church produced no outstanding technological innovation after the seventh century. Its professed duty was to inner prayer, renunciation, and the fashioning of such cloisters as this haven. Consider the desert ascetic, idling away his crusty years in stern devotion to a concept, implanted beside cactus. Compare this with White Benedictine monks in Europe who vanguarded new power sources such as water wheels, which they entrusted to the virgin of their church. She had not been endowed with so much energy in millennia. The Biblical interest in technology derives from Solomon's fascination with the natural sciences (Kings 1:13), from the farmer's disposition in Isaac (Genesis 25:12), and from those rampant archaicisms in Hebrew – verbs such as *rada* and *kabas*, meaning to "tread out" as with a wine press and to "have dominion"; verbs of conquest and construction. As for Nimrod, the "mighty hunter," throughout Genesis he is only a peripheral character, having failed to build his Tower of Babel, emasculated on the very pulpit in progress. His earlier, Paleolithic grandeur had been converted to a more anemic, static agenda, namely, "the sons of men [Psalms 115:6]." Genesis assures them of "order, regularity, and permanence in nature." The Lord gives man everything: "Every moving thing that lives shall be food for you [Genesis 9:3]." And "Thou hast given him dominion over the works of thy hands; thou hast put all things under his feet [Psalms 8:6]."

The power granted in the Bible was the power of religious discipline overcoming the wilderness into which all struggling evil and dutiful penance had been rudely cast. Transforming that wilderness into the ideal of heaven became a spiritual obsession with early Christianity. The starker the environment, the greater the challenge. These original desert fathers were not seeking technological pacification, but difficult enlightenment.

The deserts of the Middle East received a rush of misfits prepared to rebuild paradise in their hearts. Many of them were remarkable fellows. Yet unlike Taoist precepts, the desert fathers' ideology was guilt-ridden, audacious. To suggest that the world was not already perfect was insanity for the Taoist. But both doctrines – Eastern and Sinaitic – converged with gusto on the *idea* of paradise. Ironically, European Christians, tired of the Sinai by the fifteenth century, would look to

China for their heaven. It was simply more luxurious. But not in the beginning.

The early Church contained an atrium, an open vestibule called paradise. The earliest monasteries had paradise gardens. The history of early asceticism in the Middle East combines all of the elements – the allure, terror, inaccessibility, sanctity – of the wilderness-paradise syndrome. The rock monasteries of Cappadocia were inhabited in the Middle Ages by as many as 40,000 monks. The Eastern Gnostic Symeon Stylites, born in 390, posted himself atop a column 40 cubits high and lived there for 37 years. His followers were entranced with height. The thirteen granite mountains of Sinai simulated the very height of heaven, figured in the icons and illuminated manuscripts, and played essential, metaphoric roles in the monastic rules. I spent several months living in a cave beneath Mount Sinai and witnessed monkish ascents at all hours, regular meditations carried out on the exacting tallus slopes. Later on, the Christian looked towards the Himalayas. El Greco's fantastic rendition of St. Catherine's Monastery merges the two latitudes, lends otherworldly prominence to the desert peaks, reflecting this altitudinal substitution.

The ascetics shut themselves up in caves for their entire lives, passed food up and down pillars by rope, stood up for unremitting years, tied stones tortuously to their groins, chewed salt, lived amid herds of mangey deer to stay warm in the snowy winters, plunged in to mosquito swamps for punishment at other times, and went everywhere naked across the stubbly land. The fourth century Era of Retreat attracted as many as 20,000 die-hards to the deserts of northeastern Egypt.

The Greek *History of the Monks of Egypt* relates a story of one Marcarius. A lover of the desert beyond all others, he explored every facet of its waste and lived at Scete, a two day's walk from Nitria. Scete is the hottest, the worst, the ugliest, the holiest of epicenters for self-mutilation in all of Arabia. A certain Brother carrying a gift of grapes trekked in to visit the wise man. Marcarius graciously accepted the fruit, which had by then rather dried up. He held on to it, eventually passing it on to another pilgrim. The grapes passed from hand to hand over decades, finally returning to the original, if wizened pilgrim. These dried-out little pits of their former selves were precisely representative of what the ascetic had in mind for himself by way of leaping the confines of this life for some better one up high.

St. Paul and St. Anthony exchanged notes at Der el Memum around 340. Both relished cave life. But a new, monastic tendency was at hand: paradise socialized. St. Basil introduced the concept on the shores of the Black Sea. St. Benedict began his own career in a cave in the Sabine Hills around 500. Disciples made his life more miserable than it already was. Frustrated, he went first to the Vicovaro Monastery and later to a retreat high on Monte Cassino. Rules of benediction and abstinence hung ponderously over each institution. Orders sprang up across Europe – Gray Capuchins, Brothers of Death, Hermits of the Order of St. Paul, *girovagi* (wandering hermits), and sarabites (those who lived in small groups of three). Lonely cave dwellers, the begging solitaire, and the city monk with his immense library and garden club, came to offer medieval Europe something of a heritage, a half-way house on the way to heaven.

From the time of Genesis, western paradise was situated on the eastern-most limit of the world, populated by a beastiary, between the known and unknown, beyond the Don River. Isidore of Seville's *Etymologiae* (seventh century) inherited centuries of myth and marvel, quoting Augustine on the fabulous races of Asia. A high barrier was said to prevent all outsiders from reaching the Asian paradise. The peaks of Sinai, and the wild cats and barbarous human apes said to scavenge on pilgrims there, satisfied this Eastern geographic lust – at first. But maps began placing Adam and Eve atop a revitalized rim, the true heaven. It was not the ascetic's private wilderness over the hill from the Nile, but an unimaginable place encircled by tumultuous seas, obscure convergences of land mass, nation state, race, and demon. The apostle Thomas' trip to India, as outlined in the Apocryphal Acts, incited the European Christian community into believing that canonical descendants were living there with Adam and Eve. In 1122 a supposed Oriental patriarch by the name of John visited Rome. Forty-three years later, a letter to the Pope addressed by this now legendary Prester John, King of the Indian Christian empire, was to touch off a 500-year determination to find the place. Henry the Navigator's brother Pedro supposedly obtained permission from Prester John to search for the Garden of Even through his kingdom. A masterful tale was woven, citing the crossing of 680 leagues of desert by camel, talkative parrots floating down the four rivers of paradise, and an insurmountable mountain range that so terrified Pedro's men that they all quit the expedition.

Translated legends of Buddha's life were Christianized into popular parable, and extensive cartographies labored over the precise whereabouts, scale, and amenities of the Indian heaven. In one geography of China, based on classical Greek heresay, Dante's own mentor Brunetto Latini referred to the Seres people, a peaceable tribe prospering beyond "immense snowy wastes." Following the Mongol defeat of Poland and Hungary in 1241, Pope Innocent IV dispatched a Polish Franciscan to seek out the Grand Khan. Friar John's subsequent *History of the Mongols* excited its incredulous readership. The Council of Lyons decided to send other emissaries to the Mongol capital of Karakorum – no easy journey, which was the whole point. Andrew of Longjumeau and William of Rubruquis made unsuccessful attempts to convert the Mongols, who were now extending their power south of the Great Wall. The Kublai established himself at Cambaluc in 1264 (Peking) and built a summer retreat a Xanadu (Shang-tu), overlooking his fabulous capital. Then the Polos arrived. They served the Mongol administration for 17 years, the young Marco actually travelling to Yunnan and regions bordering Tibet. His *Description of the World* became a minor best-seller before his death in 1324, with 119 manuscript versions today extant. All of them are different, heaping ever-increasing marvels on the legend, making Cambaluc out to be paradise itself. Polo's reference to Japan ("so rich that no one can tell their wealth . . . immense quantities of gold, golden islands") further confounded the Asian legacy. Oddly enough, the first Mongolian legation to Rome, Paris, and Gascony, headed by the Chinese Nestorian Bar Sauma, failed to gain a common understanding either with the Pope or King Philip IV and Edward I. A second Mongol mission of sixteen persons came to Avignon in 1338 and petitioned the Pope again. Benedict XII responded by sending four Franciscans to Peking. Franciscan *Annals*, the writings of Jordan of Severac, Odoric of Pordenone (the first Westerner to enter Lhasa), and Ibn Batutah (who traveled 75,000 miles before his thirtieth year and spent 8 years in India and China) added to the chaos of descriptive fantasy, legend, and observation. By the fourteenth century, enough content had been ascertained to establish other, ulterior motives for finding paradise, namely silk products, for which northern Italy constituted a voracious market. Chinese brocades and damarks satisfied that hunger. By 1400, some 200 Asian slaves – half of them Chinese – had been sold to Florentine princes, Prester John's Kingdom removed to Africa, and the terrestrial paradise confusedly relocated deep in India.[6]

Four hundred years later, responding to the industrial revolution's mania with dominating Asia, Voltaire said that "if the Indians had remained unknown to the Tartars and to us, they would have been the happiest people in the world."

The Arab world preserved Levantian trade routes with India, China, and the east coast of Africa, their geographers compiling the records of exploration we have today. But it was Portuguese and Spanish exploration that firmly established the modernist passion for such voyages. Following Prince Henry's patronage, compilations of geographical data were begun in earnest. After extensive journeys along Africa's west coast and the penetration of Sudan in the east, this king foresaw valuable trade with India and renamed the Cape of Storms after the Cape of Good Hope. In 1498 Vasco de Gama made it to Mombasa and on to India.

In 1474, Paolo de Pozzo Toscanelli pointed out that the east coast of Asia might be more easily reached by sailing across the Atlantic. It was a mind-boggling suggestion. And the Genoan, Columbus – who had already been inspired by that fateful meeting between the Norseman

Othar of Helgeland and King Alfred the Great, a friendship that succeeded in introducing to Western literature the midnight sun – sailed to Iceland and then pursued the phantom that would exhaust his life. With Spanish patronage he would make it to the Azores, then on to the West Indies. Following the Pope's suggestion a meridian line running north-south through the middle of the Atlantic would be fixed as a treaty line between Spain and Portugal, Spaniards staying west, Portuguese east.

A small, lovely Jamaican cove was about as close as Columbus ever got to paradise. Not bad. But he was after much more. Columbus wanted gold, empire, cannibal slaves (who Prince Henry had liberated on all of his own journeys), and the four rivers of paradise. More than anything else, he wanted a western route to Cathay, following the hyperbolic calculations of apocryphal documents and their frenzied cartographers. After 8 years of bureaucratic wrangling, Christopher had succeeded. He loaded the *Santa Maria* and two caravels with heavy granite balls and lead projectiles, plenty of men and food, and one Luis de Torres, a Jew fluent in Biblical languages – in the event they should land in Paradise, which was precisely the plan. On his first voyage – on all four of them for that matter – disaster was narrowly averted time and again by omens that drove Columbus harder and harder. During his third voyage, in the Gulf of Paria, women wearing pearl necklaces reconfirmed for him their proximity to the Orient. He sailed southward, came within 4 days of discovering the source of the Amazon, observed a powerful current of fresh water, and believed it to be the mouth of one of the four rivers of paradise atop the pear-shaped planet. On his fourth voyage, having overcome yet another mutiny of his disgruntled crew, he plied the coast of Honduras, traversing his imaginary map, writing to Ferdinand and Isabella that the Ganges was but 10 days away. On June 23, 1504, he found himself half-dead in this cove and wrote the most evocative letter of his peripatetic career. His dreams, hopes, wanderings, and delusions were all set down in the style of his beloved Isaiah, whose apposite passage (11:10–12) had first triggered Columbus' prophetic fabrications. But Columbus was not alone in his readings of the Bible, or of Toscanelli, Ptolemy, and Marco Polo's *Book*, as it was called. All of Europe's monarchies were stirred by the mercantilist expectations of the East, and with the East, of paradise. "There gold is obtained in great plenty, huge elephants abound, with wild trees of all sorts, and ebony; and the men are taller, handsomer, and longer lived than anywhere else," wrote Herodotus 2000 years ago.[7]

In his celebrated tenth-century translation of Boethius' *Consolation of Philosophy*, King Alfred of Britain had written, "Men followed the path of nature in strict measure. They ate but once in the day. The fruits of trees they ate and roots. They always slept out of doors in the shade of the trees; pure spring water was their drink." It was perfect, the life of natural man in paradise. Rousseau would adopt this precise line of reasoning. Throughout medieval Europe the earliest inhabitants – Adam and Eve, then Elias, Enoch, some elect souls, and the good thief Luke – spoke directly to the yearnings of the rural poor who swelled the cities in search of their own pot at the end of the rainbow, eventually breaking the back of feudalism. They wanted more out of life and were willing to stretch for it. The monks and artists had at first prompted that impetus in the *Hexaemera* and various *Canticles, chansons,* troubadourian tales, the picaresque epic. Marco Polo delivered more than mere gun powder. He brought the idea of abundance and fresh air to a teeming, beleaguered continent that was dire for a new horizon.

In 1510 Vasco Nunez de Balboa escaped creditors in Haiti, plowed on to the Columbian shoreline, and bushwacked to the Isthmian coast. In the Darien region of eastern Panama, he and forty-one survivors of a former Spanish frontier community in San Sebastian fortified the new Santa Maria de la Antiqua, the first stable settlement on continental America. In December of that year, King Ferdinand V appointed Balboa as interim governor of Darien. During subsequent explorations inland, Balboa became convinced of the "other sea." Warned of an impending successor, he set out with 190 Spaniards and 800 Indians, crossed the legendary Tierra Firme (Panamanian Isthmus) and gained a hillcrest near the Gulf of San Miguel, September 25, 1513, from where he caught a momentous glimpse of the frothy Pacific. He took possession of the South Sea, as he called it, for the Kings of Castile. Lured into a trap by his rival governor Pedrarias, the discoverer of the blue, full-bodied Pacific was condemned on ludicrous charges and beheaded with dispatch. There were other setbacks in this era. Hoping to find China via a pathway to the north, Jacques Carter found Montreal instead.

For a millennium the cartographical debate over paradise held sway. The word itself, *paradeisos*, was first intoned in classical Greek by Xenophon in *Anabasis* and used to describe the vivarium of hunted animals kept by King Cyrus.[8] It derives from the Persian *faradis*, to form around, an enclosed park. For the Greeks, Elysium, eutopia (good place) and Arcadia merged with the Hesperdes on the far banks of the Oceanus river, where life was perfect. Rabbinic and Christian commentary had perverted the mellow quadrant into a place of tense ressurection, of multitides awaiting judgment day, eyes tilted upward, while busy angels commuted back and forth along their ladders in preparation for the grand turning-of-one's-back on the Earth.

Samuel Johnson's *History of Rasselas* defined the urging to build pyramids as that "hunger of imagination which preys incessantly upon life." For John Ruskin the longing for infinity was subsumed in "no more than such a mere luminous distant point as may give to the feelings a species of escape from all the finite objects about them." The Renaissance was rent between legend, human credulity, a passion for numbers, and greed. The Greeks had similar ambiguities to contend with. Between 2500 and 1500 B.C., the Middle Eastern height of the Bronze Age, the world west of the Ganges Valley and of the Urals composed a commercial unity of trade – amber, glazed quartz faience, tin, and copper. But with the discovery of new smelting techniques in northeastern Turkey around 1400, iron came into being, offering the prospect of self-sufficiency to much of Europe. By the time of Homer, most trading and the memory of nearly 2 millennia of commerical voyages were erased. Crete was destroyed, the remnants of its people dispersed onto the Hellenic Greek mainland. By the time of Herodotus, the known world consisted of Asia Minor, Egypt, Greece, Palestine, Syria, peripheral Mesopotamia, and the frontiers of Bulgaria and Persia. The Homeric legend of Jason and the Argonauts was more likely the description of one of the last late Bronze Age trading ventures, and "golden Rhadamanthos at the world's end where all existence is a dream of ease" merely some primeval town of great wealth where merchants converged.

Every Medieval and Renaissance writer of any salt added his own opinion to the problem of location and contents thereof, but Paradise eluded them. The burden of proof was either lacking or misleading. A hot contest of mathematical theorems, providential illogic (if man survived such a plunge, the Fall, perhaps it was fun, might not he make a habit, a sport of it?), and weathered fable raged on. John Salkeld's *Treatise on Paradise* (1617) summarized the barrage of evidence, paraphrasing learned treatises by such celebrities as Sir John de Maundeville (whose fictional *Travels* of 1370 had been accepted as gospel), Sir Walter Raleigh, St. Ambrose, Moses, and Bonaventura. The location of paradise was a question of geometry: Would not the moon crash into the high Himalayas? A question of climatology: Did humans survive in the tropics? (Aristotle said not); A crisis of oxygen: Could anyone breathe at the top? Asked Salkeld:

> *What may be the reason why Paradise was never found? Why, it was the hugeness and insuperable height of the mountains, which are betwixt us and Paradise, secondly, for that there be mightie wilderness full of all kind of most venemous serpents and wild beasts. Thirdly, because there is no way but through large regions of most pestiferous aire in which no man can live.*[9]

John Prest has added fascinating botanical data to the conundrum that Eden offered the Renaissance imagination.[10] If Amerigo Vespucci's anticipation of finding the eternal homeland was roused by the discovery of certain plants that adhered to the Biblical *hortus botanica* and herbals, Portuguese sailors were not so euphoric. Since its discovery by Pedro

Cabral in 1500, Brazil had not met the zealous expectations of paradise practitioners stranded on its shores. Prest relates that by 1540 Tupinamba tribes were leaving Brazil for the West, with dreams of their own heaven. And what were commentators to make of the new profusion of plants and animal being discovered, creatures unrecorded in the Bible? Were they secretly atop Mount Ararat? Vanilla, turkeys, the anaconda, Indian beans? There seemed at first glance, to be little in common between the Indians and the Jews. What about a second Adam? Not about to be sobered, other first arrivals – the English in North Carolina, for example – were carried away by their encounters with the local aborigines whom they described in glowing terms. But a papal bull pronouncing such people to be human after all only hastened their own genocide. Sir Walter Raleigh went to Guiana himself to verify the spate of suppositions innundating Britain. His observations were not so kindly. Though he left England assured that Eden must exist in the tropics, he was not prepared for "horrible and frequent earthquakes, the dangerous diseases, the multitude of venimous beasts and wormes."[11] In short, thumbs down. Granted, Raleigh was an atheist. But others were raising the spectre of existentialism, of man without the comfort of paradise.

It was left to the *garden* to re-create what was not there. Emblems culled from throughout the world – an Asian camel, an African leopard, American parrots – and plants reproducing the medical benefits of life in paradise – balms, herbs, mints – were gathered in systems of parterres and pulvillus, units organized according to the Biblical maxims of order and diversity. At Uppsala, Oxford, Paris, Leyden, and Padua these botanical collections aroused tremendous interest. Most important was the banana tree, not easily transplanted in London, but considered the tree of good and evil nonetheless.

Most explorers sought the East via the West, but Richard Chancellor, Martin Frobisher, and John Davis were going at it via a theorized northwest passage. There was only one problem with this methodology. Instead of finding a tropical Cathay, they were stranding themselves against ice-choked inlets, cold and rainy, no palm tree in sight; not in Labrador, anyway. From 1577 to 1580, Francis Drake made a second circumnavigation (after Magellan), asserting that the Atlantic and Pacific oceans indeed met up. On his return via the Philippines, feeling that same old discouragement for having missed Cathay, he headed southwest rather than continuing north as he should have. Sir Thomas Cavendish made the same error a decade later, upon completing the third circumnavigation. The Dutch were now embedded in India but had not given up on the northwest passage, searching in vain for an easy access to China. Willem Barents, after discovering Spitsbergen, died icebound in Greenland. Hudson and Baffin (in 1607 and 1616, respectively) reached

77° latitude, 45° north – nowhere, in other words.

Explorers were having equally frustrating times searching for the southern route. Magellan had believed Tierra del Fuego to be connected with Antarctica. In 1605 Pedro de Quiros and Luis de Torres were sent by the viceroy of Peru to take possession of the southern continent. Just like that. They got as far as the New Hebrides, called them Australia, and poked around New Guinea on their return. Very simple. The seventeenth century saw Dutchmen flocking to Micronesia. By the early 1700s, all of the coasts of Australia had been scoped out. Imagine the first desert aborigines to look up from their frolicksome cockroach clambake, a spouse picking lice from her adorable wild baby's brain, when suddenly, here comes the *Glory of Amsterdam*, sails full blown, scurvy-ridden soldiers yelling land-ho! If recent environmental perception tests on preindustrial peoples tell us anything, the aborigine father probably threw a rock at the boat, thinking it some deformed sea gull riding the waves.

The history of technology and money shaping the materialistic biases of exploration are relevant here, but we must venture back toward its beginnings. By the middle of the sixth century, the old Roman scratching plow had been replaced by an innovative Slavic device – a heavier plow with wheels and double blades. A mouldboard would turn the sod. But because the new plow required a team of eight oxen, which most peasants could not afford on their own, cooperative farming came into being, setting the groundwork, as it were, for the feudal adaptation of labor. Eventually, a three-field rotation system, utilizing horses rather than oxen, accelerated the productivity of Medieval farms. The horse collar, horseshoes and tandem harnessing had made such agriculture feasible.[12]

By the eleventh century there were at least 5624 water mills in England alone. Mill reclamation drained marshes and converted wastelands throughout the Low Countries. In 1260, Roger Bacon was speculating on the future, and his vision included flying machines and perpetual motion. By the time of Leonardo da Vinci, water-driven hammer gorges and bellows enabled blast furnaces to produce immense quantities of iron, which opened up mining in the Alps and Scandinavia. The Plague drastically changed the balance between arable land and pasture. Sheep won out, plucked down whole towns, as Thomas More put it. With the expanding woolen industry came land enclosures and dispossession of the farmer. For every sheep put behind fences, seven people went without bread. Inflation bridled the destitute, compelling Elizabeth I to pass an Act in 1601 forcing landowners to pay rates for the maintenance of the poor. Thomas Malthus would later castigate this action, anticipating the life-boat ethic. But the sheep were overindulged: Domestic wool markets declined. The only alternative was to export. But where? A series of voyages (many of those aforementioned) were, in part, inspired by the British wool industry's hope to sell its surplus in China, Japan, and India. In 1600, the East India Company entered into serious competition with the Portuguese and Dutch for spices of the Malay Archipelago. But their underlying motive was the sale of British textiles in central Asia and the Far East. When Vasco de Gama offered domestic items such as hats, strings of coral, and domestic fineries to the King of Calicut, the king's officers merely smirked. The king already had such things and more: a wash-basin of solid gold. In Japan, it was noted with some dismay that a picture of a bird was worth more to the locals than a master's portrait of St. Paul. The British found no market and ended up having to pay for their desired Oriental goods with Indian-cultivated opium. So feverish was the sixteenth-century trading in Asia that some eighty different dialects competed at the harbor in Malacca.[13]

The preceding decades spawned rural insurrections throughout Europe: The revolt of the Jacquerie swept France in 1358, with subsequent demonstrations adding to the continuing debacle of the Hundred Years War. In England, John Ball led the peasants in a rather upbeat revolt (1381), out of which came demands for the end of serfdom and the alleviation of excess taxes. Urban coppersmiths in Rouen and dyers of cloth in Florence rose up, preparing the political backdrop for the Italian monk Girolamo Savonarola's later overthrow of the Florentine Medicis, the breakdown of guilds, and the emergence of big business in Flanders and Italy. With the banking empires of Jacques Coeur of Bourges and the Fugger family of Augsburg, it was clear that money — fluid capital — was going to replace land as a *bankable* medium of power and exchange. The taste of it would lead the Hapsburgs, Bourbons, the Republican Dutch and Lutheran Swedes into bloody trial. It served for reinvestment in the nascent technologies — from bituminous coal smelting to coastal reclamation; pumps and mills. More money could be spent on enlightened agricultural practices: 4-year rotation in the Netherlands, wheat, clover, grass, and turnips; animal manure given back to the earth. Money tended to uproot locals, but they were already on their way from fiefdom to the big cities. The population bomb had been engendered.

A young man at Oxford at the beginning of the sixteenth century had the whole world at his feet. He touched the live-wire of technology, exploration, and urban adjustment: of geographical and religious foment. He was also fully insulated from the realities of peasant life and thus able to speculate with some impunity. Sir Thomas More spent 2 years at Oxford, at the height of the Renaissance, where his acerbic talent was coddled by the diatribe of Erasmus and Colet, among others. He would leave school for a promising legal career. In 1513, he read Machiavelli's *The Prince*, with its overpowering insistence on expediency and autocracy. More, not about to settle for so grim a picture of the Italian city state, settled into his own dream of what democracy might be all about. By 1516, he had completed his book *Utopia*, which sparkled with clues written in Greek for the learned: Utopia ("no-place") with its capital city, Amanote ("dream town") and a narrator named Raphael Hythlodae ("dispenser of nonsense").

Hythlodae recounts his harrowing experiences aboard Amerigo Vespucci's ship. It is a narrative bolstered in the reader's mind by Vespucci's own books, *New World* and *Four Voyages*, both seething with monsters, savages, and intimations of paradise, and published but a decade before. A series of utopian paintings, beginning with those of the 1518 Basle edition of More's book, Raphael's "School of Athens," Lorenzetti's "The City of Good Government," and Piero della Francesca's Umbrian ideal city states further charged the pictorial anticipation that More had to exploit. More, Hythlodae, and More's friend Gilles begin a conversation about the ideal urban setting and economy that would ignite much of Europe in the years to come. Paradise, recall, was on the verge of paying off. Gold would begin flowing into Spanish and Portuguese coffers. We know, for example, that the eastern seaboard of America would yield 181 tons of it between 1500 and 1650. Giovanni Bellini's "Saint Francis in the Wilderness" (1485) marshalls the first congratulatory impulse. There is the Assisi patron hallowed in the humanized light of day, his wilderness cross-bred with just the right amount of civilizing wealth. The light is golden, oriented to utopia. There is even a drainage pipe. John II of Portugal had sent Alfonso de Parva and Pedro de Covilha to mysterious Abyssinia in search of Eden. They didn't find it. They didn't have to: The gold they recovered was enough.

This frenzy to find paradise had driven Marco Polo. Then, like a strange malady, it compelled Magellan through the cold, blue straights of Tierro del Fuego, abetted by crude astrolabes, too little fruit, shipwreck, piracy, and cannibalism across the cold abyss of the unknown. The painters were quick to seize on its religious implications. Bathed in the new light, Titian, Tintoretto, Lorenzo Lotto, Leonardo, Dürer, Grunewald, Bosch, Rembrandt, and, finally, Breughel, all took to the new worlds—great drama in it, to be sure. For hundreds of years the landscape had been locked up behind cloistered walls where dutiful scholastics tottered about with farming tools, stopping every few minutes to scarf their sweat and poke into Augustine's Confessions, as Petrarch had done beneath Mount Ventoux. At the other extreme, peasant life was carried on under brutal conditions. The explorations of the Renaissance created the myth of everlasting replenishment, as identified in the accessibility of China, America, and deep Africa. If such lands did not conform, expressly, to the Genesis passages, they came recommended on the basis of apparent infinity, riches, and beauty. The details would all be filled in. In his *Treatise on Painting*, Leonardo provided the techniques for making sense of such details, observing stains on walls, fire embers, the uneven color of stones. From his first known drawing of a cliff-entrenched village in the Arne Valley to the later women in his life, Leonardo submersed his unconscious urgings in the details of scientific inquiry to unleash a wholly new seduction of distance. Breughel's appeal was yet more blatant because his subjects were polarized by the everyday, and most markedly so in "Winter, A Dark Day," painted in 1565. The *distance* would broaden. By the late sixteenth and early seventeenth centuries, large numbers of emigrants were leaving Europe. While Monteverdi was describing religious infinity in his choral work, and Galileo fixing an equal storehouse in the spheres, a Zacharias Janssen, Dutch spectaclemaker, was on to the microscope, and then to micro-organisms. All of the world was being sought after.

Such were the energies catapulting the inquisitive More, who, in conversation with his invented storyteller, Hythlodae, developed a highly creative scenario for European city life. Utopia was an island, "somewhere." Its fifty-four cities each supported a population of 80,000, England's entirety in More's day. Expert gardeners, they shared everything, wore woolen cloaks and white linen underwear; the family was ruled with iron will by the father, who was monogamous. A second adultery offense was punishable with death. But Hythlodae condemned the use of capital punishment in mere cases of theft, citing England's desperate peasant population, which, under subsistence standards, threatened by land enclosure and the erosive properties of sheep farming, and by deforestation for the iron industries, was out of work and forced to steal for its livelihood. More introduces a new agricultural economy, cleaning up Medieval democracy, diffusing the power of monarchs, and modeling his populous on a system of constant attentiveness, a prototype that resembles the earliest Communist agendas in fifteenth century Czechoslovakia:

> *There's never any excuse for idleness. There are also no wine-taverns, no ale houses, no brothels, no opportunities for seduction, no secret meeting-places. Everyone has his eye on you, so you're practically forced to get on with your job and make some proper use of your spare time.*[14]

Slaves are imported to do the dirty work. Condemned criminals within Utopia or from other countries are also used. More proposes a natural religion, with priest and priestess. No Christ. The Utopians, under the rigorous leadership of the Syphogrant council, engage in no war. They use psychology to demoralize potential adversaries or buy them off. If all else fails, they hire mercenaries. Incorruptible, without material ambitions, the Utopians offer their international peace-keeping services to other monarchies.

As lord chancellor, More did not intervene to halt executions during his regency, despite whatever he had said about clemency and capital punishment in his scholarly satire. Embroiled in the well-known trials with Henry VIII and Katharine of Aragon and with Anne Boleyn and Elizabeth Barton, More ended up in the Tower of London, charged with treason. In 1535 he was beheaded. As for Utopia, it was never to be. Many emulated More – Antonio Doni [*Mondi* (1552)], Tommaso Campanella (*The City of The Sun* (1623], Gerrard Winstanely [*The Law of Freedom* (1652)] – but it was only fantasy. Domination of nineteenth-century social thought by utopian writers, politicians, and engineers indeed restored certain laws and primacy to the individual and helped to foster a new understanding of urban aesthetics, labor laws, and cooperative living – in the experiments of the Amana Society and its community of True Inspiration (Buffalo, New York, 1843), Father George Rapp's Community of Equality (western Pennsylvania, 1803), Robert Owen's New Lanark (Scotland, 1800) – but not during the Renaissance. The great dream had been sustained from the depths of Chinese prehistory, delivered unto Marco Polo, and brought to Europe. The plague first instilled the message of overpopulation, as bodies littered streets. It was clear that more space, cleaner horizons were needed, an escape. The Bible was interpreted according to utopian game plans that endorsed autarkic, absolutist societies with high density condominium living, botanical verification, and exotic trading enterprises. No European landscape could do without some comparison to its higher form. China, India, America, even Abyssinia had been skirted, in some cases combed. But where was paradise? More to the point, *what* was paradise? But there was a mysterious middle ground – a remarkable fusion of China and Europe, perhaps the very last glimpse of paradise. It was commissioned in the Persian city of Tabriz in the 1520s.

In 1501 the young Shah Isma'il took over the Safavid dynasty, and, modeling himself after Alexander, conquered all of Iran. He entrenched his court in Tabriz at a time when Chinese and Mongol culture had flooded the whole of Persia. The Shah's son, Prince Tahmasp, was sent to Herat for 10 years. When he returned home, he was a man. In celebration, the Shah commissioned the greatest painters of his time to illustrate Iran's most famous book, the *Shah-nama* (the book of kings). Among those artists called upon was Sultan-Muhammad.

We know very little about the painter. But we obtain some idea for Sultan-Muhammad's divergent wants by examining his work, "Worldly and Otherworldly Drunkenness," which merges street chicanery, slapstick, saints and transcendence, a painting characteristic of Shah Isma'il's own temper. Compare it with other, contemporary illustrations, like those by Muhammad's compatriot Shaykh Zadeh who illuminated the *Diwan*. There is traditional arabesque, mystical passion, and plants in wild effusion. But the spirit is not there, not with the grace and enigma of Muhammad. How much time Muhammad spent on the *Shah-nama* is unknown. But sometime between 1522 and 1527 he completed a work that is, without controversy, the greatest painting ever to emerge from the Near East, and one of the most startling landscapes ever envisioned, anywhere.

Muhammad's "Court of Gayumarth," the gradations, detail, and overall vision strike the resonant chord of all art, all nature. In design and color, there is the Chinese Buddhist ethos. In style, geometry, and brush stroke, an abundance of fused Asian and Western strains. But the painting eludes both atmospheres. It is deeply Persian, reaching some other tone and time and place never before encountered.

Gayumarth was the legendary first Shah of Iran, and his idyllic reign was not to last, which is no surprise. A demon, Ahriman, plotted against him, introduced evil into the world. The angel Surush, pinioned on the upper left of the illumination, warns Gayumarth. But the Shah is trusting. His son Siyamak is slain in battle by the demon's own son, the Black Div. The painting humanized paradise, capturing the beauty, distance, and subtleties of wilderness.

The Shah died, his son reverted to extreme orthodoxy, outlawed most pleasures, and changed the course of Persian art. Sultan-Muhammad's own son, the painter Mirza 'Ali would caricature the outmoded ways of his father, who by that time had been silenced by the royal court. Muhammad's last known painting, "The Ascent of the Prophet to Heaven," says farewell to his half-demonic, long vision. Angels view the ascent through a rising cloud. The prophet rides the human-headed Buraq upward and away. Ironically, it was the very ascent, known as Mi'raj, which had inspired Brunetto Latini 300 years earlier during his studies in Spain. Latini in turn imparted the legend to Dante, whose own image of the spiralling mosque-like ascent, under auspices of the lovely Beatrice, had taken the European imagination back to China and to the Elysian fields in Greek mythology. In Sultan-Muhammad we are witness to the final incandescent memory of what those fields must have really been like.

NOTES

[1] Paul Panish, "Hsieh Ling-yün's 'Poetical Essay on My Mountain Dwelling,' an Annotated Translation," Oriental Language Master's Thesis, University of California-Berkeley, 1973.

[2] J.D. Frodsham, *The Murmuring Stream*, 2 vols. (Kuala Lumpur: University of Maylaya Press, 1967).

[3] Han-shan, "Six Poems," Translated by Burton Watson, in *The Mountain Spirit*, ed. by Michael Tobias and Harold Drasdo (New York: Overlook-Viking Press, 1979).

[4] Yi-Fu Tuan, *Topophilia: A Study of Environmental Perception, Attitudes, and Values* (Englewood Cliffs, NJ: Prentice Hall, 1974).

[5] Ch'ao-ting Chi, *Key Economic Areas in Chinese History*, New York, 1963, quoted in Yi-Fu Tuan, "Discrepancies Between Environmental Attitude and Behavior," in *Ecology and Religion in History*, ed. David and Eileen Spring (New York: Harper and Row, 1974).

[6] Donald Lach, *Asia in the Making of Europe*, 2 vols. (Chicago: University of Chicago Press, 1965).

[7] Herodotus, *History* 3.114.

[8] Xenophon, *The Anabasis* 1.11,7.

[9] John Salkeld, A Treatise of Paradise and the Principle Contents Thereof (London, 1617).

[10] John Prest, *The Garden of Eden: The Botanic Garden and the Re-Creation of Paradise* (New Haven: Yale University Press, 1981).

[11] ibid.

[12] Lynn T. White, Jr., *Medieval Technology and Social Change* (Oxford: Oxford University Press, 1962).

[13] Claude A. Alvares, *Homo Faber: Technology and Culture in India, China, and the West* (New Delhi: Allied Publishers, 1979).

[14] Ian Todd and Michael Wheeler, *Utopia* (New York: Harmony, 1978).

Chapter 5

The Embattled Lotus

During the traditional Japanese tea ceremony, samurai, peasant, and Emperor would stoop humbly to enter the tea hut and there pay group devotion to that which is obscure, moon-enchanted, and simple in nature. All weapons were to remain outside. In an era when the samurai kept his sword on his person at all times, such ceremonial separation from the cutting edge of steel was remarkable.

I walk through the main gate at the Higashiyama-dono in Kyoto, a magnificent temple with gardens known as the Ginkaku-ji or Silver Pavilion; past the pine, camelia, bamboo, and green bush; along the elegantly cropped corridors to the main wooden edifice and frontal garden. Before me rises a miniature Mount Fuji, crafted of gleaming white sand. It is the month of June 1483, early Edo period in Japan.

The eighth Ashikaga Shogun, Yoshimasa (1434–1490), has retired to this place, his Eastern Hill palace, having survived the worst decade of civil war in Japan's long history. The surrounding city of Miyako, capital of peace and tranquility, also known as Heian, city of purple mountains and crystal streams (only much later called Kyoto) has been destroyed by a new class of combatant: *ashigaru*, foot soldiers. They have leveled the magnificent city, reducing its population from nearly 0.5 million to a scavenging 40,000.[1] The Onin War was long in brewing. Yoshimasa's wife Tomi had given birth to a baby. One camp of nobles wanted this lovely child to succeed the apolitical Shogun. But Yoshimasa himself had already promised the appointment to his brother, formerly a Buddhist monk. The wolves conspired at every gate, ready to fight for their own faction. The city consequently sank into perpetual military readiness. Weapons-and-strategy experts were brought in by the competing *daimyo* (local lords). In the wake of samurai destitution (from overexpenditures on military preparedness and engagement) the Shogun has issued debt cancellations, further exciting the anger of residents and war hunger of the fighters.[2]

The streets swirl in battle, heads clanking to the cobblestones. Monasteries are burned and looted, whole marketways reduced to fume and ash. But somehow Ginkaku-ji is spared. And, oddly, the war in general is not bad for Japan's economy: This very day a ship has exported 37,000 swords to China and will be returning with silk, books, porcelain, paintings, and copper money.[3] Desperate food needs in Kyoto have stimulated surrounding rural agriculture, and with it, useful tax yields. Yoshimasa is a shy man in his late 40s; an artist, connoisseur, and foremost lover of tea. Departing from his long lineage of power, he has dedicated his life to aesthetics. And it is to this blatant disregard of political duties that we must admit a profound debt: he alone will pass on a legacy after all of the war embers have gone out.[4]

I continue walking past the Silver Pavilion garden to the Togu-do tea hut and from a discrete distance watch a tea ceremony in progress. The tea master and Shogun have invited three people to the *cha-no-yu* (way of tea) gathering, among them the famous Noami, the Shogun's own art collector. They are all dressed in kimono and hakama.

To live without wrestling angels in the soul's dark night, but rather to turn to one's friend and say, have a cup of tea with me! Yoshimasa relished the many aspects of the tea ceremony, from its rock artistry, moss gardens, and stylized calm to the requisite lotus pond and moon reflections; the flowing cascades and raked sand gardens. The Togu-do, where the ceremony is occurring, refers to a very ancient, specialized kind of garden, meaning "the search from the east for Amida's western paradise." Its origins, like tea itself, are Chinese. But here, now, Buddha (Amida) can be found in every grain of sand.[5]

The ceremony is underway. The participants have assembled at a covered bench. The signal is given, they follow the path through the garden, stoop down to enter the teahouse, and silently take their seats. The great tea master Shuko assumes his spot before the hearth. With a bamboo spoon he gracefully scoops up some powdered green tea leaf from the caddy and places it in the tea bowl. With a wooden dipper he pours boiling water over the tea, whisks it into a froth, and provides cakes for nibbling so as to harmonize the tea's brilliant bitterness. The guests have each a small piece of silk brocade on which to rest the tea bowl. Murata Shuko (1421–1502), a Zen priest, has introduced the idea of *wabi* tea, the word meaning chill, lean, withered, rustic. Together, these terms apply to the notion of appreciating insufficiency. Amid so much shogunate splendor, it is refreshing to come upon the modesty of tea.

Consider the ceremony. There are 100 tea styles, variations about the size of the teahouse, location of hearth, arrangement and wiping of utensils, warming of the whisk and bowl, the making of thick or thin tea (3.5 grams for thick, half that for thin; and always 15–20 strokes of the whisk); variations in water temperature, dress, etiquette, conversation (*"O saki ni;" "O shoban sasete itadaki masu!" "O temae o chodai itashimasu."* Ask a Japanese friend what it means). And the delineations continue with regard to flowers, lacquer stands, alcoves, portable braziers, the entrance, stone lanterns, the gong, the type of charcoal, of water basin, the season, the hour, the moon's position, a breeze against the cheek, an animal howling in the hills. Everything is considered! And all swords are left *outside*. In an era when the samurai kept his sword on his person at all times, even during lovemaking, this ceremonial separation from the cutting edge of steel is unprecedented.[6]

Urbanity, purity, courtesy, and imperturbability are the qualities after which Shuko aspired in the tea ceremony. This Amida garden exterior reveals the integration of forces; where warrior, shogun, peasant, and Zen monk share simplicity together before all-embracing nature. The very posture of tea service connotes a stooped respect. The tea master must act unceremoniously with those above and with sincere courtesy for those beneath; he must consider all things earnestly, love neatness from his heart, start the tea kettle an hour before the guests arrive, show discretion, and properly attune the ritual in all ways. The meal must include vegetables, sake, raw fish, and fowl broth; a flower and a painting must be arranged, along with the right incense. One hundred tiny cups of tea are served, and time must be allowed for urination in a private privy. This business of piss – most important! It suggests the Taoist passage of all things in the universe – from clay to camellia, from lips to bladder. The Buddhist avows a higher reality, and tea is the supreme method. Furthermore, tea sustains fine health, which was the purpose of importing it to Japan by the Zen priest Eisai who travelled to China during Sung times and retrieved the leaves for his fellow monks. It could revitalize a tired heart. Such reverence for life-through-nature during the rampant wars and irreverance reveals a sure and delicate pattern to human reckonings wound around the vine and impetus of all-healing nature.[7] Such prescience typifies the basis for all of Asian metaphysics: an astute pragmatism melded to otherworldly preoccupation and forged in the garden. Exhorts the Zen trickster, "You carry home the fish in the basket. Don't cook the basket!"[8]

The guests at the ceremony admire the utensils and examine closely the painting in the alcove. The work is unsigned but painted by Tensho Shubun, Kyoto's darling, teacher of Sesshu, Japan's greatest landscapist. The painting in black and white ink portrays a temple high on a mountain, drenched in mist, defended by steeple ramparts of aerial cliff. All is indistinct, elusive. Below, a tiny bridge spans the atmospherics, and a man crosses. The temple is of T'ang architecture, subtly composed. The painting hangs beside a single morning glory. This image – painting, flower – is the Pure Land, Amida's paradise. The tea ceremony is the recreation of paradise, a middle ground; and our own access to understanding several key features of Chinese and Japanese history: the Silk Route, aboriginal hunting and agricultural worship, two centuries of Tokugawa Shogunate renunciation of European firearms, the evolution of nature poetry, mountain worship, flower arrangement, and of a culture's grand effort to cope with contradiction – good and evil, order and chaos, nature and civilization. A mere beverage, this tea? Hardly. It is here at Ginkaku-ji that Kannon, the androgynous goddess of mercy (*Avalokitesvara* in India) is enshrined. This *is* paradise. To find its original source, we must make our way into the central Asian deserts and go back many centuries.[9]

But First, we visit the west wall of Horyu-ji Monastery in Nara, not far from Kyoto. The painting portrays Kannon and Seishi (a second Bodhissatva disciple of Buddha) flanking Amida himself who is making with his hands a gesture, or mudra, to the wheel of law, the cosmic embrace of life and death. It is perhaps the most subtle and emotional of all existing portrayals. This mudra refers to a special contemplation of the Pure Land Buddhist sect, a Buddhism based entirely on the notion of a wilderness paradise known as *saiho-jodo*. Where is it? The Chinese looked to five mystical islands, eastward; the Japanese sought the very same islands, but in the opposite direction. The original images of the place, called *henso*, derive their form from a smattering of caves along the two major silk routes through northern Tibet: cosmographies of howling wind and dust.[10]

We come to a second painting in Nara, at the Gango-ji Paradise Hall, a close-up of the Chiko Mandala. In the eighth century a monk named Chiko dreamed of being reunited with his deceased friend in the pure land of Amida. In the dream, Chiko meets Buddha, who discerns Chiko's inability to concentrate with enough power on the Western paradise. True to character, Buddha offers the monk an alternative: In the palm of his right hand he is to meditate on the pure land. That is enough. And from it, Oriental culture derives its tradition of the miniature. William

Blake seized on a similar revelation, "eternity in the palm." Two late sixteenth-century tea masters Rikyu and Oribe offered opposing formulations of reduction and aggrandizement in the ratio of stepping stones along a tea garden path; six practical to four aesthetic stones, suggested Rikyu. No, just the reverse, argued his student. Not merely a frivolous dispute, the ratio determines the Japanese conception of *to know what is enough*. To *flow* along the steps is to be nature's own transmutation. Similarly, the mandala painting joins the viewer to Western paradise — if the painter's equations and the depth of his feelings are correct. Buddhism is filled with entry points.[11]

A windswept plateau, Sinkiang Province, China: Shrine 1X at the Bazaklik-Turfan caves is 80 miles southeast of modern Ürümqi. Inside the cave is an early eighth-century painting called "Musicians and Mourners." It pictures Arabs, Iranians, Chinese, and Tibetans, gathered to see the Buddha pass away. The communion of races asserts the very itinerary of Buddhism: from the jungles of southern Nepal, throughout northern India, up into Tibet, and then east to China. From a 53-meter statue of Buddha at Bamiya, northwest of Kabul, to the "Cave of the Coffered Ceiling" at Qïzïl near Kuca in western Sinkiang, wherein is seen a spectacular Tibetan immortal, past the third-century cave paintings at Mirân, and then on to Tun-huang, where Lo Tsun saw a vision of 1000 Buddhas later to be cast in stone, we apprehend the Buddhist landscape — Himalayan desert. At Cave 19 of Turfan Bazaklik we behold a dragon leaping out of water; behind it, a fantastic mountain range, weeping willows, a cherry tree in bloom.[12]

Between nomadic marauders, long hard winters, and the incessant wind, this resplendent outreach, clinging to the other world, burst colorfully from a watershed of war, along a refugee corridor marked by the spread of Manichaeism, the toppling of King Yazdagard III, and the dissolution of the Sassanian Empire. From islands in the Atlantic, from Byzantium, Syria, Persia, and India, a super-charged influx of artistic styles and religious quarrels merged in these caves, as the Near Eastern Kushan Empire battled to force commercial contact across the rugged Tarim Basin with the later Han Dynasty, creating the two Silk Routes of travel. Sericulture, like tea, is embedded in the soul of humanity. Here, in the high deserts of central Asia, it worked an iconography of startling freshness. It can be seen in the mask of Gautama's vision, with his illumined almond eyes and azure countenance — all Apollonian. The central Asian artwork reveals gold patinas on stone, goatherd lads — hapless bystanders paying mute testimony to events that had wings, which moved relentlessly toward the gardens and inner depths of southern China; moved upon Bactrian camels, were swept up in wars, apparitions, into dark caves, and open courtyards. The beneficiaries of this migration were the T'o-pa Tartars who ruled north China from 398 to

557. Known as the Wei rulers, they adopted – and became immensely wealthy by – Buddhism.[13]

Blessed with ebullient vegetation, northern China produced a religion of the Earth. The Chinese character for rest and peace, connate with that of *park*, portrayed a man sitting under a tree. This image merged with that of the blue water lily, basic to Nepalese and Indian Buddhism, symbolizing the ascent of the soul out of elemental mud. Chinese ethics were mightily enriched by its ecosystem, particularly by the bowl-leafed nelumbium lotus. The cult of Amida was taken up by a coterie of court savants, renegade poets, and Taoist wayfarers centered in the Lu-shan Mountains. They also founded the White Lotus Society. All of the earliest champions of *shan-shui* landscape painting and elegy – Ku K'ai-chih, Tao Yuan-ming, Tsung Ping, and Hseih Ling-yün – joined its ranks. Meanwhile, Han Dynasty emperors were busily employing hundreds of thousands of laborers to divert rivers, build up paradise islands, even embarking on perilous expeditions in search of the rumored Taoist–Buddhist immortals said to inhabit the misty five isles off the Coast of Shantung. Near Lo-Yang, the Emperor Sui Yang Ti engaged one million workers to construct his western park. Endowed with so myriad, so luscious a landscape, why all the effort to seek it elsewhere, to build it? In this discrepancy of logic lies a clue to the arrogance and dissatisfaction of power. True Taoists were quite contented, we may assume. Only the wealthy, the political ingenues, fretted and schemed to insure their religious good-standing. This disparity between political hubris and the urgings of nature, epiphany, paradise, and the everyday stand out beyond the T'ang Dynasty. The accumulation of Taoist and Buddhist sentiment regarding landscape and the Earth organism (gaia) reaches a crescendo during the Sung Dynasty.[14]

Ch'an monks of the Sung Dynasty powdered and whipped their tea drink in a ceremony honoring their founder, the Indian Bodhidharma, twenty-eighth Buddhist patriarch, discoverer of tea; a man who meditated wordlessly before a large stone boulder for 9 years. Such austerity helped shape the greatest painter of the early Sung, Li Ch'eng, whose one verifiable work, "A Buddhist Temple in the Mountains," sets the tone and pace for all future Oriental landscape art.[15] The serenity competes with turbulence; there are worlds within worlds. These two maxims consummate the wilderness – paradise scenario. All else is incidental. Li Ch'eng's immediate successors – Fan K'uan, Li T'ang and Kuo Hsi – deified the painter's tenets, equating them with the quintessence of Zen. Kuo Hsi commands our interest. His *Essay on Landscape Painting (Shan Shui Hsun)* provides remarkable insights into his own masterpiece, "Early Spring," done in 1072. There has never been a more lustrous paragon of Taoist ecology in action. Of immense size, taking on the whole world as its subject, with plunging cataracts of gold, a cornucopia of plants and motion and intimation ravages the viewer. Today we must journey to the National Museum of Taiwan to see it.[16] In his treatise, Kuo Hsi provides an account of mankind's place in nature, a moral echoing of energy and presence and purpose. He discerns four types of landscapes: those in which one can travel; those that can be gazed upon; those for dwelling; and those for mere rambling.[17] In other words, the imagination, rooted to experience, is advised to don a rucksack, step into the canvass, and spend a pleasant evening in a rustic retreat (discernible in the foreground) before setting off the next day into the heart of the painting, where mist and mountain converge in dazzling labyrinths and clean outlines. Only true observation can engender such a sacred space for action, exhorts Kuo Hsi. Ink gradations, the dimensions of rock, perceptions of distance, various atmospheres: The details of reverence are carefully delineated by Kuo Hsi to his son, with whom he wanders, relating the various theorems for the painter. Kuo Hsi abhorred stability, setting all of life in turbulent motion. His objects are depicted in progressively lighter tones as they recede into depth. Localized clumps of mist accentuate height by masking the bases. A profusion of miniature worlds compound design. Hypnotic, reductionist, transcendent, allegorical, Kuo Hsi's work gives us a new sense of space intensity (as had Li Ch'eng's) not rivalled until Albrecht Altdorfer's sprawling portrayal of the aerial "Battle of Alexander Defeating Darius" in the early sixteenth century.[18]

By the time Jurchen Tartars ransacked the capital of China in 1126, a sufficiently indoctrinated cavalcade of artists could refocus its energies at the new southern Sung capital in Hangchow. The Jurchens managed to exact a heavy price for peace. But such tribute is meaningless compared to the lasting triumphs of the southern artists, whose principle masters were Ma Yüan and Hsia Kuei. The city's luxuriant terrain provided an intriguing contrast of tropics to the temperate river plains of the North. Thousands of paintings in the manner of the "Ma-Hsia" school were carried abroad by merchants to Japan, eventually on to Europe. Ma's major work, a signed album leaf, was entitled "Walking on A Mountain Path in Spring." A scholar stops to admire two orioles in a wind-blown scene. A verse couplet treating the dance of flowers adjoins the work. With the surest economy the painting becomes *renga* (a type of Japanese poetry), relying on emotional pathways of evocation. Surrounding emptiness is charged with human longing. Such simplification is actually quite complex, a literal science. Details are reduced to mere likenesses, faint shadows. Graded washes fulfill the inclination of the eye. The Ch'an master painters of the southern Sung held out in monasteries surrounding Hangchow. Liang Kai, the greatest of these painter-monks, executed his "Sakya Coming out of His Mountain Retreat" in the early thirteenth century. It served as a model for later Japanese artists who wished to explore the perplexing emotion of perspective itself. Here the Buddha is portrayed looking into the future, to the agricultural plains of population centers. The figure casts an anachronistic visage as if anticipating some crisis of conscience, beautifully realized in the Japanese Kato Moriuke's 1683 rendition.[19]

Kuo Hsi had believed in a living force animating the inanimate. In his *Essay*, he wrote that "Water-courses are the arteries of a mountain; grass and trees its hair; mist and haze its complexion." His term for mountain was *ku fa*, backbone; and he continually borrowed from anatomy to describe this organism for life. Liang Kai's painting intuits conflict and the destruction of nature at a timely crossroad of culture. For it was during his era (thirteenth century) that the Japanese Zen monks flocked to China (Dogen among them), implementing trade negotiations and superficially converting the Kamakura bureaucrats. And so it is that we return to Japan.[20]

From the Moth of Clouds Trail, above Kyoto on the highest peak of Mount Hiei, we can see all the way to the Pacific. But when the wind drives mist through the cryptomeria trees, there is not much of a view – only introspection. The Japanese language does not yield a view, either. That is to say, it is a poet's language; it is subtle, subjective, full of nuance, shades, mists, elipsis, double negatives, innumerable verb forms, dramaturgy, finesse, and occult grace. We have to search for outrightness. Mount Hiei is one of the most important religious sites in Japan, center of the Tendai sect of Buddhism that adopted central Asian Pure Land Amida.[21] The tenth and eleventh century Fujiwaras, like the Florentine Medicis, cultivated the religious arts, readily accepted the doctrine of Pure Land, leaving the military arts to the Taira and Minamoto clans. This led to Fujiwara decline, though nobles of the ancient house still live on in Japan today. But for purposes of power the Minamoto family won out and created a shogunate in Kamakura under the leadership of one Yoritomo. An early emperor would say that he was powerless in relation to three things: the flooding of the river Kamo through Kyoto, the winning at dice, and the monks of Mount Hiei. By the time of the Kamakura shogunate, pitched battles between armed monks was common. The very word *osho* in ancient Japanese, meaning priest, also meant spear teacher; blades in sure supply were traditionally kept outside under temple eaves and heavily relied upon to acquire religious *strength*.[22] Monasteries literally conducted warfare. It was no coincidence that the militarization of Japan under Yoritomo coincided with the influx of Zen, Nichiren, and Amida Buddhism. Zen, particularly, contained elements philosophically attuned to the samurai spirit. The art of metal fittings, sword making, and tempering of edges; the doubling, welding, and forging of armor were all dominated by Zen artisans of some fame.[23]

Meanwhile, at Kamakura, warrior exploits were being recounted in painted screens and prose tales: sweeping descriptions of the Taira–Minamoto wars, scrolls illustrating the Mongol invasion of north China. When Yoritomo assumed power, he acted to form a schism between his rule and that of the imperial family's in Kyoto. This resulting dualism of Japanese allegiance between the Shinto-derived Emperor and the new insolent shoguns, bent upon force of arms, produced an irreparable gulf in the Japanese psyche. The shogun clearly had the decisive advantage – a standing army. The patronage of Buddhism was only so expedient, as was the Zen solicitation of political sanction. But these conciliatory gestures – the shogun looking to Buddha with one eye and to a would-be adversary with the other – led to inferno. In fact, by 1571, not a single living creature was alive on Mount Hiei. In a single day an infuriated shogun (Nobunaga) massacred thousands. Why? The monks challenged his authority. The shogun was, after all, a dictator; no real irony in this. Religious sects should have stayed out of military business to begin with. What prompted the Buddhist will-to-power is no real mystery. Wrangling, proselytizing administrators were at the helm.[24]

From aboriginal times the Japanese worshipped *kami*, resident deities whose symbols – mirrors, swords, jewels and oral myth – were incorporated into all of the temples *(jinja)*. To this day a child is taken to his precinct's *jinja* within 30 days of his or her birth, there to become a child of the local divinity. He will pay special obeisance to it throughout his life, and in so doing respect his village and homeland. The sun goddess Amaterasu Omikami oversees this bond under the name of *Shinto*. Japanese Buddhism consistently accommodated this native creed, erecting inner tabernacles for the resident kami within the Buddhist temple. Numerous cults emerged at the juncture of the inner and outer shrines, and these were the more powerful doctrines. There are more than 80,000 Shinto locales in Japan.[25] Invariably, a stone, tree, pillar, or flower situates the point of belief. At Mount Yoshino, water kami control irrigation. The cult is named after the watershed of the region. On Mount Mikasa north of Nara, the Fujiwara clan worshipped five agricultural deities. There is no shrine: The mountain itself is sacred. Japanese self-awareness, in concert with Pure Land Buddhism, came to perceive *all* of Japan as a sacred land, a divine nation. Like the Jews, the Japanese adopted a *chosen-one* ideology. Beneath Mount Hiei is a second, secret mountain – Hachioji. Mandalas of this peak accord the viewer special privileges, again, a high form of power. But the most important meeting of native cults occurred at Kumano. There, at the southern part of the Kii Peninsula, naturalistic folk belief attached to Amaterasu's brother Mikoto, the underworld figure who taught human beings forest conservation. A religious movement in late Heian times promulgating the Buddha Amida believed the Pure Land western paradise to exist right there, above Kumano. Shugendo mountain wizards were said to possess its secret. An old woman was apparently delivered unto Buddha atop the Nachi Waterfall; yamabushi priests practiced their religious adherence in ascetic alpine retreats, and were accorded the gift of flight. As a substitute for military strenth, the peasantry of Japan expressed tremendous belief in its own power. That the imperial family attached itself to these energies there is no doubt. In fact, generations left the Imperial Palace in Kyoto on pilgrimage to Kumano – no light matter considering the 500 miles involved. The emperor, family, and staff dressed in white robes and sandals and reciting the *Nembutsu* chants to grant them safe passage into paradise, took such religious observance quite seriously. Inasmuch as hunting, agricultural, and Buddhist deities were merged, we can only admire the practical zeal of the emperor, otherwise branded an apolitical effete. But his approach was that of peace. The shogun Yoritomo, on the other hand, paid to these cults tribute by building a garden at Eifuku-ji. His chronicler was careful to record that it matched the beauty of Amida's own paradise. To construct the garden, samurai were hired to move the heaviest rocks into place, the bulky soldiers competing to see who could lift the most. While Eifuku-ji is no more, to this day the emperor plants rice seedlings to the aboriginal deities in the Imperial Palace garden of Kyoto every May to mark the new season.[26]

Which leads us back to the Ashikaga Shogunate, to the hidden gardens of Yoshimasa outside the late fifteenth-century ruins of Kyoto. The city's magic, understood by a cursory inventory of its place names, had been raped. The Temple of the Absolute, of Gratitude, Pure Foundations, of Serene Quietude, Western Fragrance, Celestial Dragon, Benevolent Harmony, Enchanting Knowledge, Calm Light, of the Quintessence of Enlightenment, and of the Blue Lotus, were all destroyed, their gardens ransacked. But today they are back, largely because of the power granted to the Zen functionaries, painters, poets, priests, and tea masters who inherited the Sung Chinese naturalism. These men endeavored on behalf of the bankrupt Ashikaga to implement a favorable balance of trade with Ming Dynasty China, allowing them in a single stroke to import the Ch'an pantheon of riches and fully entrench themselves in Japanese government. They came to dominate advisory positions and maintained the only existing educational institutes in Japan.

The smooth sweep of a samurai's sword – in the gestures of *kendo* – and the spartan styles of the monochrome palette – of Sesshu, the brothers Ami, of Shubun and the Kano School; the *shoin-zukuri* architecture noted for its low rooms and friezes, its painted sliding screens, unpainted woodworks, unfinished mud plaster walls, its windows of oiled paper and bamboo, natural stones, and total lighting effect; and the wide proliferation of gorgeous, naturally hewn ceramics and earthenware tea accoutrements deliberately blemished to emulate the Earth's own soil — all were cultivated to satisfy the military rulers' craving for certain, token austerities. The dictators of Japanese unification each constructed flamboyant, lavish palaces for themselves: Oda Nobunaga's Azuchi; Toyotomi Hideyoshi's Mansion of Assembled Pleasures, Osaka and Fushimi Castle; and Tokugawa Ieyasu's Nijo Castle and funeral mound above Nikko. Zen served the shoguns by counterbalancing their extremes, if only symbolically. Nobunaga acquired a fine taste for ceramics, but only to impress the merchants of Sakai who were in control of Portuguese weapons imports. For Nobunaga to make a gift of an Ashikaga-owned utensil to one of his officers implied just the requisite touch of culture. Could we expect any more from him? Hideyoshi built his own tea room inside Fushimi Castle, where he spent his last reclusive years. Yet he was to have his revered tea master Sen no Rikyu commit suicide. The Shogun Ieyasu would similarly force his son's tea master Oribe to commit suicide. Fits of anger, ultimate irreverance, sham learnedness: such devices contest whatever lasting impact the Zen aesthetic may have had on medieval politics in Japan.

At the same time that Yoshimasa was completing his silver pavilion, in 1488, Masamoto, the son of Lord Hosokawa, was rebuilding the Daiju-in gardens at the other end of Kyoto. This Heian villa was in rubble, its former resident monks having taken up evacuation quarters at a Zen center. Masamoto appointed the Zen priest Tokuho Zengetsu to reconstruct the temple. But we are uncertain who designed the famous Ryoan-ji raked sand garden. Needless to say, it constitutes the finest hour of Zen mysticism. Its fifteen rocks, in five clusters after the legendary paradise isles, mark the meeting of simplicity and abstraction. The uncanny intercourse provokes an earnest, unadorned balance. To this day, Japanese students of architectural landscape and aesthetics are unable to re-create the energies these precisely configured natural rock surfaces unleash. The eye is staggered, caught in a cross-fire of force fields that scintillate over the white sand, obtaining elaborate rhythms of the Tao. On the back of one of the stones is carved the names of two men who worked on the garden — Kotaro and Hikojiro. The late Loraine Kuck pointed out that this may be the only signed garden in the world. The same impulsive, egomaniac, Hideyoshi, who compelled his guru to kill himself, managed to spare Ryoan-ji between his many rampages. He is said to have been greatly moved by the place, ordering his subalterns to protect it. There is little reason to tax ourselves with theorizing about political personality and its fluctuations. Hideyoshi mourned the loss of Rikyu within a day of the tea-master's *seppuku* (self-execution). But what is it worth? History copes, somehow, with its contradictions. Hideyoshi could thrill to Zen simplicity while erecting the most outrageous estate up until that time in all Japan, Fushimi at Peach Hill (Momoyama).[27]

Since the first introduction of Portuguese matchlocks at Tanegashima in 1543, the regents had watched the deterioration among fighting men of the samurai spirit. With the new weapon any moron could bring down a warlord. Such tactless success both seduced and infuriated the samurai. Hideyoshi determined to stop it by controlling the distribution of weapons. He announced the construction of a massive statue of Buddha that would require the iron from all civilian weapons. The edict failed, the Japanese military continued massive production of arms (even overtaking European levels), and the supposedly peace-loving Buddhist shogun invaded Korea with 160,000 thousand samurai, 40,000 of them gun-toting. But Hideyoshi failed to take Korea, let alone China and the Philippines, which were also on his mind. The chief reason lay in impasse, the bizarre "constriction of . . . technology."[28] Though Hideyoshi's forces had strong gunner ranks, routing the Koreans for the first two years of battle, Chinese gunners came in and bolstered their ally. Neither adversary would risk the graceful heroics that dominated true hand-to-hand combat among samurai, not with gun barrels pointed at them. As a result, both sides entrenched themselves and waited, and waited. Nothing happened. The outcome was deferral, sullen alliance, and the frustrated search for other, more rewarding trials. The sentiment of winning relied less on the efficiency of killing, which guns indeed satisfied, and more upon the traditional artfulness of bodily engagement — *bushido,* the samurai code of ethic aimed at the perfection of the body and soul. Bushido was Teddy Roosevelt's banner in 1898. More importantly, it was a doctrine of violence basic to Japan. The rise of the samurai class accompanies the downfall of the Fujiwaras and the first public executions in 350 years in Kyoto. The Heian Masakado and Sumitomo rebellions; the eleventh-century Former Nine, and Later Three Years Wars — these bloody contestations fueled legends of "patriotic gore," emotional hooks that were resoundingly intimate and concerned with *issho kemmei* ("desperate holding onto land") in a country notorious for its geographical constraints and dense population. The sagas of particularly brutal, or cunning, or successful warriors cannot be separated from an ancient sense of topophilia, or land love.[29]

We have seen how deeply immersed the peasant psyche was in the annals of Shinto. Similarly, the warring communion seems to suggest a further allegiance to territory, to the extent that one is willing to fight for one's lord and daimyo and shogun in the name of property. Territorial conflict is perhaps the most basic conflict in any country (or animal population). But Japan's record of conflict and abuse over land is exaggerated because of that conflict's particularly keen religious connotations and its unique feudalism. One may note the celebrated incidents of warfare memorized by school children in Japan: Benkei, who fought 30,000 men alone; Hatakeyama Shigetada crossing the Uji River; Kumagi's beheading of the boy Atsumori, as retold in the *Tale of the*

Heike; the loyalist Kusunoki Masashige who emerged in the battles of the Ashikaga Takauji campaign; the forty-seven ronin who avenged their Lord Asano by killing Kira (as recounted in *Chushingura [Treasury of Loyal Hearts]* a 14-hour play); and finally the invincible Minamoto Yoshiie who was spurred on by such clan loyalty as to terrorize all the barbarians who fled rather than face him in his wrath.[30]

And it was the very "barbarian-suppressing" generals, the bakufu of Shogun Yoritomo, that first spawned the two centuries of war, divided all inheritance, diminished property holdings, and aggravated a land scarcity. Only a feudal order could manage under such disintegration. Even the term daimyo is rooted in political ecology, myo meaning rice fields. Land was measured in the *koku* of rice it could produce. Prior to shogunate times, the earliest codes of law had insured a transferrance of public property (all property) every 6 years; a redistribution under the Fujiwara meant to insure that every family was able to lease a viable amount for growing rice. *Dai*, of daimyo, refers to great name. When a warrior plunged into battle, he "read the family roll," enumerating his lineage and shouting his exploits and those of his daimyo across the land.[31]

Along with legendary gallantry and feudal allegiance, the warrior had adopted Zen, as witnessed in the fine art of *kendo*, the way of strategy, equivalent to the way of tea, of archery, of healing. It was *the* way, and all ways were the same. Sword play aims at higher realms than mere death, namely *satori*, Zen enlightenment. Hideyoshi's tea master Rikyu, faced with seppuku, by which he disemboweled himself with a sword without so much as a whimper or squirm, wrote, "Oh, sword of enlightenment. . . This very moment, Into heaven I throw high, This only sword of my armor."[32]

In his *Book of Five Rings*, Miyamoto Mushashi, a younger contemporary of Rikyu's, outlined the Zen of wielding a samurai sword and in the process orienting oneself to the cosmos. Musashi killed his first man at the age of thirteen and went on to defeat sixty more individuals before he turned thirty. His last two years were spent in a cave where he payed ultimate homage to heaven, kneeling before Kannon and Buddha. Mushashi's book explores every facet of contact with an opponent, as if the death-look were a love embrace, the total effect given to choreography. He details stance, gaze, footwork, sweep, and the no-attitude attitude; timings, the method of becoming one's enemy. Sen Shoshitsu, the fifteenth tea master, would speak of the "opportunity to share a bowl of tea" with another person as occurring only once in a lifetime. Mushashi similarly lent intense and solumn meaning to the battle. He described a sword dance, meant for combat, but intrinsically a part of other fine and erotic Japanese arts and wholly indebted to the spirit of Buddhism and the journey toward paradise. The play of swords is defined according to philosophical sword*cuts*: the cut of red leaves, of the void, of the rock, of soak, confusion, crush, the mingling, and finally, of the ox's neck. This latter stroke is especially significant in Japanese Buddhism because of the famous ox-herding Zen pictures, in which there persists an empty space between the ox arriving and the ox leaving. Similarly, the tea master used many references to the abode of the void, the asymmetrical, and applied them to the teahouse itself, a place in keeping with the Buddhist concept of *sunyata*, of rigorous austerity. This litany portrays the ultimate Zen art, a reduction of outward violence to inward principle, which explodes back outward and kills. The metaphor of death gives nothing away of itself. Like the hidden view, it promulgates stoic resolve, attraction to finality, Nirvana, to Amida's pure land. One treats one's enemy as an honoured guest.[33]

The samurai practiced the "twin way" of literary and military arts. Country lords became ardent practitioners of linked verse, *renga*, which again exploited the artistic powers inherent in communalism. The first Tokugawa codification of Laws Governing the Military Households made clear that "from of old the rule has been to practice 'the arts of peace on the left hand, and the arts of war on the right'; both must be mastered." It comes as no surprise, then, that the samurai endeavored to become refined tea ceremonialists, amature painters, ceramics dilettantes. This integration could even be discerned in the three manners of wielding the tea ladle during the ceremony, each motion meant to simulate the shooting of an arrow in combat.

Oda Nobunaga, who had engaged the Japanese navy in its first fated gun battle against the Buddhist monks of Ishiyama Honganji monastery (where dwelt the shogun's own tea master Jo-o, a samurai) died appropriately by the sword at age forty nine. It was then Hideyoshi who rose through the ranks to replace him and drive Japan into its first imperialistic madness abroad. When he died in 1598, nearly 10 percent of Japan's population were samurai, meaning a standing army of some 2 million men with a kamikaze's zeal. When compared with the 30,000-member total force of England at that time, we begin to appreciate why the Portuguese never once considered laying siege to Japan; and how easily the Japanese defeated the Spanish in Siam in the 1620s; and why the missionary St. Francis Xavier would remark — upon a two year's stint in Japan during the mid-sixteenth century — that she was the fiercest, the most war-hungry of all nations.[34] A man moving in graceful accord with the Zen principles of kendo, with religious will and concentration and a dancer's style, both hands on the *katana* curved sword, provided a far richer intimidation than he who held a harquebus. Musashi would say that guns were only useful from inside a fortification, or on the battlefield prior to the gut clash of ranks. But once swords were crossed, the guns were useless. This was both a technical and spiritual warning, one hailed at the beginning of the sixteenth century in Europe when Altdorfer's "St. George" (1511) defeated a dragon with his sword.

Azabu, outside Yedo, 1618. A group of the shogun's engineers have accompanied the Dutch weapons experts to a testing ground. Disturbed by the advent of superficial alien matchlocks over the symbolically infamous sword, the Tokugawa Ieyasu began a systematic policy of renouncing guns, and with them Christians and foreigners generally. One of the last instances of known gun celebration occurred at Azabu, when 300 coolies unloaded huge mortars, constructed by Dutch casting experts at Hirado. The Shogun's top engineers joined in to view the exhibition of explosions. The spectacle delighted all those present, in spite of the occasional misfire that bludgeoned not a few. Houses burst, blazes were set, sparks flew from massive detonations that may have been clumsily off their mark but no less impressive. The Japanese were recorded to have clapped their hands gleefully, like children at play. And why not? Many of the daimyos were eager to purchase the guns and cannons for their personal sport while they were still allowed to do so.[35]

Noel Perrin, in a lucent exposé of Ieyasu's reversion, demonstrates without doubt Edwin Reischauer's own conclusion that the two and a quarter centuries of the Tokugawa reign, dating from the final suppression of the Christian-motivated peasant revolt at Shimabara in 1638, "was probably the longest period of complete peace and political stability that any sizeable body of people has ever enjoyed."[36] Guns were indeed used to massacre 20,000 men who had taken over Shimabara castle. But gunsmiths were put out of business from then on, save for four families, and even they were subsidized with cash in lieu of production. After 1636 Japan was closed to all outsiders. Japanese caught abroad by the closure were not allowed to return home. Remaining Christians were crucified. Not the happiest advent for a period of peace, but then this paradox was characteristic. We are reminded that the Silver Pavilion of Yoshimasa was erected atop the gutted ruins of an earlier temple. When Henry Heusken, translator to the first U.S. Ambassador to Japan, experienced a devastating typhoon, he was amazed to see how matter of factly the Japanese took the destruction of one of their cities, setting right away to rebuild it.[37] Perrin outlines many of the advances witnessed under the Tokugawa's fifteen shoguns, from the development of a purely Japanese mathematics well ahead of European calculus, to the largest canal in the world, the first tractates on mining technique, mail service, packaged grocery goods with price labels, and the world's first anesthetic. While Europeans suffered from scurvy, widespread skin ailments, vitamin deficiencies, lice and flea infestations, the Japanese revelled in the application of certain medicinal and religious prophylactics. Lice were ground up and taken with winter grain (a remedy for epilepsy); fleas were consumed as a remedy for frostbite; insect larvae—maggots, dusted moths, even silkworms—were consumed as delicacies. Herbs were mandatory in any meal. While the British were wolfing down huge quantities of flesh and drinking equally massive volumes of ale (the Queen said to consume one gallon with brakfast), the Japanese drank tea.[38] In addition to their medicaments, the Japanese excelled their European contemporaries in one further nutritional way: rice. Unplowed wild fields could yield 1300 pounds of it, in addition to 1300 winter grain pounds per quarter acre, enough to support five to ten people each working but one hour a day in cultivation. This maximization of space was in tune with *za*, the communal priority afforded villages within the commanding Shinto framework; the deities, recall, were rice and irrigation deities. The Christian gods could not compete, had nothing to do with agriculture. The British ate salted meat in the winter months while the Japanese dined on fresh fish, thus avoiding many of the bladder and kidney problems so rampant in Europe. The *za* legacy is undoubtedly medical. The breakdown in the traditional Japanese collective among Japanese-Americans has produced a dramatic, documented increase in their heart attacks and hypertensive rate.

The first generation of Westerners to enter Japan provided moving testimony to what the Tokugawa leadership had accomplished. When Commander Rodgers landed on Tanegashima in 1855, his armed party was stunned to realize that the islanders were ignorant of guns; descendants, no less, of the very islanders who had first imported them three centuries before. As in the United States at the time, 75 percent of the Japanese were farmers (or 26 million people). Her population growth was under 1 percent annually. If ever there was a steady-state economy, it was Japan's. For more than 2 centuries she experienced no future shock.[39]

Wrote Townsend Harris, U.S. ambassador, "It is more like the golden age of simplicity and honesty than I have ever seen in any other country." Harris' assistant Henry Heusken kept a diary and had this to say of his first trip to the Imperial Palace.

> *Not one diamond sparkled among the crowd of courtiers. A small gold ornament on the handles of their swords was hardly visible. . . . The simplicity of the Court of Edo, the noble and dignified bearing of the courtiers, their polished manners which would do honor to the most illustrious court, cast a more dazzling splendor than all the diamonds of the Indies. . . . Not one bayonet follows them; they carry no other weapon than two swords. . . . I fear, Oh, my God, that this scene of happiness is coming to an end and that the Occidental people will bring here their fatal vices.*[40]

So refined were the Japanese that a Jesuit chronicler could not resist relishing this one detail: The Europeans picked their noses with a thumb, but the Japanese used the little finger.[41]

Heusken, while in his late 20s, was killed by a samurai. The Japanese government paid the boy's mother $10,000 in compensation. The murder is a measure of the pain and indecisiveness, the angry antipodes of opinion regarding westernization through which the Emperor had to steer a course. Under the Tokugawa Shogunate, with no battles to fight, the nobel samurai became dispossessed and impoverished. Under the new Emperor Mutsuhito and his inner circle of young advisors, the samurai were ordered to get rid of their swords, retaining no legal privileges other than a token stipend. The new Meiji might have used this immense and ready force, but did not. They relied instead on the rising peasant class to flesh out their nascent military and take up the old Chinese adage of "rich country, strong military." Mirroring the Western approach (five blatantly disadvantageous treaties forced on Japan by the Western powers after 1853), she forced her own trade agreement with Korea in 1876, the same year that the samurai were disenfranchised. Within a few years, Japan had major ammunition factories in Tokyo and Osaka, had 138 naval vessels, and had fortified herself for battles with Russia and China. By 1900 her gross national product had doubled. When General Nogi accepted the Russian surrender in 1905, ending the Russian–Japanese War, Japan's population had reached 45 million. She had become a creditor nation and a military presence in the world.

Eloquent tea celebrant and art historian Kakuzo Okakura indicted Japan for renouncing her past and castigated the West for obliviously intruding. Okakura was in both worlds, having assumed the curator's post for Oriental art at the Boston Museum of Fine Arts. "When will the West understand, or try to understand, the East?" wrote Okakura.

> *We Asiatics are often appalled by the curious web of facts and fancies which has been woven concerning us. We are pictured as living on the perfume of the lotus, if not on mice and cockroaches. It is either impotent fanaticism or else abject voluptuousness. Indian spirituality has been derided as ignorance, Chinese sobriety as stupidity, Japanese patriotism as the result of fatalism. It has been said that we are less sensible to pain and wounds on account of the callousness of our nervous organisation! Why not amuse yourself at our expense? Asia returns the compliment . . . You may laugh at us for having 'too much tea,' but may we not suspect that you of the West have 'no tea' in your constitution? Strangely enough, humanity has so far met in the tea-cup. It is the only Asiatic ceremonial which commands universal esteem. Let us stop the*

continents from hurling epigrams at each other, and be sadder if not wiser by the mutual gain of half a hemisphere.[42]

Okakura mourned the collapse of serious tea ritual, urged its reflowering, and drew incisively from historical Asian sources to pinpoint the roots of Japanese refinement. It is *cha-no-yu*, the way of tea, born of the Sung–Zen conjunction in Kamakura times, and taken to its renaissance during the Ashikaga Shogunate, when Kyoto was the artistic epicenter of all Asia.

Yoshimasa's mountain villa was ruined in wartime, then reconstructed by Toyomori Miyagi. The late fifteenth century must mark the beginning of the modern world, of a paradoxical conscience, one fraught with the unchecked impulse to annihilate and the ever-redeeming act of delicate penance. Japanese balance grew quietly out of turmoil. Wending, fondling inclines, steadied with two-dozen species of moss, incremental seedlings, virescence that bathes the hillsides in edible life – this vision in humane proportion always borrowed from heaven a formula propitious to human limits and limitless dreams. Curiously, the word for *dreaming* was sacred in Japan and could be used only once in 100 lines of renga poetry. This dream – of a sacred space embodying all of Japan, of a pure land, of war and peace and simplicity of a cup of tea – became the predominant métier of the Japanese, and one which throws up the cry of solace in the face of that which would tear it down. There is a biogeography, if you will, a humanist's testimonial to terrain that has given birth to cultural ambiguity: Moghul gardens, mountain retreats in Lu-shan, Etruscan villas, the Ginkaku-jis. These hopes for resurrection imply an outer victory of the soul, a directional talent that must reflect a neural configuration. This "talent" is the same burial-mound reverence found in Shanidar, of neatly arrayed crania in Choukoutien; of those moments, in other words, throughout our history, when the soul has spoken to stamp meaning on illusion. And what of Japanese illusion?

The foot plods lovingly through monkey-infested highland on Mount Hiei; the eye reflects off blue mist. Deep in the shade of the Shinto shrine is the thud of drum through endless high copse. A rice festival is in full play. Here, civilization has grown up. And only a few 100 kilometers away, the first atom bomb was detonated. Mankind seems bent on this contradiction: create, then destroy. Japan's terrible dilemma is its total embrace of a pattern, now additionally strained, which carried each generation farther from some beginning in these forests; where religion maturated in the woodcutter's pile and snowy yard and in the censer's fragrance. Clay gable shingles vent the rain throughout Kyoto, nourishing the Earth. Classical *kanji* characters enshrine the poet's soul in an animistic language that endeavors to hold on to such beginnings but is hard-pressed to do so. Tokyo speech and action is all *tokyu*, express, given to an absolutism, microchips, the *shinkansen* (a 177-kilometer per hour commuter train), every camera-angle denuded, every convenience made over in the new sun and steel. We can only speculate on the troubles besetting the late Yukio Mishima, the last of the samurai-philosopher poets to commit seppuku, in the early 1970s. I have not been able to find any record of a samurai who ever killed himself with a gun. Yukio cut out his stomach before a crowd.

There is still civility in the surviving Fujiwara countenance, indisputable royal blood that gleams from the past. And at Koya-san, the "Mother of Japanese Civilization," Kukai still sits meditating over the rear cemetary.

Gentle giggles permeate the woman-folk in rural villages. Hills (some aborted to chaos, concrete overpasses) make up 88 percent of the country, their deep forests still refulgent, intact, because of Japan's policy of importing most of its timber. These hills gave birth to yamabushi, and to the first great wave of religious tolerance, of syncretism; and here, in Kyoto — the very city laid down according to cosmic principles earlier embodied at Ch'ang-an — salesgirls gesturing that all the world should buy Amita Damascene, erotic silk kimonos, paper fans, imari and kutani, painted screens, lanterns, tea whisks. The ascetic cults, haiku in the northern hills, fond curvatures, steaming volcanos, smooth eyes of the old mottled folk; and knowing, merciful wisdom shuffling silently up backstreets in glazed slippers; the atmospheric briskness, the hundreds of monasteries and temples: Here the terror of militarism confronts the sovereign passion of a community ensconsed from time immemorial in nature.

Since Ginkaku-ji, since Meiji, since World War II, Japan has tasted the contradiction that now bears down on everyone. Kyoto is indeed a model city. Like Old Jerusalem (or Calcutta or San Francisco) it contains its own addulcent blend of vitality in the centuries of art, religion and continuous greenery. It all might have been destroyed if not for the saving efforts of one Harvard professor, Serge Elisséev, who prevailed on the U.S. military to spare this one city, at least, when every other city in Japan was being fire bombed.

America's recent pressure on Japan to again militarize herself against the illusory window of vulnerability, and Japan's acceptance, is but one more example of the West's historical blindness to what Asia has to teach us. We have abandoned the notion of the miniature, so magnificently preserved in one "curiously trained pine tree" after another. A Boston scientist by the name of Edward Morse, one of the first to visit Meiji Japan and later to become the president of the American Academy for the Advancement of Science, put it this way:

> *A foreigner, after remaining a few months in Japan, slowly begins to realize that, whereas he thought he could teach the*

Japanese everything, he finds to his amazement and chagrin, that those virtues or attributes which under the name of humanity are the burden of our moral teaching at home, the Japanese seem to have been born with.[43]

Kyoto, the historical vortex of profoundly embattled sentiments, has engendered a stable image, a lotus of great pertinance to a troubled world. Not that Kyoto herself remained free of tragedy. Yet something speaks eternally to the soul from Kyoto; lends clarity to the churning of human contradiction – that penchant for paradox that is our genius and terribly unclear destiny.

These days I wander Kyoto's streets in the bliss of so much hidden meaning, discovering every secret garden as the epitome of someone's private dream, gardens that blush with solutions, that cannot help yielding up their every solace and devout peace to the accepting heart.

Kyoto! With your monkeys on Mount Hiei. Caught off guard am I. For even now a Shinto sanctuary feels obliged to keep caged two tormented monkeys so as to ensure the propitious eye of a nearby god, said to be incarnate in the monkeys, who might otherwise, if freed, disappear into the mountains, forsaking the temple. Religion is not without its hypocrisy. Westerners are more than familiar with the subject. But a greater reality presses home the true substance of Kyoto; in the quiet sentience and proliferation of holy men in their colored robes, praying beneath their wooden shrines, washing hands in stone basins, and all within the deep cyprus glens. Those almond eyes from central Asia's Pure Land Buddhist rock art, Modigliani-like, cobalt, haunting; and the sinuous stride that comports the total sensation.

Sampling every delicacy, every vantage, all that paradise that humankind has ever been capable of fostering in the space of a few bamboo mats and surrounding humble plot of evergreen grove.

Ah, Ginkaku-ji!
My hut in the Spring.
Indeed, there's nothing in it –
This teahouse full of moonlight![44]

NOTES

[1] Paul Varley (with Ivan Morris and Nabuko Morris), *Samurai* (New York: Delacorte, 1970).

[2] S. R. Turnbull, *The Samurai: A Military History* (New York: Macmillan, 1977).

[3] Edwin O. Reischauer and Albert M. Craig, *Japan, Tradition & Transformation* (Cambridge: Harvard University Press, 1978).

[4] Ruth Benedict, *The Chrysanthemum and the Sword* (Boston: Houston Mifflin Company, 1946).

[5] Abraham Kaplan, "Synthesis of Contradictions," (Aspen Design Conference, Summer, 1979).

[6] J. M. Scott, *The Tea Story* (London: 1964).

[7] Tanikawa Tetsuzo, "The Esthetics of Chanoyu, Part 4," *Chanoyu Quarterly*, no. 27, 1981.

[8] A. Kaplan, *op cit.*

[9] Michael Sullivan, "Pictorial Art and the Attitude Toward Nature in Ancient China," *Art Bulletin*, March, 1954; and Yoshiaki Shimizu, Carolyn Wheelwright, eds., *Japanese Ink Paintings: From American Collections, The Muromachi Period* (Princeton: Princeton University Press, 1976).

[10] Haruki Kageyama, *The Arts of Shinto* (Tokyo: Weatherhill, 1973).

[11] Sen'o Tanaka, *The Tea Ceremony* (Tokyo: Kodansha International, 1973).

[13] Mario Bussagli, *Central Asian Painting from Afghanistan to Sinkiang* (New York: Rizzoli International, 1979).

[13] Bussagli, *ibid.*; and Sen XV Soshitsu, *Tea Life, Tea Mind* (Tokyo: Weatherhill, 1981). See also, Shoshitsu, "Understanding Chanoyu" *Chanoyu Quarterly*, Spring, 1970.

[14] Loraine Kuck, *The World of the Japanese Garden* (Tokyo: Weatherhill, 1980).

[15] James Cahill, *Chinese Painting* (New York: Rizzoli International, 1970); and Joji Okazaki, *Pure Land Buddhist Painting* (Tokyo: Kodansha International, 1969). See also Peter C. Swann, *Chinese Painting* (Paris, 1958).

[16] Taichung, National Palace Museum, *Three Hundred Masterpieces of Chinese Painting in the Palace Museum*, 6 vols. (Tokyo: 1959).

[17] Esther Jacobson Leong, "Place and Passage in the Chinese Arts: Visual Images and Poetic Analogues," *Critical Inquiry* (Winter 1976): 345–368.

[18] Michael Sullivan, *Birth of Landscape Painting in China* (Berkeley: University of California Press, 1962); and George Rowley, *Principles of Chinese Painting* (Princeton: Princeton University Press, 1974). See also J. Prim-Moller, *Chinese Buddhist Monasteries* (New York: Oxford University Press, 1937). See also John Major, *Topography and Cosmology in Early Han Thought: Chapter Four of the Huai-nan-tzu*, Harvard University Dissertation in Chinese Religion.

[19] J. Cahill, *op cit.*

[20] Kuo Hsi, *An Essay on Landscape Painting*, trans. by S. Sakanishi (London: Wisdom of the East Series, 1936). See also Benjamin March, "Linear Perspective in Chinese Painting," *Eastern Art*, III, 1931; and Alan Priest, "Southern Sung Landscapes: The Horizontal Scrolls," *Metropolitan Museum of Art Bulletin*, March, 1950; Osvald Siren, *Chinese Painting: Leading Masters and Principles*, 7 vols. (London: 1956–1958); Saburo Ienaga, *Japanese Art: A Cultural Appreciation* (Tokyo: Weatherhill/Heibonsha, 1979); Kojiro Tomita, *Portfolio of Chinese Paintings in the Boston Museum of Fine Arts*, 2nd ed. (Cambridge: 1938); Noritake Tsuda, *Handbook of Japanese Art*, (Tokyo: Charles Tuttle Company, 1981); W. Liebenthal, "Immortality of the Soul in Chinese Thought," *Momumenta Nipponica*, VIII, 1952, p. 378; Ko Hung, *Alchemy, Medicine and Religion in the China of AD 320: the Nei P'ien of Ko Hung -Pao-p'u tzu-*, trans. and ed. by James R. Ware (Cambridge: MIT Press, 1966); J. D. Frodsham, "The Origins of Chinese Nature Poetry," *Asia Major*, VIII, 1960: pp. 68-103; Henri Dore, M. Kennelly, *et al.*, *Researches into Chinese Superstitions* (Taipei: Ch'eng-Wen Publishers, 1968).

[21] Fosco Maraini, *Meeting with Japan*, trans. Eric Mosbacher (London: Hutchinson, 1959).

[11] Henry James Coleridge, *The Life and Letters of St. Francis Xavier*, 2 vols. (London: Burn and Oates, 1980).
[7] Varley, *op cit.*
[24] Albert Craig and Donal Shively, eds., *Personality in Japanese History* (Berkeley: University of California Press, 1970).
[25] Allen Grapard, "Flying Mountains and Walkers of Emptiness: Toward a Definition of Sacred Space in Japanese Religions," *History of Religions* 20 (1982).
[26] L. Kuck, *op cit.*
[27] Walter Dening, *The Life of Toyotomi Hideyoshi* (Tokyo: Hokuseido, 1955).
[28] Noel Perrin, *Giving Up the Gun: Japan's Reversion to the Sword, 1543–1879* (Boston: David Godine, 1979) – relied upon extensively throughout.
[29] Varley, *op cit.*; and Reischauer and Craig, *op cit.*
[30] Kenneth Butler, "The Heike Monogatari and the Japanese Warrior Ethic," Harvard Journal of Asiatic Studies, vol. 29, pp. 93–108.
[31] Peter Duus, *Feudalism in Japan*, 2nd ed. (New York: Alfred Knopf, 1976).
[32] W. Dening, *op cit.*
[33] Miyamoto Musashi, *A Book of Five Rings*, trans. Victor Harris (New York: Overlook Press, 1974).
[34] Perrin, *Giving Up the Gun, op cit.*
[35] C. R. Boxer, *Jan Compagnie in Japan 1600*–1817 (The Hague, 1936).
[37] N. Perrin, *op cit.*
[37] Henry Heusken, *Japan Journal 1855–1861*, trans. Corput and Wilson (New Brunswick, NJ: Rutgers University Press, 1964).
[38] Henry Smith, ed., *Learning from Shogun: Japanese History and Western Fantasy* (Santa Barbara, CA: University of California Program in Asian Studies, 1980).
[39] N. Perrin, *op cit.*
[40] H. Heusken, *op cit.*
[41] H. Smith, *op cit.*
[42] Kakuzo Okakura, *The Book of Tea* (New York: Dover, 1964), p. 3.
[43] Quoted in H. Smith, *op cit.*
[44] Haiku by M. Tobias.

Chapter 6

A Tale of Two Centuries

Steam, hedgerow, barbed-wire fences, and electricity transformed the icon of wilderness into the billboards of democracy. Between Wordsworth and Dickens, Jefferson and Velben, this enormous change over has reconditioned the human perception of nature. Though we may question the speed at which the drum is beaten, we can not but acknowledge that the drum has always been within us.

ROUSHAM

> *[I]t was the season of light, it was the season of darkness, it was the spring of hope, it was the winter of despair, we had everything before us, we had nothing before us, we were all going direct to heaven, we were all going direct the other way.*
>
> Charles Dickens, *A Tale of Two Cities*

Jesus was buried there. St. Francis plucked its flowers. Edmund Spencer wrote sonnets to its baying lambs, to its stars, manger, and virgin. The Japanese called it "borrowed scenery." But to Westerners this original garden was no mere symbol. Rather, it had become the world itself, much to the credit of topographical painters of Flanders and the

Netherlands and of southern Italy, who introduced the Biblical garden theme to the British upper class. If the madding crowd ran riot with "belittling influences" (Johann Schiller's phrase), there was always this refuge to escape to. Byron's Childe Harold toured its turbulent perimeters, sanctioning savagery, upholding a new, revitalized poetic — the wilderness. By the 1830s in London, Parliament was concerned to appease the public's need for nature; plans were adopted for the first city park, Birkenhead in 1843. In America, Frederick Law Olmsted was busy drawing up his own plans for paradise in New York, Central Park. These little tributes of human organization date back to Nilotic garden estates in pre-dynastic Egypt, to fervent historical crossroads in Persia, but — without belaboring it — we can trace the idea of the *pure landscape* to the work of William Kent.

Kent's garden, Rousham, laid out in 1740 at Buckinghamshire, had, according to Horace Walpole, "leapt the fence and saw that all nature was a garden." It was Kent's style to keep all evidence of humanity from view. "Nature abhores a straight line," he said. Rousseau was enamored of Kent and had his own island tomb at Ermenonville created according to his tenets. Jean Jacques based his notions of ecological contract on the intellectualized and dubious evidence of ethnographic data that he read in journals about Central American aborigines. But Kent strove headlong into local thickets, there reinventing nature and offering its application to the gentry. Lake poets fell under its trance. Kent had determined to rid his domesticated wilderness of stereotypic elements, the nestling brook and cleared greens and requisite ungulates. He wanted the beyond, and Buckinghamshire was as good a place as any to fashion it. The land was suffused in legacy, established with agriculture since the time of the Norman Conquest. Milton had lived in a cottage nearby, and Thomas Gray had composed his own "Elegy" on a tombstone in the country church. Forested hills could be scene in the south, the fertile Vale of Aylesbury to the north. This was the Britain that everyone had known for 700 years. Roman, even megalithic, ruins attested to a far greater acquaintance. Kent crossed the various components, left lawns unmowed, trees unpruned, and inspired the first *rustic* retreat. He died in 1748. Nor was his untamed England to survive.

STEAM

Following the patent in 1698 for Thomas Savary's "fire engine," steam power enabled miners to drain water from underground coal mines. In fact as early as the fourteenth century, the English had started using coal. The switch from wood had been implemented by Henry II for the people of Newcastle. During a journey to Scotland in the fifteenth century, Pope Pius II commented on his surprise at seeing people in rags lining up at church doors to "receive for alms pieces of black stone." With deep coal mining made possible, heat could be generated in the crucible of a developing iron industry. Abraham Derby succeeded in forging the metal from coke. Iron and steel would replace wood in England's navy; and coal-fired brick would change the style of British home insulation. At the height of England's poetic renaissance, the transforming effects of steam were also underway. Following the invention of Savary, James Watt's firm sold more than 500 engines, each device consisting of two power strokes per cycle. Flour, rolling, and textile mills could employ more labor as a result of more work, though this ratio would be reversed. But in the beginning, the expanding operations began an era of capital appropriation, the very pivot upon which the modern West got so rich, and Marx so vituperative. In 1800, the absence of industrial competition for steam power enabled capital to create its own market. The downside of labor-to-technology had largely to do with the dispossession of the rural poor. In 1820, for example, the Duchess of Sutherland sent 15,000 tenants packing in a single command, replacing them with 131,000 sheep. Pasture land had become more valuable than tillage land.

What fully revolutionized the steam source of energy was Oliver Evans' engine for converting high-pressure coal into energy at a 1 percent efficiency. Today's steam engines function at a 40 percent-plus rate. Evans' upgraded model found jubilant applicability in the transportation sector. A steam locomotive called the *Rocket* managed to cover twelve momentous miles in a mere 53 minutes. This infant train was the rage, and just in time: by 1800, nearly 14 percent of British and Welsh countryside was taken up to grow feed for horses. An alternate means of travel was deemed imperative. The founder of thermodynamic principles Sadi Carnot would write in his memoirs:

> *The study of these machines is of the greatest interest, for their importance is enormous and their use increases day by day. Already the heat engine works our mines, propels our ships, deepens our harbors and rivers, forges iron, fashions wood, grinds our grain, spins and weaves our fabrics, pulls the heaviest loads, etc. . . . Perhaps there is no single factory in England whose existence is not founded on the use of these elements and in which they are not abundantly applied.*[1]

The European population would increase by 200 million in the nineteenth century, fueled by seemingly innocuous steam. London's population rose 20 percent in the 1840s alone, and with it a phenomenon now called the suburb.

Water power replaced human labor for the first time in the operation of the Arkwright spinning frame in 1768. Now, labor dispossession afflicted urban as well as rural populations. Machine brakers arose throughout central England, following the example of one outraged Ned Ludd. By the 1820s the labor unions eclipsd the Luddites, even though Lord Byron and others defended the brakers' cause before the Parliament, and Samuel Butler (in his novel *Erewhon*), warned of the coming replacement of the human soul by machines. Such qualms were squelched with big profits for the rich and continued poverty for the poor. England believed thoroughly in the trickle-down notion and went after money with single-minded zeal. By the 1830s, for example, she was importing *skeletons* from all over Europe and Africa to use as fertilizer. The chemist Justin von Liebig wrote that the Industrial Revolution was stealing the fertility, the very "heart-blood," from the rest of the civilized world. Engel correctly predicted that the steam engine (capitalism) would smash the wheelbarrow (traditional livelihood). By Victorian times, utopia would be wrought in iron. Renaissance work shifts had been the standard 8 hours, but nineteenth century British days consisted of 16 hours. The heat was on.

To understand England's chilling changeover from a rural to a technological society, we need only compare two literary effusions separated by some 50 years, but by much more. Listen to Charles Dickens' often-cited portrayal of Coketown, from his novel *Hard Times*:

> *A town of red brick, or of brick that would have been red if the smoke and ashes had allowed it; but as matters stood it was a town of unnatural red like the painted face of a savage . . . serpents of smoke trailed themselves for ever and ever . . . It has a black canal in it and a river that ran purple with ill-smelling dye, and vast piles of building full of windows where there was a rattling and a trembling all day long . . . melancholy madness . . . people equally like one another, who all went in and out at the same hours, with the same sound upon the same pavement to do the same work, and to whom every day was the same as yesterday and tomorrow, and every year the counterpart of the last and the next.*[2]

Dickens, who'd been likened to Shakespeare while still in his twenties, was riding the crest of unbelievable fame. But he was not entirely happy, as such passages demonstrate. Industry had swallowed up the garden of Dickens' childhood romping grounds in Kent. The slate-roofed, thatched-covered cottage— traditional Britain of Edwardian and Victorian times, the very stuff of "Tintern Abbey," Hardy's Wessen, of *Wuthering Heights* and George Eliot's Warwickshire—had become but a forlorn daydream. In *Little Dorrit* (1857), and *Bleak House* (1853), Dickens chronicled the pace and ardor of demolition across the English psyche. The American chapters of *Martin Chuzzlewit* (1844) and

American Notes (1842) were no less damning, full of murderers and braggarts, of a wilderness demoralized.

Just fifty years before Coketown, Wordsworth was still free to reflect on his "Solitary Reaper."

> *Whate'er the theme, the Maiden sang*
> *As if her song could have no ending;*
> *I saw her singing at her work,*
> *And o'er the sickle bending —*
> *I listened, motionless and still;*
> *And, as I mounted up the hill,*
> *The music in my heart I bore,*
> *Long after it was heard no more.*[3]

Wordsworth died in 1850, having heard the shrill, grinding song of industry to cast him from the heavens of his muse.

JEFFERSON

Thomas Jefferson's vision was very much a common sense orthodoxy. Its passion was its pragmatism, a legendary breadth of inquiry, and Jefferson's insistent belief in the rights of the individual within the state of nature. In his concern with big business he anticipated the Marxian critique of exploitation. His *Declaration of Independence*, written the same year as Adam Smith's *Wealth of Nations*, provided the first political context for man in the environment, holding both entities inalienably joined.

Eighteenth-century Americans had made it clear that they were prepared to engage nature. The American Philosophical Society addressed issues pertaining to wilderness through the dedicated voices of botanists William Bartram, William Byrd, Ben Franklin, and Jefferson himself. Franklin had helped found the society. He taught himself to swim, recommended the strange pastime to others, and flirted with the wilderness on an unprecedented scale in the invention of a lightning rod. His "Pennsylvania fireplace," now called the Franklin stove, was devised to economize on wood. In Paris, with Jefferson, the two men encouraged the international exchange of seeds and horticultural ideas. Jefferson would later experiment at Monticello. His colleague, Dr. Nicholas Collin, delivered a famous "Essay on Natural Philosphy and its Relationship to the Development of the New World," the first ecological credo in America. Collin called for moderation in the axing of forests, which he designated "a natural treasure." Another Jefferson contemporary, Charles Peale, started the first natural history museum in the United States in 1785. Named the "great school of Nature," a branch of the enterprise opened in Baltimore 10 years later, then in Cincinnati. Throughout this early environmental movement, Jefferson stands out. To the president of Harvard he had written that students must be apprised of the potential for research in such unexplored topics as minerology and natural history. And during his brief government retirement, he wrote his one book, *Notes on Virginia*, which remains a precious, important statement, both infuriating – when he predicates a racist sociobiology – and rhapsodic when waxing on the benefits of true republicanism. Jefferson's book was a milestone in mankind's understanding of the physical world, an understanding imbued with more than mere science. In his description of the natural bridge formation, he prescribed a psychological vantage point. Acquainted with the literary conceits of sublimity, Jefferson could describe the Potomac breaking through the Blue Ridge, and conclude that the scene was worth a voyage across the Atlantic.[4]

THE FRONTIER

"I wish to have rural strength and religion for my children . . . and I wish city facility and polish. I find with chagrin that I cannot have both," wrote Ralph Waldo Emerson.[5] But in three days a man could reach any frontier settlement and get himself just the right mixture – and a comfortable cabin, no less. The wild prairie grasses would feed his livestock while he grew his first crop, which was invariably corn, the first year's yield known as sod. Seed was merely dropped into broken-open turf, cleaved with an axe, the soil not mellow enough to take a harrow. Homesteading might cost $1.25 per acre, as dictated by the U.S. Land Office, but breaking the prairie demanded another $2 an acre. A quarter section could cost $1000 in net expences, and the possibility of grassland fire wiping out investment was omnipresent, except during winter when the winds drove the sleepless farmer to insanity and the snowpack threatened to bury the whole farm. Compare that $1000 expenditure in the wilderness to Thoreau's expenses of $61.99 his first year at Walden Pond. Was Thoreau so remarkably frugal? No. By 1845, Concord, Massachusetts was no wilderness. The Midwest was. And wilderness had become an expensive undertaking.

After he settled, the coming summer would see the farmer sowing wheat between the previous year's corn stalks. Crops then were not even half the size they are today. Agricultural mowing and threshing machines would not come into use until 1846. The settler ate cornmeal, hominy, potatoes, pork, and wheat flour. The first grist mill was established to supply these farmers in 1830 at Dayton, Ohio. It was at this time that the American landscape painter George Catlin (1796–1872) was busy in the Great Plains painting *natural man*. He pleaded for the preservation of these plains as a permanent home for Indians, "a nation's park containing man and beast, in all the wildness . . . of their nature's beauty."[7] Washington Irving, Francis Parkman, James Fennimore Cooper, Timothy Dwight (president of Yale) – all rallied around that entreaty. But De Tocqueville was more realistic: "In Europe people talk a great deal of the wilds of America, but Americans themselves never think about them; their eyes are fixed upon another sight, the march across these wilds, draining swamps, turning the course of rivers, peopling solitudes and subduing nature."[8] Indeed, the soil was largely ill-drained, contributing to an abundance of stagnant pools throughout the midwest, breeding rampant agues. River towns experienced outbreaks of malaria.

The Pennroyal was the first important body of grassland encountered in trans-Appalachian settlement. Pioneers called them "the barrens." Soon the term was replaced by the local French word, *prarie,* a word Americans did not know when Kentucky was settled. Surrounding these open lands, between Louisville and the Tennessee line, the remaining forests were scarred: testimony to the indigenous Indian practice of burning out the dense stands, a practice that had ended some 50 years prior to the intruding white settlers. All of America was once bound in thick glen. The first paleolithic inhabitants of this land used fire to capture game. It was they who actually altered the ecosystem. But the Puritans had similar effects on the forests of New England. Nearly 50 percent of all trees fell beneath their axe. By the 1850s New England had been evacuated. The two industries that previously brought any income to the region – lumber and potash for soap – had been devastated by overcutting. But it was not industrial greed entirely that spawned the blitz. Puritan children suffered from chronic bronchial infections, and the parents attributed this "tree sickness" to the dank thickets. Claustrophobia set in. They needed to clear out, to head West where there was rumored open space, horizons. We can actually trace the indignance and impetuosity of the earliest settlers to this Puritan psychosis.

In 1672, the Puritan John Josselyn described Sugarloaf Mountain in Connecticut as "a rude heap of massive stones, daunting, terrible, full of rocky hills as thick as mole hills in a meadow"; in 1679 Father Louis Hennepin, the first white man to report on Niagara Falls, called it "a horrible precipice . . . hideous . . . dismal." Cotton Mather's ethic made clear that "what is not useful is vicious."[9] When Eleazar Wheelock founded Dartmouth College for the Indians, the school's motto became *"Vox clamantis in deserto,"* a voice cries in the wilderness. But solace was on the way.

With the support of the first canals and railroads, the last great wave of settlers established themselves throughout all of Iowa and Missouri. Horses replaced oxen. Farms were divided into fields of alternate corn, wheat, oat, clover and grass; barns were erected (those salubrious old red ones) to store grain and husband lifestock. And, finally, fences marked the true character of American land tenure. To this day, in most other nonindustrialized countries, there is no such thing as a fence. It went against everything Jefferson had promoted, in fact. Capitalist to the very bone, the fence was synonymous with the word accumulation, which was Hamilton's own motto and the basis for all wealth and big business. Daniel Webster, another Dartmouth lad, had shared Jefferson's sympathies. In 1820 he wrote that "the freest government, if it could exist, would not be long acceptable if the tendency of the laws was to create a rapid accumulation of property in few hands, and to render the great mass of the population dependent and penniless." Clearly, at least in the beginning, each farmer was entitled to his share and had rather egalitarian chances of getting his foot in the door. The transition from that diffusity of privilege to the period of monopoly is of course our story, with its woes and successes; the story of land grabbing and the rising ecological movement. Even before his presidency, Washington decried the speculator's land rage. William Wyant has keenly assessed the history of this free-for-all, and it is to his work that I am indebted.[10]

Robert Morris, who helped finance the American Revolution, was foremost among the earliest real estate professionals, and when he died he owed 12 million dollars in unpaid land-seizure debts. The Northwest Ordinance of 1787 offered mile-square sections to the farmer, a size greatly reduced to 160 "quarter section" acres in the Homestead Act a century later. While the Indians were seeing their territory stolen from under them, the federal government was overflowing with the 233 million acres ceded to it by seven of the original colonies; colonies which had, themselves, obtained or stolen land by conquest, crown rights, or paltry sums (New York for under 30 dollars for a starter). Jefferson more than doubled the original amount by purchasing from Napoleon the Louisiana Territory at a cost of $23,213,000. In 1819 Spain gave up Florida to Monroe; in 1848 President Polk wrung 334 million acres from Mexico; Franklin Pierce extracted another 18 million Mexican acres 5 years later. And finally, Seward's "Folly" added an additional 365,481,600 acres from Alaska in 1867, at the shrewd price of $7 million. Two years after the Alaskan purchase, the Central and Union Pacific railroads met at Promontory Point, Utah. America was up for grabs, and had been for nearly a century. The railroad companies enjoyed windfall loans and grants exceeding two Colorado's worth of land. Millions of acres were given away to logging and cattle interests, but only token efforts (unsuccessful at that) were legislated on behalf of the freed slaves. In Texas, any head of a family, excluding Negroes and Indians, could come by 4600 acres of land. So common was the giveaway ingrained in the economic spirit that a Senator Ralph Cameron from Arizona could openly boast of his ability to make a bundle off the Grand Canyon, if only Congress would let him. Congress *did* designate all natural resources as "leasable" with the General Mining Law of 1872, a crude concept still on the books. Land was perceived as cheap enough to absorb all manner of fraud, which that particular law inspired, courting the ambiguity later inscribed in the 1934 Taylor Grazing Act, the 1960 Multiple Use and Sustained Yield Act, and the Wilderness Act of 1964 which defined wilderness, ironically, as that place where human beings "do not remain." But in fact the democratization of wilderness in America has always demanded the satisfaction of conflicting interests, be they fish and game, recreation, range, timber and watershed, industrial and aesthetic. Whatever legal or perceptual pretensions, the American landscape has been completely penned in, circumscribed by satellite, and culturally converted.

Culture, the invasion of immigrants, came alongside white pine, the primary lumber of the frontier that was shipped from the Great Lakes through the newly opened Michigan Canal (1848). Wells were dug, fields drained, and the General Land Survey established to divy up its now-too-recognizable pattern of rectangular land sections for public domain, Four families per square mile was the recognized optimum. Railroads offered ready-cut houses, the first prefabs. Mason jars could store fruits and pickling vegetables. State fairs grew up, as did family orchards for making apple butter and cider. Even nurseries, featuring ornamental trees and shrubs, facilitated otherwise sterile, often deadly homesteads. Carl Sauer has provided an intimate record of these droves.[11] In 1849, a Lieutenant Simpson, passing through the midwest would speak with loathing of the "ever-recurring sameness of plain and mountain, plant and living thing."[12] In that same year, Thoreau first visited Cape Cod and addressed these lines to Columbus:

Quit, now, full of heart and comfort;
These rude shores, they are earth;
Where the rosy clouds are parting,
Where the blessed isles loom forth.[13]

MARX

We go back to Dickens' London, to that nerve of memory in which is stored the impulses of past aspirations – pharoahs, warring dynasties, painters, archaeologists, cowrie shells. Cowrie shells? We come to the British Museum, with its examples of primitive barter – cowrie shells, stone-age economics. That is where Karl Marx began his studies, from there proceeding to absorb the whole history of mankind. In the 1850s he camped out there, like a hermit bedding down in voluminous stacks, reading, brooding, waiting. Marx's *Economic and Philosophic Manu-*

scripts of 1844 came at a desperate time for Britain. A disastrous fungal infestation in Ireland and Western Scotland, potato blight, had caused a widespread famine. Potatoes had been planted singularly as a monocrop. When they rotted, Ireland had no alternate food source. One million people died, 2 million emigrated. The potato had fostered expansive population growth, from 3 to 8 million. Ireland was in ruins. The Corn Law of 1815 granted subsidies to farmers. Now, in 1846, Prime Minister Benjamin Disraeli let pass the law's repeal, thereby instituting free trade in corn with the hopes that such hubbub would free competition and make for a healthier economy all around. For 30 years Disraeli had believed in agriculture as England's future imperative. But popular agitation, stemming from a deep rift between the rural agricultural poor and the city elite, won over the Parliament. As a result, in absence of traditional subsidies, more and more of the farmers closed up shop and moved into the cities. Britain's food base was disappearing. Its effects took 30 years to be truly felt. But by 1881 England had 100,000 fewer farm laborers than even a decade before. It meant that England was increasingly dependent on imports, for which its needed capital could only come from factory workers, themselves dispossessed farmers: a vicious cycle. And it was this burgeoning, oppressed constituency about whom Marx genuinely worried.

In his 1844 manuscripts, Marx defined humanity and nature as a necessary conjunction. Later existentialists would actually echo the same sentiment of struggle as its most fundamental level — that of pure existence. Social Darwinians reaffirmed the connection as the rudiment of all growth and consciousness, of civilization. For Marx, it was capitalism that alienated man from the soil, and — all nature being the same — from himself. In his major analyses of surplus value, set down in *Das Kapital*, he relocated the ironies that had nearly destroyed Jefferson in his dispute with Hamilton: namely, the capitalist's appropriation of the means of production, which in turn signalled the emergence of consolidated power. "By means of his work the labourer creates new value which does not belong to him, but to the capitalist," wrote Marx.[14] A man works for his wages. But he works for more than his wages, obviously, if the capitalist makes a profit by him. For Marx, the laborer by definition could make no profit but was working for his family's mere subsistence. And this was the preferred state of nature. It was the capitalist who went beyond that condition, forced additional sweat from his workers, and, in the form of capital, reinvested that excess for new value which he alone commanded and enjoyed. Thus there was a surplus in society, dominated by an elite bourgeois, the property owners. Land was victimized, equally exploited. The very biology of nature would rise up, and in its name, class struggle would become invevitable. Marx himself struggled all his life.

Marx was not appreciated in the United States until the early twentieth century, but he had pinpointed the American style: "Capitalist production disturbs the circulation of matter between man and soil . . . it develops technology, and the combining together of various processes into a social whole, only by sapping the original sources of all wealth — soil and the labourer."[15] But elsewhere, in his *Grundrisse*, he had perceived nature as the laborer's own proving ground, and this contradiction was later seized upon by Engels, who in 1876 wrote an essay drawing on the work of George Marsh in Italy, condemning man's presumption of superiority over nature, citing Spanish planters in Cuba who were slashing and burning their way to bedrock within a single generation. Neither friend could fully grasp the coming consequences of the capitalist accumulation. Nobody could.

George Marsh, author of two important works, *Man and Nature* (1864) and *The Earth as Modified by Human Action* (1874), had been a Vermont naturalist and farmer prior to his appointment, under Lincoln, as U.S. Ambassador to Italy, where he spent a year with very little to do other than to write. Marsh warned of America's downfall, basing his conclusions on the same syndrome that destroyed Rome's forests and Rome itself. We had become a nomadic people, he said, and lost the means of insuring permanence of "relations" to the earth, "the meadows and the pastures, to the rain and the dews of heaven, to the springs and rivulets with which waters water the earth."[16] His term "relations" is curiously placed. Marx had constantly used the word to indict the capitalist skewing of nature, "contradictions" in the very root of all production and society. Marsh and Marx usher in two distinct themes during the latter half of the nineteenth century, and we will pursue both of them. In Marsh was the grand spokesman for the coming era of conservationism that would reach its zenith under John Muir. But Marsh's inspiration had come almost verbatim from Thoreau. And in Marx we may perceive the technocratic enormity which was overtaking Britain and the United States.

ENERGY AND MONEY

American inventiveness led ineluctably to England's decline as a producer, but to her rise in banking stardom. Multiple innovations in medicine, agriculture, and transportation dramatically altered demographic adjustments, catering to competition. But competition kills competition. Rockefeller in oil, Hill and Harriman in railways, Carnegie and Morgan in steel — these were the beasts that resulted, despite the Sherman Antitrust Act (1890). More and more immigrants arrived at New York's back door. The United States grew from 4 million in 1790 to 23 million in 1850; by 1900 we had tripled our population to nearly 70 million.

No other industry had such an effect as did steel during the late 1800s. Sir Henry Bessemer's converter, which cleansed molten iron of carbon impurities, enabled the United States to replace England as the world's greatest producer of the metal. After 1873 the North American railway system was fashioned of such steel. It was the railway – financed with nearly 3 billion dollars of British investment capital – that set the pace for America's economy and her exploitation of water, gas, mines, labor, and product. As Britain lent out more and more sterling, she instituted the loan, an economic arrangement allowing for the forwarding of imperialist motives under the appealing guise of material promise. But her role as international moneylender had its most profound effect on the developing world, where she came to rely on favorable balances – in Turkey, Japan, and India -- to settle her American deficits. Cotton mills were no longer the big money makers but, rather, metallurgy and energy. In effect, Great Britain had outsmarted herself by lending the capital and knowhow to the Americans who were, within a matter of decades, able to manufacture on their own. By 1898 the cost of steel rails in the United States had dropped nearly 100 percent. Sixteen dollars a ton was the going rate. Protectionist tariffs were implemented, first by Bismarck in 1879 (Germany had seen half of its populous going urban) and then in France and Russia. But England kept open her free trade and enjoyed continuous annual surplus, primarily through her banking and the unleashing of textiles on the Asian market.

But the trend was set: England was isolated, with an expanding population, decreasing agricultual base, and decreasing share in the world market. It was only a matter of time before she would have to turn to the Common Market in our own era. Her supremacy had been lost precisely through the highest participation in capitalism. This pattern suggests a physical property to competition: that a center must always emerge to the relative erosion of peripheral participants. The center was always traditionally determined by its land. More fundamentally, we can now see how Europe's colonialism had engendered massive world imbalances in consumption, the brunt of all trade benefits going to England. With America, and the new self-generating technologies in the field, England was forced to recognize finite resources. In that respect, she was ahead of the game. As for the Americans, the word *finite* was a century away from any working vocabulary.

In every discipline the nineteenth century witnessed spectacular insights. Justus Liebig, a German chemist at the University of Giessen made massive headway in the development of artificial fertilizer. Sir John Lawes, in experimenting with manures, then patented a process for producing superphosphate synthetics. By the 1870s, 40,000 tons were being produced in London annually. With the discovery of bacteria, the world of agriculture was on the brink of a new age. In medicine, certain toxins from pathogenic microorganisms were fabricated; all modern antibiotics trace their origin from the studies of Pasteur, Joseph Lister, and Paul Ehrlich. Alfred Nobel had invented dynamite in the 1860s, and H.Y. Castner had devised a cheap method for the manufacture of sodium, allowing for commercial aluminum production. In fuel technologies, cheap oil exerted growing competitive pressure on traditional coal and wood stocks. From its original tests in 1854 by the Pennsylvania Rock Oil Company at Titusville to Standard Oil's unleashing of 60¢ a barrel Lima crude in the 1880s, oil had become a persuasive, cost-effective alternative and appeared to be unlimited in supply. But the one truly revolutionary development, one that had built up in stages for many hundreds of years and would turn the fundamental ionic structure of all matter into easy use, was electricity. With a flick of the switch a child could illuminate a palace. There was an Übermensch in such power.

Holborn Viaduct, Pearl Street Station, New York, 1882: A grand experiment was underway, financed by Cornelius Vanderbilt and J. Pierpont Morgan and supervised by the most formidably practical scientist of his time, Thomas Edison. Four boilers were on the ground floor, and six 200-horsepower dynamo generators – capable of lighting 1200 lamps apiece – were installed on the second floor. For months workmen had dug trenches and laid the cable. It was the first commercial venture to use electricity, and would serve as the model for all others. But Edison's reciprocating steam engine, with 7.5 kilowatt unit efficiency, would be greatly "steamlined." The same year at the Paris Exhibition, Nikola Tesla's alternating generator was displayed. Within a decade, after Tesla's uniting forces with George Westinghouse, alternating current electrical motors mechanized the home throughout America. Washing machines, dishwashers, electric refrigerators, vacuum cleaners – all such gadgets owe their existence to the obscure Tesla.

Electricity changed the face of Chicago and New York. E.G. Otis' 1854 design of a hydraulic lift, in concert with the silent electric elevator, made buildings of the skyscraper species habitable. Fan ventilation – crucial in the New York summers – could not have come about but for alternating current. Light and ventilation, artificial heat, telephones, elevator, and flush toilet – these were the requirements of the skyline. At the turn of the century Chicago's twenty-one-story Masonic Temple was the largest existing building in the world, one of the seven wonders. Within a very few years, fifteen other high buildings would grace the city, surpassed only by New York, with twenty-nine skyscrapers.

In 1870 Nikolaus Otto developed a four-stroke gas fuel engine that compressed before ignition. It was to be the forerunner of all internal combustion engines. Karl Benz patented it for the automobile. Ransom Eli Olds switched from steam to internal combustion in 1900, and Henry Ford followed 3 years later. Using the idea of overhead tracks borrowed from slaughterhouses, Ford engineered the first mass production lines. His Model T would sell more than 15 million vehicles by 1928, when the line was superseded. All the while, Rudolf Diesel was working furiously to maximize heat-engine efficiencies. Elsewhere in the economy, coal and natural gas production had doubled, and petroleum quadrupled. With the Great Auto Race on, readers learned – over months of daily coverage – that the auto could go anywhere, even across Siberia. In 3 years, from 1908 to 1911, there were over 400,000 sales. And with such sales, every other industry was flying. The GNP was increasing by 50 percent annually.

DEPRESSION

It took much more than World War I to awaken Americans to the inherent instability of so much growth. But the droughts of 1934, 1936, and 1937, coming at the height of the Depression, drove the message home. Even so, during 1933 at the Chicago World's Fair, it was scientific technology that had been assembled in glittering showcases. The fair's theme was "A Century of Progress." Henry Adams, and John Ruskin before him, had warned of technology's dangerous sides. Having visited the Paris World's Fair in 1900, and seen the stellar workings of steam and electricity, Adams had written in his *Education* that such advents had utterly eliminated the biological and spiritual bases of human life. For him, "the cult of the dynamo had replaced the cult of the Virgin."[16] It was Adams, in his "Letter to American Teachers of History," who suggested that the human mind was itself vulnerable to entropy. The Chicago World's Fair exhorted mankind to fall into step with industry and in the Hall of Science projected its optimism in the sculpture of a robot, imploringly extending its either hand to man and woman. The cantankerous academic who predicted the Depression and the self-destructiveness of over-consumption, a mordant wit, professor of psychological economics, trenchant eccentric, graduate of Yale in 1884, was Thorstein Veblen.

Harold Laski called him "the most distinguished speculative mind in the field of social philosophy that America has produced since Thomas Jefferson."[18] It was Veblen's genius to expose the inordinate privileges that had accrued for business in the false guise of America's religious – that is to say, financial – backbone. Any challenge from labor unions would appear to upset the very balance of Democracy, which is right where industry wanted itself to be identified, charry of future bartering. If De Tocqueville had seen the greatest tyranny to lie in that autocratic covenant, Veblen extended the metaphor: He prophesied the disappearance of the small farmer, the abolishing of a gold standard, criminal abuse of limited natural resources stemming from the inefficiency of alienated, careless workers and their unresolved enmity with employers, and finally, "that critical point of chronic derangement in the aggregate beyond which a continued pursuit of the same strategy on the same business-like principles will result in a progressively widening margin of deficiency in the aggregate material output, and a progressive shrinking of the available means of life."[19] Seeing the coming Second World War and knowing its genesis to be as much a question of profits as the fear of despotism, Veblen stirred his younger contemporaries into restive action and called for the immediate recognition of an interdependent world. What had happened to England would happen in the United States before long, he shrilly insisted. For too many years it had been assumed in America that the feverish energies of the super rich would ooze into the destitute melting pot, infusing the bottom rungs of immigrant working class aspirants with the same luck and potential. It was Veblen who first questioned this supply-side thesis, doubting strongly whether any selfish hoarding of wealth could ever benefit a civilization that was, first and foremost, bound to natural laws. The Dust Bowl resulted from over-use of the soil. Money could similarly overuse its extraction sources.

A University of Chicago economics professor, Veblen's annual salary was $520 in 1899 when he first published his *Theory of the Leisure Class*, a book filled with what Lewis Mumford hailed as "desperately accurate circumlocutions." His *Theory of Business Enterprise* was so abrasively perfect that the trustees of the university fired him. His wife left him. He dressed like a slob, stayed potently unkempt, lashed out at the university system in America in a book he nearly titled *A Study in Total Depravity*. During the First World War he published a dry but cogent plea for world peace, translated old Icelandic epics, edited the liberal *Dial* magazine, lectured at the New School for Social Research, and finally retired in his sickly 70s to a cabin retreat in California. He died in 1929 and left a will that read, "It is my wish . . . that no tombstone, slab, epitaph, effigy, tablet, inscription, or monument of any name or nature, be set up to my memory or name in any place or at any time," etc., etc. Wish it or not, Veblen was in fact ignored by the obituary columnists, who made, if any, but light mention of the "false prophet." Though he shared a basic belief with Marx that capitalism resulted in war, Veblen did not ignore what he perceived to be the proletariate's envy of the rich. There would be no class struggle, only covetuousness. What Veblen really looked toward was clean industry, nonpolluting, austere, effective and egalitarian commerce that would enable workers to enjoy their labor and be elevated by it, reunited with their own nature.

During the Depression, it was the paleontologist Lester Ward who suggested that the human mind accounted for the schism between man and nature; that society changed more rapidly than the environment, appeared to be invulnerable to it, damaging. He assailed individualism and capitalist economy on that basis by calling for a pragmatic collectivism. This appeal resounded throughout the New Deal policies. Because human society was apparently overstepping nature's own limits, the only real safeguard against total collapse was some emergency boosting of surplus to counteract the lessons of Ricardo, of diminishing returns. Henry Wallace, the secretary of agriculture during Depression years, expounded on these callings in his own book *New Frontiers* (1934). Out of his and Roosevelt's efforts, America was able to fight a war and the Tennessee Valley Authority was transformed from its intended purpose of catalyzing harmony between man and nature into the manufacture of atomic bombs at the Oak Ridge facilities. Following the Depression, the new modus operandi had been established: The great dams were built – Hoover, Bonneville, Grand Coulee. The Civilian Conservation Corps, originally established to return unemployed single young men into work that would benefit America's natural resources, was turned over to the Federal Security Agency, in whose precinct those same men worked on defense projects.

No product of keen social inquiry better illustrates the change that occurred in the American psyche between the late 1800s and the New Deal than Professor and Mrs. Lynd's *Middletown* and *Middletown in Transition*. When a laborer had confided to Mrs. Lynd that he had stopped trying to understand the New Deal and the complexities of America in general, the reader was given the fullest realization of cultural hypertrophy. Muncie, Indiana, before and after the Depression, evidenced wrinkle and weakness. The incorruptible glow of democratic idealism was washed up. The Lynds attempted in their second book, published in 1937, to portray the unsullied yearnings of Jefferson and Thoreau against those lesser, uniform goals of men like Sinclair Lewis' Mr. Babbitt – stout, chirpily resolute in his full mediocrity, depressed but proud of pettiness, and going nowhere; subject to the expensive grip of convention, propaganda and routine, and living the very life of "quiet desperation" that had come remarkably to suit capitalist intentions. It was the American Romantic movement that best documented this second theme of dualism that we have encountered before; a tension that earlier fuled the Marxian critique of appropriations. It is the apotheosis of wilderness as an antidote to social despair that provoked so lyrical an enthusiasm in the nineteenth century. And to that realm we now turn.

TRANSCENDENTALISM

"No idea is so soaring but it will readily put forth roots," wrote Thoreau. "No thought but is connected as strictly as a flower, with the earth. The mind flashes not so far on one side, but its rootlets, its spongelets, find their way instantly to the other side into a moist darkness, uterine."[20] This binary appreciation perfectly encapsulated the Transcendentalist conviction of balance in the world. Thoreau was something of an oddball, spoiled, for all of his lonely vigils. He was stuck, like most of us, in a middle ground of ascetic ideals and exasperating comfort. He was, finally, seduced by the charms of moderation.

The poet spent from July 1845 to September 1847 at Walden Pond, outside Concord, Massachusetts, living in a house that cost him $28.12½ by his figurings. He was not a wealthy man. In 1849 he finished his book *Walden* published in 1854 at his own expense. Four years later, it had managed to sell 300 copies. He earned emergency funds by selling pencils, that is when he was not bushwacking. Our lives "should be as impressive to behold as objects in the desert," he wrote, after spending a week on the Concord and Merrimack Rivers. And in his essay "Civil Disobedience," he urged Americans to *live* their opinions, not just voice them.

Thoreau estimated his costs of living beside Walden Pond at 4¢ for food and 8¢ for utilities each day. No special training had prepared him for this kind of modesty. In fact, reflecting on his education he would say, "To my astonishment I was informed on leaving college that I had studied navigation. . . . Even the poor student studies and is taught only political economy . . . while he is reading Adam Smith, he runs his father in debt irretrievably."[21] But Thoreau was not, as myth likes to recount, the total renunciator. Nearly everyday he would stroll into nearby Concord and lazy away the afternoon. Sometimes he would visit his mother. Once he lectured at Harvard. And then there was his little industry, pawning pencils.[22] But he was America's first celebrated solitaire, a symbol as convincing as a sex star. "Why should I feel lonely," he says. "Is not our planet in the Milky Way?"[23] Later, he bemoans the fact that he cannot find a blank page to write in, they being all ruled for dollars and cents. The train passed near the end of the lake each day. Thoreau surprisingly found comfort in its whistle. His copious notebooks reveal an unimaginably beautific New England. If the mountains he climbed are still there, the rivers are filthy. Thoreau's principal anger stemmed from the economic realities he witnessed, from the Mexican War and the Fugitive Slave laws. He believed strongly that the "opportunities of living are diminished in proportion as what are called the means are increased." It was a Spartan, Buddhistic equation and might be the very pull Thoreau is said to have exerted on both Gandhi and Tolstoy. If he was sometimes contradictory, as students of Thoreau constantly carp, he was extraordinarily human, shot through with storm and sunlight, truly of the wilderness. His *Journal* (1937-1861) is the most prolific and passionate body of ecological prose in the English language.

Walden became the Bible of American Transcendentalist nature philosophy. But it had a companion, and that was *Leaves of Grass* (1855). Walt Whitman, its author, a strapping, cocky, 28-year-old adventurer, called himself "a rough, an American a kosmos to himself." (His use of the German *k* was his tribute to von Humbolt's own book *Kosmos*) The poet wanted to live with animals, afoot with heart-stomping, rib-splitting vision. "No shuttered room or school can commune with me," he rallied. But to roll and froth and spasm in the grass. "The scent of these armpits is aroma finer than prayer," he went on. "Song to Myself," part of the larger *Leaves of Grass*, may be the greatest American poem. One cannot sympathetically read Whitman without taking on his own sentience, the hulk of blue, vast ever-extending America. Nathaniel Hawthorne said that he could not identify with a landscape any larger than his native New England. But Walt took on the whole nation in his rucksack, lyrically bounding, doubling over with riotous, sometimes political, always transcendent poignance, nonsense, journey after journey, madcap autobiography across the windswept altar of the frontier. Together, Whitman and Thoreau unleashed powerful criticism of those specifically American disorders, the cult of Mammon, the dislocation of spirit and body resulting in a characteristic syndrome: where people most want to be, they cannot. The adage is not simply oriented to topography. The record of American manufacturing is one of upbeat frenzy. It relied upon a narcissistic calculus of gain, meant to stimulate for the sake of stimulation. The *idea* of wealth exceeded its requirements. This run-away symbolism – the self-made millionaire, like auto-eroticism – was its own manic transcendentalism.

ALEXANDER VON HUMBOLDT

It was Humboldt whose massive explorations to exotic lands, and equally voluminous descriptive tomes, incited to poetic frenzy an eclectic, wayfaring cabal of artists and backpackers throughout the European and American nineteenth century. It is a mouthful, but that was Humboldt. With his friend Aimé Bonpland he first sailed to Tenerife, went down to Caracas, followed the Orinoco River and in 4 months traversed nearly 2000 miles of uninhabited country; he went on to Cuba, then back to South America where he climbed Andean peaks from Columbia to Peru. He made the first ascent of Chimborazo (5759 m), which was an altitude record for 36 years. He conducted a detour to the sources of the Amazon and spent a year in Mexico exploring plants. Finally returning to Europe, he wrote up his 23-volume *Voyage*. During his journeys Humboldt investigated every facet of the natural sciences and collected more than 60,000 specimens, including 1000 new species and genera. He introduced new knowledge in magnetism, volcanism, seismology, tectonics, meterology, zoology, and ethnography. Humboldt later completed a five-volume masterpiece, *Kosmos,* one of the world's greatest scientific speculations, in which he characterized *all* of the known facts — about everything! — into a single,. compelling theory of nature.[24] Like his contemporary Johann Herder, who theorized on the powerful "interdependence" of all living creation, Humboldt sought to identify the effect of plant life on the imagination, on art. He introduced a paradise factor into human geography by explaining the Andean Indian's preference for bare, exiguous highland habitats over the fertile tropics by reminding us that — whatever the inhospitality — the Indians *loved* the mountains.[25]

Humboldt completed an isothermal world map based on 25 weeks of wandering across Russia, and spent thirty years of his life engaged in various data surveys. He died the year Darwin first published his *Origin of Species,* a work indebted to Humboldt's spirit. When Darwin postured the "complex relations of all animals and plants to each other" he was voicing Humboldt's revelation of a monotheistic biosphere.

Humboldt stood for a breed of late-eighteenth and early-nineteenth century explorer. His predecessor was James Cook, whose use of the chronometer to determine longitudes, and whose provisioning of proper diet for his men in the Antarctic (thereby proving that yellow fever was not a function of homesickness) lent for the first time a massive erudition, empathy and contagion to the assimilation of new species, vistas, and stars. The possibilities for such science were inexhaustible, the Earth pregnant with meaning. Humboldt, like Cook before him, managed to convey the excitement throughout the Western world. Publishers and lay readers lapped up his itineraries and aesthetic intoxications.

By the early nineteenth century, interior Australia, Africa, the Arctic and Antarctic, as well as most of the Himalayas were still largely unknown. Despairing of conflicting rumors from Arab traders who had gone into Central Africa, the French cartographer d'Anville simply blanked out the map, calling it unexplored territory. A resulting wave of interest for the dark continent was thusly spawned. James Bruce traveled the Blue Nile from its Abyssinian source to the White Nile. John Ledyard crossed the Sudan. Mungo Park traced the Upper Niger. In 1827 Dixon Denharn and Hugh Clapperton explored the Sahara and Lake Chad. Twenty years later, the missionary David Livingston probed the Kalahari, spent most of his life researching the puzzling hydrography of Central Africa, and in 1855 traced the course of the Zambezi River, later pushing northward to Lakes Nyasa and Tanganyika. At the time of Livingstone's death in 1873, he was following the Cualaba River, hoping to find its source in the Nile.

In America it was Humboldt's effect on the painters that interests us. In the way that John Constable was greatly affected in his cloudscapes by

Luke Howard's classification of condensed vapors in 1803, or in the study of optics that so enamored the French plein-air palettes of Barbizon and Impressionist masters, Humboldt's measurements and myriad universal taxonomies, the life canvass of his travels, grabbed the painters Frederic Church and Albert Bierstadt. Both men painted the American wilderness as it had never been seen, and in so doing helped to popularize nearly every scenic spot on the continent, codifying imagination as Congress codified land grants and windfall profits to industry. The record of American Romanticism tragically fixes in our minds the pristine expanse, the vulnerable air, water and soil, the unwitting Indian whose very life – like a state of grace – most clearly symbolized the rigorous beauty of America.[26]

Church, with a geologist's fanaticism, had read everything by Humboldt and actually revisited his itinerary, embarking on several journeys throughout South America, the Arctic, Caribbean, and Mediterranean. His success was paralleled by that of his friend Bierstadt. Both men were explorers, each artist managing to collect unheard of fees for his work. The White House purchased their paintings. Through specially illuminated exhibitions and picture post cards. These in turn came to exert a spiritual power upon Americans.[27]

After six months in Columbia, Equador, and Peru, Church arrived back in New York by steamer from Panama with dried palm trees entire, an Indian boy, a newly adopted pet jaguar, cages of macaws, and a straw hat on his head. During his second romp he trekked around Chimborazo and painted his "Heart of the Andes," which was hailed as the most compelling landscape painting in nineteenth-century America. Thousands of visitors came to view it at special theatrical exhibitions. Neither Samuel Clemens nor Emerson could get over its magnitude. As for Church, he went by one theory only: The artist must restore to nature what was there at the creation. But undoubtedly his most engaging work – in terms of his own theory – is that enduring sunset scene in the Adirondacks, painted in 1851 "Twilight In The Wilderness." It went beyond realism into ether and gave its very calibre and glint to Transcendentalism.[29]

Meanwhile, Albert Bierstadt, in 1859, had joined Lander's expedition west under President James Buchanan. Their mission was to scout the Overland Trail. The train trip ended at St. Joseph, Missouri, and the huge caravan then took off for the Rockies. The discovery of mountains in America – which Bierstadt best depicted – occurred in the context of amateur ethnography and the U.S. Government-sponsored surveys. The Lewis and Clark, Powell, Kearney, Major Long, and Captain Steward expeditions were each acccompanied by painters who documented the grandeur of place and the exotica – or brilliance – of Indians. Some men, like Alfred Miller, the Swiss Karl Bodmer, C. R. Leslie, George Catlin, and R. W. Weir, brilliantly empathized with writings on the wall – the coming extinction of the Indians, and with them, of the wilderness itself.

But Bierstadt's paintings were not exactly documentary in restraint; he preferred visionary belief over reality. He conveyed the Indian in all of his religious splendor, used every cliff escarpment at sunset to promote the second coming – not a Christ figure, but the *radiance* of American light as it struck the plains and Swiss-like glaciers. Back in New York, Bierstadt viewed the first exhibition of wet-plate photographs taken by the explorer C. W. Watkins. It was among them that Bierstadt first glimpsed Yosemite. He left nearly immediately and arrived in Carson City, Nevada, with fingers numb against his pistol.[29] If outdoor photographs competed with paintings, they also fostered *belief* in canvasses so dramatic that some few were indignant over what they took for fabrication. But the mass photographic editions of William Jackson, Edward Muybridge, Timothy Sullivan, and Carleton Watkins found a voracious market for their landscapes. Americans who had begun to taste the real industrial era were panting after refreshment. Jack London, and the various publicized forays of Teddy Roosevelt made an institution of the American quest for alternatives. As for the settlers, they were either fabulously rich or wasted. Gertrude Stein no doubt voiced the complaint of those who lost heart – when you're there, there's no there.

Bierstadt painted scenes all over the high Sierras. Some were as beautiful as life, which is saying a lot. Easterners had their white and green and Applachian Mountains, but they had nothing like Bierstadt's America. Incredulous, many made the arduous trek to get there. Transcendentalism made the way for tourism.

YOSEMITE

The Valley of Yosemite was originally inhabited by grizzly bears, and is to America's shame that it did not remain that way. This is wishful thinking, of course. The southern Sierra Miwok Indians took their local name from the animals, *Uzu-mati*, grizzly. Yosemite was the white man's way of pronouncing it. The Uzu-mati worshipped no outside celestial entities – only the cliffs, sequoia, and waterfalls of the valley itself. Of the linguistic wealth employed by the Miwok, only the word *Awahnee* remains, meaning "deep green valley." I was married in that valley, have spent a good year soloing its walls, bathing in its many hidden pools, sleeping soundly in its caves. I am, like so many of the five million visitors a year, heartsick and confused over its democratization.

The first soldiers to view the valley from above were moved to tears. Others searched for gold. Then the editor of *California Magazine* exhorted the visitation of tourists, built a hotel, and published bleary-eyed religious awakenings in newspapers throughout the country. In

1864 Senator Conness managed to get Lincoln's signature on a bill granting Yosemite to a California board of commissioners. Bierstadt was already living in the Valley. John Muir was to arrive within 5 years. By that time, thousands of vacationers had been drawn there over the rugged coach road and stone trails. By the 1870s, forest within the valley was being cut back to allow tourists better views from the convenience of hotel rooms. Federal troops had to guard Yosemite from vandals and poachers. At Yellowstone, where other troops were stationed, a frustrated acting superintendent quoted John Muir: "The smallest reserve, and the first ever heard of, was in the Garden of Eden, and though its boundaries were drawn by the Lord, and embraced only one tree, yet the rules were violated by the only two settlers that were permitted . . . to live on it."[30]

Muir began writing his environmental essays, enraged by the "improvements" that had been allowed to stand without any apparent opposition from the public. In 1900 the first automobile entered Yosemite. Under newly established Department of Interior jurisdiction, two regiments of the army assumed permanent residency: A power plant and telephone lines were installed, and a railroad track was laid to the entrance of the park. Under the National Park Service's first director Stephen Mather, automobiles were officially commended:

> *Our chief duty is to make our National Parks comfortable for all the people. . . . [I]t is the right of the hundred million owners of the Yosemite to visit that incomparable shrine in as much luxury as each can afford, and it is the duty of the Government representing these millions to provide each one who comes as nearly as possible, all things considered, the degree of comfort even of luxury that he requires. . . . The Yosemite National Park, as a whole, will become in a few years the choicest playground of this nation, and, in connection with another experimental development at Mount Rainier, the model upon which the National Park Service will use to administer and develop its National Parks.*[31]

Four years previous to this pronouncement, Secretary of the Interior Franklin K. Lane had paved the way of park policy by officially deciding to allow all automobiles into Yosemite, "to make our parks as accessible as possible to the great mass of people."[30] By 1920 not a single Indian was left in Yosemite. And by 1928, nearly 1.5 million automobiles had come into the sanctuary of sanctuaries.

John Muir's story is familiar enough. Born in 1838 in Dunbar, Scotland, he came first to Wisconsin then settled in California in 1868. Writing primarily for *Century Magazine*, then edited by his friend Robert Underwood, the bearded, shy figure came to be regarded by

Emerson as the noblest spirit in America. When invited to Massachusetts to study and discourse with the venerable older celebrity, Muir declined, emphasizing his own wilderness priorities. "Insist on yourself," he repeated. In 1892 Muir became president of the newly formed Sierra Club. Theodore Roosevelt would come for a 3-day backpack with the forest sage and press legislation for an expanded park system as a result of their séance. But Muir did not anticipate a Stephen Mather.

In 1879 Muir had kayaked to Alaska and lived for a year on the terminus of the glacier he discovered in what is now Glacier Bay National Park. Exactly one century later, I was chased down by a gorgeous grizzly bear at that ice terminus and had a month to contemplate Muir's journal notes:

> *Here . . . I saw the glacial rocks and traces which seemed yet more fresh and telling in all that relates to the action of the ice-sheet — its course, the way it deposits the so-called glacial drift, excavates harbors and fiords, and brings landscape features in general into relief. Yet, strange to say, man, with his reason, builds his houses, grades streets, tills the glacial soil on the ground prepared by this mighty agent, where the phenomena are so strange and so striking as to attract and arrest the attention of animals, without once attracting his. So truly blind is lord man; so pathetically employed in his little jobs of town-building, church-building, bread-getting, the study of spirits and heaven, etc., that he can see nothing of the heaven he is in.*[32]

That was typical, unapologetic Muir. And it fairly sums up the romantic–conservationist response to nineteenth- and twentieth-century American industry and mass culture. Muir spent his last years in the San Francisco Bay area, on a farm in little Martinez. He died the year we entered World War I.

NOTES

[1] Saudi Carnot, quoted in Wilson Clark, *Fnergy for Survival — The Alternative to Extinction* (Garden City: Anchor Books, 1974), p. 10.

[2] Charles Dickens, *Hard Times* (London: Macmillan Publishers, 1983).

[3] William Wordsworth, "The Solitary Reaper," in *The Norton Anthology of English Literature*, Vol. 2, M.H. Abrams, General Editor (New York: W.W. Norton, 1968), p. 156.

[4] Hans Huth, *Nature And The American — Three Centuries of Changing Attitudes* (Lincoln: University of Nebraska Press, 1972), pp. 14–22.

[5] Remarked by Andrew Burnaby during a tour of the Pawtucket River, 1768, in Huth, ibid., p. 27.

[6] Arthur A. Ekirch, Jr., *Man And Nature In America* (Lincoln: University of Nebraska Press, 1973), pp. 36–40.
[7] See William Truettner, *The Natural Man Observed: A Study of Catlin's Indian Gallery* (Washington D.C., Smithsonian Institution Press: 1979).
[8] Alexis de Tocqueville, quoted in Roderick Nash, *Wilderness and the American Mind*, 3rd ed. (New Haven, Yale University Press: 1982), pp. 85–95.
[9] Huth, op.cit., pp. 5–9.
[10] William Wyant, *Westward in Eden: The Public Lands and the Conservation Movement* (Berkeley: University of California Press, 1982).
[11] Carl Ortwin Sauer, *Land and Life: A Selection from the Writings of Carl Ortwin Sauer*, ed. John Leighly (Berkeley: University of California Press, 1923); and Frederick Jackson Turner, *Frontier and Section* (Englewood Cliffs: Prentice-Hall, 1961).
[12] Quoted in Sauer, ibid.
[13] See *The Journal of Henry D. Thoreau*, ed. by Bradford Torrey and Francis H. Allen (New York: Dover Publications, 1962), Vol. II, p. 11.
[14] Karl Marx, *Capital*, 4 volumes, trans. J. Cohen and W. Ryazanskaya (Moscow: Progress Press, 1954).
[15] Ibid.
[16] George Marsh, *Man and Nature: Or a Physical Geography as Modified by Human Action* (New York: Charles Scribner, 1871).
[17] Henry Adams, *The Education of Henry Adams*, quoted in Wilson Clark, op.cit., pp. 567–568.
[18] Harold Laski, *The American Democracy: A Commentary And an Interpretation* (New York: Viking Press, 1948).
[19] Thorsten Veblen, *The Theory of the Leisure Class* (Boston: Houghton Mifflin, 1973).
[20] Barbara Novak, *Nature and Culture: American Landscape Painting 1825–1875* (New York: Oxford University Press, 1980), p. 116.
[21] Thoreau, *Walden and Other Writings*, ed. by Joseph Wood Krutch (New York: Bantam Books, 1977), pp. 143–144.
[22] Thoreau, ibid., p. 356.
[23] Thoreau, ibid., p. 203.
[24] Alexander von Humboldt, *Cosmos: A Sketch of a Physical Description of the Universe*, 4 vols., trans. E. Otte (New York: Harper & Bros., 1844).
[25] Clarence Glacken, *Traces on the Rhodian Shore: Nature And Culture In Western Thought From Ancient Times To The End Of The Eighteenth Century* (Berkeley: University of California Press, 1973).
[26] Barbara Lassiter, *American Wilderness – The Hudson River School of Painting* (New York: Doubleday, 1978).
[27] David Huntington, *The Landscapes of Frederic Edwin Church* (New York: George Braziller, 1966).
[28] *West As Art: Changing Perceptions of Western Art In California Collections*, Palm Springs Desert Museum, 1982.
[29] B. Lassiter, op.cit.; and Weston Naef and Hames Wood, *Era of Exploration: The Rise of Landscape Photography In the American West, 1860–1885*, Albright-Knox Art Gallery, and Metropolitan Museum of Art, 1975.
[30] William Wyant, op.cit.
[31] Stephen Mather, *Outlook Magazine*, January 1917.
[32] John Muir, *Travels in Alaska* (Boston, 1915).

PART 3

Proportions of the Human Shadow

UNITED NATIONS

Chapter 7

Ground Truth

Politicians and economists of every persuasion look toward a world that will have adopted the American city as its model. The impoverished children of Earth know too well what such a model implies: overconsumption, famine, the destruction of forests, and watersheds, the pollution of the biosphere. This geography of despair is readily discernible, even from a satellite.

> *The fact that Arcadia was the code-name for a Washington conference on joint strategy when the United States entered the Second World War, might serve as a clinching illustration of the truism that men cannot escape from history.*
>
> Harry Levin, *The Myth of the Golden Age in the Renaissance*, 1969

CONQUEST AND THE AMERICAN DREAM: THE PREMISE

That billion dollar photograph, our first glimpse from the moon, back to the lovely floating sphere, deep in its fragile azure, like an infant's eyes, and clouds as silken as gossamer, and the budding continents adrift murmuring seas. A paradise unrivalled in all of the galaxies. To some other sensitive form of life in space, reaching out inquisitively towards

Earth, certain immediate blemishes would meet its own untamed look — the Houston astrodome, the Great Wall of China and the electrical generating plant in the Four Corners region of Arizona and Utah, whose sulfur dioxide effluent is indeed visible from the moon. An electron-capture detector yields the suspicion that some form of life is busy at work on Earth, reconstituting inorganic accumulations from deep below its surface. On closer inspection, it discerns fervent bacterial activity in human armpits, and throughout the botanical Rhizosphere. It notes chimpanzees copulating for seven seconds, Great Apes for fifteen minutes, human beings for several hours. Who is this latter species? Its eye zeroes in over the teeming Amazon Basin. It follows the sensuous curves of coastline into the Antarctic ice sheets, passes the symphonic outcry of millions of species furiously at work in jungle, highlands and air. The visitor from outer space notices a minute but curious clue along an ice cliff at the back of a maroon fjord, a tell tale caveat: two hundred years of animal reorganizations and industrial revolution have left their record of grimy particulate in the ice. Again, the unusual beacon of human beings, even where there are none to be seen.

Our creature is delighted with Earth. It basks in the ever-present sunshine, throws his face into the warm rain, sloshes through bogs, dives pell-mell after mountain goat, dances with the bumblebees, nestles into burrows, feasts on abundant plant life. In short, he loses his mind; he beams, shrieks, turns circles in ecstasy. No interstellar sojourner could have anticipated so inviting a planet amid so much darkness and void. The extraterrestrial dons a set of clothes, merges into a crowd, assumes the look, purpose, and posture of *Homo sapiens*, whose electrical impulses give off a desirable rush of energy unique in the animal world. Now we've lost him: He's ended up in New York, at the United Nations Plaza. He's got himself a job, a wife, even sired some kids. You'd think he'd grown up here, evolved over millions of years out in the wilderness of life and then — like so many others the world over — had come into the dazzling city, ready for action.

The cities: they encompass some of our very best and New York is the ultimate champion. The cities are the culmination of a purposeful migration. Eight generations have responded to the compelling call for jobs, material luxury, and culture. There are now 1.8 billion residents in cities. Urban life must be doing something right.

The United Nations building was designed in 1951 by a group of the world's best architects. Its final statement bears the inimitable stamp of Charles Jenneret, better known as Le Corbusier. A Conduit Weathermaster System was installed to maximize a comfortable internal temperature; glass prisms secured the structure against sunlight. The building, like most of the glass facades that followed its initiative, is the hallowed zenith of comfort to which human beings have adapted. The building's energy and structural costs, massiveness, and insulation reflect the international body of forces that the United Nations (and cities in general) have come to symbolize. The Earth has been geared to accommodate two centuries of consumption. Is such consumption natural? The Earth is not unfamiliar with greed. Is there self-indulgence in the animal world that exceeds the requirements of a successful survival strategem? Absolutely. Let us examine a popular species of logic favoring untempered growth, profit, and consumption.

We have the technology. We have seen where such technology came from. Chemicals, fertilizer, coal, steel, dams, irrigation, liquid naural gas; where do we begin to unsort the energy of our greatness? Simple reasoning tells us that everything is connected to everything else; the mechanism is moving, and there is no way to stop it. And why, after all, would we want to? I mean look at New York, look at it! A great town, a fabulous town. What about English toffee? Huh? English toffee. We can start there. Like tea, the toffee – in its baroque old world wrapper and tin container – represents technology, has about it something particularly edifying and aristocratic that nearly everyone (in the West) can afford; a battery, a radio, in fact, almost every commodity in our life depends on an unnerving array of *other* commodities. Energy, freedom, hard work, and money links them together. We are healthy, strong; we are meat-eaters. Lions eat meat. We eat meat. No tofu burgers for us! Our evolution confers dignity, complexity, and genius.

Twenty-three percent of the earth's land surface, or 3.1 billion hectares, supports more than 3 billion ruminants. Beef and dairy cattle represent the number-one farm income for nearly half of the states in America. Water buffalo, sheep, and goats compose an overwhelming resource system throughout the world, without which we would suffer. Cellulose is the stock animal's main food and, fortunately we appear to have plenty of cellulose. Where it is insufficient, we have plenty of fertilizer to boost the soil's deficit. Where the soil itself is thinning, why there's still a bit more arable land to convert (4 percent). American dairy and agricultural produce is the world's godsend. Our materialism is our religion. It saves lives, bolsters dreams. It is good.

And then there's the *New York Times*. Every Sunday edition consumes 70 acres of Canadian hardwood forest. Sounds impressive! But North America has those forests. We are on an even keel thanks to the ingenuity and capital of American enterprise. Maybe that sounds excessively patronizing, square; but it's true. Throughout the world, 80 cubic meters of wood are cultivated per capita each year. By the turn of the twentieth century, the total annual harvest of wood is projected to be much more than double what it is today. We hear a lot of sour grapes about the lumber industry. And, sure, the redwoods are great to look at and should be strictly preserved. But the fact is, 59 percent of all American forests are privately owned and nearly another 30 percent are in the public domain, including those wilderness areas and parks. Lumber folk aren't greedy; they're scientists. Now when we hear about wood-burning stoves you know we're talking ecology, Ben Franklin. Ecologists will tell you that wood burning is better than nuclear energy. But the point is this: Cutting down a few forests and applying sophisticated reforestation techniques seems to make perfect sense. Furthermore, nearly two-thirds of all tropical forests abroad are economically accessible. Already some 190 million hectares are being cleared every year for shifting agriculture, which is contributing greatly to the expansion of farmland and the battle against hunger.

When we talk about hunger, we're into big debate. Well-fed scientists argue that Guyana, Jamaica, Panama, Costa Rica, the Barbados, India, and the sub-Saharan nations have massive malnutrition problems. Well, I've been to great hotels in Jamaica and India, and the life expectancy in all of those countries is pretty damned good. The United Nations World Food Council estimates that .5 billion people are hungry. But other independent researchers claim that such figures are grossly misleading. The United Nations International Children's Educational Fund (UNICEF) has published statistics citing more than 30 million children under the age of five as dying of starvation every year. Whereas a leading demographer with the Overseas Development Council concluded that there are only half that many deaths. OK, 15 million is still horrible. But who do you believe; where's the credibility? Who is really in charge of these statistics?

If we take an average of estimates and agree that some 22 million children will die of starvation next year, while Americans are enjoying all of their prosperity, why not just ship our excess grain to the poor instead of selling it to the Russians? After all, 3 million tons of grain a year would enable 100 million malnourished people to have the right kind of diet. Maybe not an American diet — but remember, we consume nearly fifty times as much as poor people in other countries. But we've *got* to. Energy expenditure, human output, is inflationary in the United States. Such inflation doesn't mean that we actually eat fifty times more, obviously (we'd all be gluttonous slobs, instead of just some of us), but it refers to our cars, our houses, my uncle's two motorhomes, his dune buggy and snowmobile, and second home in Winnemucca. But if we didn't churn out all that steel and plastic and whatnot, well, there simply wouldn't be those 112 million tons of food exports that the United States sells to nearly 100 countries now dependent on us for their nutritional supplements. I've read the figures. I know what I'm talking about. I don't like arguing about these things. But you just wait until those Organization of Petroleum Exporting Countries (OPEC) sonsabitches come begging at our door! We're in demand, you better believe it. We consume as a function of our production. Fair is fair. What's more, we fought good and hard for it, saved the Jews from Hitler, made a valiant if blundered stand against the Reds in Vietnam and kept America as a beacon for all oppressed peoples. You think those poor folk in totalitarian, deforested Haiti would have risked their lives in smelly boats to get here if all they had to look forward to were tofu burgers? The world's depending on us and we need to be out there producing. "Produce, produce, produce!" says former Secretary of Energy James Edwards. That's his answer to the energy problem. Maybe he's right. He may have started out a contented dentist, but don't kid yourself: Ronald Reagan didn't hire him for nothing! He has that rare and sintillating pragmatism that embodies the American dream – conquest. That's the very premise of democracy.

And if that doesn't squelch any doubts, then listen to Pope John XXIII's own encyclical of 1967, urging all good souls to heed the Genesis passage (1: 28–29) that orders mankind to subdue and dominate nature. It's God's will.

We began talking about hunger by referring to the cutting down of trees in developing countries. Now we are debating about the supposed "greenhouse effect," the global warming in the atmosphere that will melt the western side of the Antarctic by the twenty-first century, raising the water level 6 meters, wiping out every KOA campsite in Florida. The bright side is that it just means more water, more fishing, more fish![1] But we have a far more compelling and respected disputant, a government man with the Department of Agriculture (DOA) who noticed something that all the highbrows missed: plants grow by converting carbon dioxide and water into carbohydrates, with the sun's energy. Sherwood B. Idso, the DOA climate specialist, believes that the greenhouse effect will actually stimulate agriculture by 50 percent throughout the world. By cutting down forests,we get two birds with one axe. We begin to get the impression that flexible science, generous statistics, a little foresight and patience are just the right antidotes to doomsday, at least in the eye of the beholder. So stop worrying. It's counter-productive.

A DREAM DENIED

I have heard countless varieties of the preceding tone-deafness; earnest beliefs from sincere, extremely well-read people across a wide political and practical spectrum. The remainder of this chapter addresses those perceptions and challenges the premise of an economic miracle.

The Landsat D Satellite orbiting 724 kilometers above Earth can focus clearly on parcels of land as small as 2.4 hectares. Landsat can determine weather movement, agricultural vitality, soil health, sand dune migration, and the likelihood of fossil fuel deposits. Central Intelligence Agency (CIA) satellite resolution, accessible to few civilians, obtains images of even greater pellucidity. Satellite remote sensing records electromagnetic radiation reflected from Earth, allowing for a variety of inferences, about mineral resources. Holographs, airborne radar systems, thermal infrared imaging techniques, seafloor sideways scanning sonar, seiscrop sections, and other nonconventional methods of scrutiny are continually assessing the Earth. But these resulting statistics — that the U.S. military consumed 2460 trillion btus of energy in 1971, for example, how many cubic kilometers of pumice and ash Mt. Saint Helens spewed forth, or that the population of China in 1982 was 1,008,175,228 — have a very nebulous value taken together. Every facet of human experience is liable for quantification, can be endlessly broken down into categories of perception, numerical translation, futurist projections, and models; and in hundreds of real and artificial languages. The modelers themselves meet every so often at obscure chalets outside of Vienna and carry high-speed printers in their briefcases. Fluency from one standard of energy, weight, and measure to another is never guaranteed. The scope for inaccuracy exceeds prediction, plays merry hell with any aggregate. Values are always weighted. The demographics, indices, and resources themselves change every second, insulting our sense of order.

The statistics may have little meaning for most of us, but satellite omniscience provides a ground truth. It is that net effect of our dreams which has shadowed once-sacred space, fouled waterholes believed to be involate, reshaped our arrogance and expectations. The last decade assailed us with the facts and figures of our own actions. Such data are usually the only responses our consumption ever incites — the only reactions we are aware of, that is. The Earth is a planned economy, but within it human beings promenade in something of a noosphere, oblivious, scratching their livelihood without much interest in the greater picture. Ferocious tombs project grim overviews, and fact-finding missions disinter the planet. Most of us try to live somewhere in the interstices of such information, denying if we can the tyranny of the *fact* Ghost slaves, ghost acreage, carboniferous legacies, subsidized prices, an essentially isolationist, power-hungering government grants us this illusory stay of execution. The Age of Exhuberance endeavors to protect us.

But on closer inspection, we see that this is not so. Let us outline the algebra of human impact at large in the world. Such pathology is like a satellite photograph; it will cater to experts. The unconverted or disinterested have little means — though plenty of reason — for assimilating such data. Where is the bottom-line relevancy? What can someone *do* with the highlights? The question has been posed repeatedly since World War II, because size, danger and complexity — the fabric of our lives — have finally overtaken native shrewdness. In the presentation of data, we erect a world picture, abstract as it may be. We struggle to take it all in, one view, a split-second apparition of consensus. The echo of our rifle-fire array may leave us bewildered, even nauseated.

Most of the following statistics, when cited, refer to the American scene, where global data is either too scant, or unavailable. It should be pointed out, however, that what is bad, unregulated, and dangerous in the U.S. is much more so in some other countries.

POPULATION

Alfred Lotka's application of physical laws to biology resulted in the characteristic Sigmoid (S) curve defining predator–prey population cycles, boom-and-bust explosions that hold for any living group, from lemmings to tsetse flies, paramecia to Easter Islanders. The Swedish botanist Carolus Linnaeus had portrayed these cycles by describing a corpse. Two flies, he reasoned, will devour it faster than a lion. The burgeoning-plunging graph embodies two intersecting pathways whose periodicity elicits the mathematician's sign for infinity, a subconscious appeal that promotes the assumption of an unending constancy to nature's biological oscillations, of a balance to which all life is always returning, despite whatever upset. There is indeed an equilibrium, but it does not remain in steady state. Rather, all of nature performs a sliding-scale gymnastic, maintaining high amplitudes, equally impressive troughs. Lemmings may multiply 500 times their normal group size during peak years. The trough ensues at three-to-five year intervals. Among the lynx populations studied by biologist Daniel Botkin, who relied on 150 years of records of fur trading left by the Hudson Bay Company, boom–bust cycles occurred in 8 to 10 year cycles, ranging in population from 4000 to 80,000. The lasiocampid moth fluctuates on an average of once in 40 years, its population increasing 10,000 times the norm. Ecological overshoot can plummet the population below its threshold for recovery. Such an event accounts for extinction. On St. Matthew Island in the Bearing Sea, colonizer reindeer devoured their surrounding plant life. Their base population — 2000 animals — exceeded the island's marginal carrying capacity. Within 3 years the herd was reduced to 46, this precipitous decline occurring in absence of any predators other than famine.

It took *Homo sapiens* several million years to reach his first billion in population, a mere 180 years ago; 2 billion in 1900; 3 billion in 1950; 4 billion in 1975. Today, mankind is 500 times more numerous than his global population at the dawn of agriculture. In the year 2000 he will number 6.35 billion, a considerable burden. The Earth's carrying capacity for human beings is not easily addressed: One man is too many if he manages to detonate nuclear weapons. Similarly, too many hunter-gatherers could not range across the trophic plateau without annihilating lower food sources. The Yokuts of 18th century Sukwutnu, Wowol (Tulare County, Southern California), numbered 20,000, which approaches a viable upper limit for the number of hunter/gatherers in one region. As for more advanced, technological populations, the balance is not known, though some scientists have voiced a meeting ground somewhere between 8.5 and 12.5 billion. The upper bound, in absence of intensified family planning, will be upon us within 50 years, short of some other ecological trough opening to engulf us.

Prior to industrialization human number remained within manageable, biological parameters, ecological sanity. Two rudimentary events challenged this traditional stability in the eighteenth century. Colin Clark believes that the climatic change in Scandinavia compelled the Norwegian gray rat (which does not carry the plague) to find passage into Western Europe, where it successfully ousted the indigenous flea-carrying black rat, thus eliminating a major threat to human population.[2] At the same time, improvements in nutrition and in agricultural technology led to massive colonization throughout the world. Again in the 1950s our population exploded in the tropics, this time with the introduction of malaria suppressants and DDT. China's overpopulation today, which has risen 45 percent in the last two decades, exemplifies all of the quintessential constraints imposed by demographic transition: food-fuel competition (500 million people suffering from a serious lack of fuel), massive deforestation and consequent flooding, large numbers of the homeless, blocked food supplies, the spread of water-borne disease; rampant, outlying malnutrition (reports vary widely); and authoritarian checks to fertility.

The Earth herself plays with populations, size, and proportion. There is a time to kill and a time to forgive. Ninety-nine percent of life is bacterial — survival machines that pollute, cannibalize, and metabolize nearly every known chemical. The remaining 1 percent of life is concerned with evolution. As a result, population has — for over 2 billion years — been the biosphere's primary site for experimentation. Lemmings and locusts are extreme examples. The 150-ton blue whale, giant sequoias, and smallest known organism, Mycoplasmas, are other test cases of size. Among most animal populations, size appears to be balanced from within the group. Starlings, howler monkeys, cicadas, hummingbirds, even gnats, appear to maintain population balance by taking an inventory of their number every day. Worms and tadpoles can secrete growth inhibitors; mammals can absorb their own fetus under stress to prevent nonviable births.

On the cellular level, 1.2 billion-year-old cytochrome *c*, the most stable and primeval protein, assures a precise number of amino acids — of energy boost and atomic polarity. This optimal chemistry affects glucose extraction, photosynthesis, and sheer numbers in every population of plant and animal, including the 77 billion children that have graced this planet and the 128 million new ones who arrive every year.

Some human groups have checked their own population growth. The monogamous Mbuti Pygmies, believing in no afterlife, exert their own modest birth controls. Man and woman refrain from intercourse for 3 years following the birth of their child. Nineteenth-century Europeans – the French, particularly – condoned infanticide. Roman Catholics go in for abortions four times more vigorously than any other Europeans, despite Papal decrees. Tibetan polyandry endorses its own form of limiting population by mating one woman to several husbands. Tibetan Buddhist Tantrism is preoccupied with premature sexual withdrawal, a type of birth control said to engender enlightenment along the way. Ironically, it was the old Silk Route through Tibet that transmitted *Pasteurella pestis*, the Black Plague, to a harbor in Sicily in 1347. Rodents account for 50 percent of all mammals, and it was the friendly rat, carrier of *Xenopsylla cheopis*, the rat flea, that caused such demolition between Iceland and India. At least 30 percent of the human population was killed.

But Europe witnessed a population revival within a century of the Plague. Thomas Malthus predicted that 70 million people would die of hunger in England by the end of the nineteenth century. He wasn't far off. Seventy million Chinese *were* afflicted with war and hunger during the Taiping Rebellion and Great Northern Famine of the latter nineteenth century.

One hundred million have died in the twentieth century from war. This, too, is the planet's recipe for balancing populations. The contradictory urgings of science, to give life and take life, characterize a level of stability often overlooked with regard to *Homo sapiens*. But whether thermonuclear war can be regarded as the ultimate stabilizer is doubtful. Though from the bacterial point of view, such destruction might prove something of a windfall; of corpses in plenty to degrade.

If human technology can eradicate aggression, convert all food sources to liquid, step up the efficiency of marine phytosynthesizers, limit all energy use to that which is solely productive of food (even recycling cadavers) our species might attain a density of 120 per square meter.[3] A total population, that is, of 60,000 million million.

Somewhere short of that scenario, architect Paolo Soleri has endeavored in the deserts of Arizona to envision and build a model city based upon the principles of miniaturization and creative complexity.

CITIES

The city is the epicenter for energy exchange. Consequently, it outreaches and fuels all human impact on the planet. In the Third World the cities set market prices, creating economics unfavorable to traditional rural farming agendas and forcing unwise husbandry practices upon those who know better. The allure of money in the cities has triggered the existence of fringe ghettoes, slums by any name – favelas, farriades, bidonvilles, bustoes, vecindades, barrios. The word *civilization* was first used just prior to the French Revolution by the Marquis de Mirabeau in an essay on population. It was not the city, said Mirabeau, but woman who made life civilized. Times haven't changed. For all of the city's utopian ideals – the Persian Persepolis, Francis Bacon's New Atlantis, Luke Howard's Garden Cities, Pierre l'Enfant's Washington D.C. – the City on the Hill has not entirely fulfilled its promise, not since Neolithic times. It has been demonstrated that evolution requires stress. Whether the city increases such stress is not known.

During Michelangelo's era, Rome and Florence each contained 50,000 residents, a number applauded by Aristotle, and the same size as Boston and Philadelphia at the time of the American Revolution. It may be the optimal congerie for creative, self-sustainability. London had one million denizens in 1820, and by 1900 ten other cities had that population range. By 1985, 273 cities will contain a million or more inhabitants, a vision acceptable to termites and no-seeums, but one that would have been appalling to dinosaurs, who ruled for 100 million years without overpopulating. In 1800, 2.5 percent of the human contingent was urban. By 1900 that sector had multiplied six times, 12 times (33%) by 1960. In the year 2000, there will be more cities of 100,000 people plus than even existed in the entire world of 1960. Seventy million additional dwellers need urban housing annually. In the year 2000, city folk will account for at least 3 billion. Eighty percent of those city populations will be confronted with little or no safe drinking water, sanitation, health, safety, or transportation. The cities are powder kegs that usurp surrounding resources the same way that Roman excess was sustained at the expense of imported slaves who accounted for more than one-third of the total population. Diffusing the cities is the basis for all future energy security. In the meantime, people already want out: 596 million visits were paid to 3851 state parks in 1980, and 300 million additional visits to national parks, just in the United States.[4]

WATER

More than 99 percent of the Earth's water is unavailable or unsuitable for human consumption. Linear extrapolations of water reserves are insufficient to fully portray the heterogeneous, fluctuating nature of supply. The Americans and Soviets withdraw the highest per capita amounts of any people. The average U.S. family uses nearly 75 percent of its 605 liters per day in the repeated flushing of toilets. Compare this extravagance with much of Africa, where per capita consumption totals 3 liters each day. The 1980s are the international drinking water supply and sanitation decade, though there is no dancing in the streets of Rio to celebrate. Sixty-six percent of the world's population currently lacks adequate waste disposal systems. More are victims of unsafe drinking water. Unequitable distribution of water resources results in areas of consistently high drought (e.g., the Sahel). In other regions, irrigation – the largest downdraw of water – wastes at least 50 percent of the resource. In the year 2000, water needs throughout the world will triple. According to the World Bank, a billion people are currently pinched for drinking water; UNICEF cites the figure at 2 billion; the World Health Organization says nearly 4 billion. It's bad. America's domestic pure drinking water – 22 percent of our total – is being steadily contaminated by salts and microbiological toxic organic and inorganic chemicals, much of these in turn arriving with the 3024 billion liters of waste dumped into septic tanks each year. Whole aquifers are percolating into industrial garbage sites.

Water is the miracle-giver on Earth. All life is defined by it. The planet's average temperature of 59 degrees caters to its liquid state. The human body requires 2 liters a day. An oyster and most trees consume 20 liters every hour. Each year the ocean recycles 456,000 cubic kilometers of water over the entire Earth. Nearly 85 percent of a living cell is composed of water. The balanced pH of 7, its perfect solvency and voracious hydrogen bonding, and its self-shielding capability (releasing oxygen molecules to form ozone) combine to make water the most essential component of the biosphere. Without it, the Earth's atmosphere would be seventy times thicker, composed principally of carbon dioxide. Eighty-three percent of human blood and 2 percent of human teeth enamel is composed of water. Water's life-fostering architecture is at work in every bloodclot, clump of jello, waterfall, and pit stop.

One hundred trillion human cells each contain a plasma membrane whose specific filtering ability ensures a balanced pH of cytoplasm, cellular water volume, nutrient stockpiling, and release of toxic compounds. This membrane was the earliest biological hospice on Earth, approximately 3.8 billion years old. Today, it is endangered from pollutants manufactured by *Homo sapiens*.

There are 325 million cubic miles of water on Earth, in a perpetual state of renewal. *Homo sapiens* consume 100 trillion gallons per day. In 15 years, this one species will have consumed 75 percent of the planet's supply. It's already getting hard to find in much of the Third World, where women and children must often walk from 16 to 32 kilometers a day in search of new water holes. States and nations have discussed rustling clouds, diverting rivers, transporting icebergs. World War III could be fought over water. And wouldn't it be ironic? The origins of life and life's extinction in the same thermos.

The National Academy of Sciences and the EPA have discovered 700 major contaminants in U.S. drinking water, twenty-two of which are suspected carcinogens. Water-borne diseases world wide result in at least 30,000 deaths every day. Eighty to ninety percent of all cancers are environmentally induced. Septic tanks, sewers, toxic waste dumps, disposal wells, waste pits – all leach into ground water. Many of the pollutants are novel. The Earth never invented them. But these new chemicals are deficient in electrons and thus eager to acquire them from other molecules. DNA and RNA have excess electrons. In this way, carcinogens infect the entire planet, every rain drop. Atoms are the oldest storytellers. They remember everything. Water is their medium.

Water has made and broken civilizations. It embodies our dreams, our aesthetics, our forays out to the unknown horizon. And it is our survival. Nine out of ten organisms on Earth live in the oceans and seas. Homer of Smyrna and Thales of Miletus looked out in Ionian times to the cutting-edge of mountain and sea and divined paradise in the water. Along the Nile, in the Fertile Crescent, across the length of the Amazon, the Yellow and Yangtze Rivers, civilizations and religions emerged explicitly indebted to the cycle of moisture, silt, flood, and fertility.

The Ganges is the lifeblood for nearly 700 million Hindus. It is the Mother Goddess, giving life, taking life, mending all conflict. The Himalayan silt load is so dense that fresh-water dolphins in the Ganges are blind. The Persians in Medieval India called the river "a hell filled with good things." Early Greek writers identified the Ganges as the central river of Eden. At Allahabad, every 12 years, millions of Indians come to this sacred river's bank to drink of the *kumbha,* the cup of the gods. Initiation, marriage, and death occur along the Ganges in India. At least 35,000 corpses are burned and deposited in the river every year – charred bones, ash, partially consumed internal organs. Ascetics trek to the river's source at Shivalinga Peak in the northwestern Himalayas; young girls mould clay phallic symbols from the river's mud to help them find a virile husband. The god Shiva is destroyer and restorer in Hinduism. He is the male, the fire; Ganga (Ganges) is the female, water. Life is born of their intercourse. But this sacred connection is losing ground throughout the world.

The Mediterranean that so inspired Homer has become the Earth's foremost toilet, fed with untreated sewage from 120 major cities. The Mississippi, that Old Man River, has become America's urethra. There is not one river in America safe for drinking in 1984.[5]

ENERGY CONSUMPTION AND WASTE

Libya's success at exploiting independent oil companies from the West after the closing of the Suez Canal triggered a multitude of demands from Persian Gulf producers under the visionary umbrella of Perez Alfonso, the Venezualan oil minister who initiated OPEC in 1959 after the Americans ignored his ideas for a hemispheric oil pact. After October 1973, Middle Eastern producers slashed production by 25 percent. Britain had withdrawn from the Gulf in 1971, and the United States sought to fill in the vacuum by installing the Shah of Iran. But Iran ended up leading the OPEC states in a 400 percent price increase. The key Straits of Hormuz and Aden and Malacca came back under the control of Arab potentates.

The various estimates for oil reserves converge upon the fact of global oil desiccation in the early 1990s, reserves that took hundreds-of-millions of years to come about – gone in two generations. From the planet's perspective, this might prove useful. We burrow deeper than any other species, and in so doing we liberate trapped carbon, thus closing the chemical circle on Earth. We have consumed fossil fuels 10,000 times faster than they can be replaced by geological forces. We consume a mean 1200 gallons per person, 6 billion barrels a year, 60 million a day worldwide; though Americans are the penultimate users. The figures reel, lose their meaning. Of the estimated trillion barrels of oil that are left to be discovered, most of it exists in deep offshore and polar regions.

Nearly 238 trillion cubic meters of natural gas is still out there, as well as 725 billion metric tons of recoverable coal. The numbers are fighting a losing battle: a 4:1 ratio of consumption over growth is necessary to sustain the American way of life. In other terms, a 20-percent compounded increase in energy consumption is equivalent to the 5 percent economic growth that politicians traditionally grapple after. With 30 percent of all industry depressed and unproductive in 1983, this ratio is somewhat modified. The big pie translates into a U.S. energy demand of between 87 and 102 quadrillion btus, 329 to 352 worldwide. Where will that energy come from? By the year 2000, shortages in the world's oil supply will amount to 15-to-20-million barrels per day. The changing projections for nuclear power in the year 2000 do not give rise to hope. In 1972 the Atomic Energy Commission (AEC) projected 1010 gigawatts of nuclear production in the coming generation. Today, the Nuclear Regulatory Commission (NRC) more soberly hopes for 145 gigawatts. As of 1984 the nuclear fission industry was experiencing its first death knells, with huge cost overruns, protracted litigation, a termination of any new facilities, and a declining consumption of electricity by 7 percent each year. The safer fusion deuterium-tritium reactors are still decades away from realization. For every quadrillion btus of electricity consumed, it took 3 quads of energy at the production end.

For those 3 quads, 9 other quads were necessary from input. And for those 9 quads, 27 quads, and so on. The 3:1 energy waste scenario, coupled with the 4:1 consumption-over-growth ratio, engenders a nightmare of abstraction: the tragedy of the commons, entropy. We are spreading ourselves thin. One American uses as much energy as three Swiss or Japanese, as sixteen Chinese, as 1072 Nepalese.[6]

SYSTEMS ENERVATION

If we exercise restraints in the following social sectors, America's GNP would double without any corresponding escalation of ecological damage: the vagaries of fashion; redundant paper services (e.g., insurance premiums); advertising, packaging and public relations; exploitation industries; crime prevention; governemnt regulations to restrain multiple abuses; government engendered inefficiency, tax loopholes — the full spectrum of environmental degradation costs.[7]

HEALTH

American health is a mixed blessing. Our life expectancy, like our chronic boredom, loneliness, alcoholism, heart disease, and cancer is high, at least one in three Americans is afflicted. Our infant mortality rate could be lower. And in between, 40 million Americans are treated for mental illness; per capita health costs are nearly $600 annually. Every day between 50 and 80 percent of all Americans swallow a prescribed drug. Per capita annual food consumption in the United States is skewed between natural and synthetic products. In fact, we eat more of the latter, taking in at least 41 kilograms of synthetics and 2500 food additives per person each year. The environmental maleffects on human health are not easily understood in terms of dollars or epidemiology. Insurance claims, missed work days, long-term studies on respiratory morbidity and mortality encompass, by themselves, hundreds-of-thousands of people and billions of dollars. America has the highest low-level radiation background of any country in the world, with 140 million tons of uranium mill tailings enshrouding our habitat in 1983.

JUNK

One hundred thousand new vehicles are produced each day, adding up to a fleet of 325 million in America as of 1980. These machines represent the overriding materials overshoot in the United States, comprising 20 percent of all steel production, 95 percent of nickel, 35 percent of zinc, 60 percent of rubber, 12 percent of aluminum, 10 percent of copper, 51 percent of lead. One in every four dollars goes into the auto, and one in six jobs are concerned with it. Every year, 7 million autos are junked.

Soon, computer hulls will begin replacing the automobile for scrap value. In addition, 8 million televisions sets, 38 billion bottles, 17 billion cans, 60 million tons of paper, and 11 million tons of iron and steel are junked annually. In his lifetime, each American consumes 1400 tons of fuel, wood, and mineral resources. Of his textiles, the American's store of synthetic fibers is 33 percent. Most of his precious minerals come from South Africa, Zimbabwe, and the USSR.[8] But they all end up in the American landscape, junked.

IMPOUNDING OUR ASPERSIONS

Four hundred and thirty thousand industrial waste receptacles in the United States receive at lest 204 billion liters of waste fluids every day (excluding home waste disposal). Seventy percent of these impoundments are unlined, and 30 percent of them adjoin groundwater (aquifer) sources. Curtailing degradation is a critical priority, but in a democracy any incremental approach risks ambient insult. Cleanup costs are staggering. Cause-and-effect analyses are insufficient to arrest the growing accumulation of wastes underfoot. The Environmental Protection Agency (EPA) estimates that there were 50,000 hazardous waste disposal sites as of 1979 in the United States. We have no global estimates. By 1990, the EPA will be lucky if it can investigate 300 of these dangerous dumps. Of the 400,000 firms notified by the EPA under the Resource Conservation and Recovery Act, only 14,659 of them ended up actually applying for permits to pollute. We have no way, as of yet, to enforce those who did not comply with the application procedure, short of hiring huge numbers of pollution enforcement officers, for which the administration has elected no funds. So what's with the administration?

GOVERNMENT CONSPIRACY

The Reagan administration Task Force on Regulatory Relief, meant to increase the number of jobs, to get America's economy really moving, ignored numerous studies indicating immense savings to be had through environmental cleanup and safeguards implementation. The AFL-CIO estimated that 600,000 new jobs could be created by the conservation industries; even the EPA itself concurred. Instead, 32,000 workers lost their mode of livelihood because of the administration's ecological renunciation, its denial of funds, its shirking of mandates. Ten top environmental administrators under Reagan were hired from previous posts that were blatantly against all environmental protection.[9] This uninformed, insensitive, mind-dulled assemblage attempted to steal America's heritage from our descendants, to strip mine the very Rose Garden. As it happened, "cemetery" was re-defined so as to enable mining companies to move in on family burial plots. In addition, the

Secretary of Interior declared that it was time to move in on the wilderness. There are only 32 million wilderness hectares in all, 23 million in Alaska. Only 1.2 percent of all American land is designated as wilderness, and only 4 percent can, under dubious law, ever be so sanctioned.[10] Oil and gas leases have already been issued for the Capitan Wilderness area in southwestern New Mexico. Other leases in Wyoming and California are pending. Elsewhere, 25,900 square kilometers of America have been irrevocably ruined by strip mining. Reclamation programs are decorative, not ecological.

Whether it be asbestos, auto safety programs, school lunches, Social Security, formaldehyde, acid rain, or food inspection regulations the administration has put on blinders, ignoring everything, even body count. Data-processing funds for the National Emissions Data System were cut by nearly 90 percent; the EPA budget for 1983 was slashed by 40 percent, to a mere $216 million, hardly adequate to carry out its too numerous legal and scientific duties to the American public. The 1983 Department of Energy's budget stipulated 47 percent for energy defense, the building of bombs, and only 1 percent for safeguards and security, 7 percent for storage and disposal of wastes. Meanwhile, Congress has entrusted to the courts the terrible responsibility of judging cases that really require the expertise of several scientific disciplines. "Conflicting interests play fiercely for enormous stakes, advocates are prolific and agile, obfuscation runs high, common sense correspondingly low, the public intent is often obscured," wrote Judge Patricia Walk in her commentary on a D.C. Circuit Court of Appeals judgment.[11] For the EPA, conspiracy was rewarded with contempt of Congress.

AGRICULTURE

With nearly one-third of all human beings chronically undernourished, the developing countries urgently require a 4 percent increase in cereals production over the next 25 years. So far, increases are around 2.5 percent, at best, compromised by population explosions, food-fuel competition, erosion, and desertification. An area the size of two Belgiums — six million hectares, the size of Maine — is becoming desert every year. By 2000, 35 percent of the ice-free Earth will be swept over by sand. In the past 20 years the Sahara has gained a 100 new kilometers. If 15 to 25 percent of all grain traded internationally were devoted to lessening the 50-million-metric-ton calorie deficit in the Third World, production in the West would simply be increased, political tensions eased, jobs created, and millions of lives saved. In addition, the government could let the market dictate more palatably a cost-beneficial price of grains, made competitive by the domestic trimming back. But to maintain price subsidies at home, grain has been sold (40 percent of all crop land for export) and no crop land idled for the first time nearly in our history. Meanwhile, our food security has been reduced to a mere 150 million metric tons, or 40 consumption days.[12]

In America today, there are 24 people per square kilometer, 4 hectares of land per person, land that comprises 1.2 hectares of forest, 1.2 hectares of pasture, 0.7 of cropland, 0.6 of marsh, desert and tundra, and 0.3 for remaining uses. The crop land sector is diminishing, fast: By the year 2000 more than 8 million hectares of arable land will be converted to urban use. One-fifth of the world's croplands have been destroyed. Part of this dilemma stems from the genetic uniformity of higher yielding varieties, making for such leaf blights as that which devastated the U.S. southern corn belt in 1970.[13]

If the world converted to the American style of agriculture – a goal implicit in the Green Revolution – 80 percent of all energy conversion would go into food production and petrochemical reserves would be exhausted within a matter of a few years. The substitution of energy for land has resulted in tremendous cereal gains in some countries; but twelve desertified nations have experienced a 40 percent drop in per capita grain output. Globally, the food balance has declined 15 percent, increasing the gap between the haves and have-nots. Agribusiness now controls between 51 and 97 percent of our domestic major staples, including eggs, vegetables, citrus fruits, and chicken. The illusion of energy-intensive agriculture is clarified when two statistics are equated: An Iowa farmer produces as many as 6000 calories for every calorie of his own expended labor. But when the net energy expenditure is calculated (gas, oil, tractor, fertilizer), his one can of corn – equal to 270 calories of product – actually required 2790 calories from his accessory technology. In other words, a 10:1 ratio. His own labor is magnitudes away from the actual workings of his calorie exchange. In this way, one person can raise 75,000 chickens or feed 5000 cattle in an automated feedlot. But at what price? Our daily intake of energy calories is actually closer to 200,000 if we include the mechanical scaffolding and pesticides.

Inorganic nitrogen fertilizer and pesticides have increased domestically by 42 and 1069 percent, respectively; but crop losses remain at about one-third of the harvest. More than 300 insect species have genetically adapted to the chemicals. In the meantime, soil fertility is greatly diminishing, as is the number of farmers. Forty percent of all Americans were farmers in 1900, 11 percent in the mid-1950s, and 3 percent today. By the year 2000, ambiguous federal subsidies will have fueled conversion of prime agricultural lands over to various urban development projects. Florida will have no farmland; Virginia will have lost 24 percent, California 16 percent. Less than a dozen corporations will own and operate the sources of most of the world's food.[14]

EROSION: THE HOURGLASS EFFECT

Two billion people depend on the watersheds of mountains throughout the world. Healthy forests maintain nutrients in the soil and naturally curtail the displacement – even during floods – by massive landslide and siltation. When the forests are impinged upon by human beings in need of fuel, when the soil's nutrients are razed by shifting cultivators who abuse the requirements of fallow cycles, or when manure is lost to the soil from its consumption as a fuel, then the soul's infrastructure is compromised. The Dust Bowl was one result. A billion metric tons of Ethiopian topsoil lost annually is another. Airborne dust actually travelled to Washington during the Depression, covered the cars of Congressmen. This was indeed a new tactic in conservationist lobbying, the modus vivendi in Ecclesiastes. But it didn't work. There is to this day virtually no legislation treating soil erosion.[15]

It takes from 300 to 1000 years to form 2.5 centimeters of soil and 33 years to lose it under natural biomass runoff conditions. The 'T-factor' relates to that point at which the natural rate of soil formation is hampered by inordinate erosive furiousness from unnatural means. In Iowa, which has the strongest state conservation soil laws of all the ten states implementing such regulations, 260 metric million tons are lost from prime farmland each year. Calibrated at 160 tons per acre inch of soil and a 20-centimeter depth, 6.4 billion tons throughout the United States vanish. Fifty percent of currently plowed land – from a base of 0.5 billion hectares – requires urgent remedial care. Regrettably, short-term profits and soil husbandry are incompatible at a 3:1 diseconomy. Every day in the United States, 3240 hectares of agricultural land are lost. The situation is worse in the USSR. The loss of organic matter, the chemical and toxic salts buildup, destruction of the soil's porousness, all contribute to the hourglass effect. In the Philippines and El Salvador, siltation is destroying costly reservoirs. In the Himalayan countries, destroyed watersheds have vented devastating floods resulting in the deaths of millions of people in this decade. The inability to enforce soil practices among private entrepreneurs (erosion control measures are actually decreasing in the United States), grossly unequal land-tenure policies in the Third World, the Green Revolution with its heavy emphasis on addulterants, has all contributed to the limiting of world agricultural production and soil disappearance. Recognizing how critical the soil is, Peruvian mountain farmers have been known to climb down to the lowlands and sniff out their own soil for transport back up to the highlands.[16]

FORESTS

The world's forests are being depleted by some 20 million hectares a year, an area one-half the size of California, or the size of Hungary. By replanting, the United States is way ahead of the crisis. Nonetheless, the domestic ratio of consumption to reforestation is 6:1. By the year 2000, Africa's closed forest will have been reduced from 180 to 146 million hectares; the Amazon basin – which accounts for 75 percent of the world's tropical rain forest – will have lost half of its groves. In the Philippines, Thailand, Indonesia, and Haiti, shifting cultivation, fast-profit opium harvesting, and watershed deterioration have devastated most of the remaining forests.

Two decades ago, forests covered 25 percent of the Earth. Today, the figure is only 20 percent. By the year 2000, that figure will be halved to a per-capita wood supply of 40 cubic meters annually, ignoring the fact of price inaccessibility in most of the Third World. Virtually no natural forest remains in Europe, where intense monocultural management promises a surplus in the year 2000, but at the price of deadened population dynamics. This refers to the animals normally inhabiting a diversified forest bioniche. A single giant casearia fruit tree in the Finca La Selva Reserve of Costa Rica, for example, may feed as many as twenty-two bird species at various times of the year, who in turn rely on many other plant species to compose their annual diet. When that particular tree is cut or replaced with commercial wood for harvest, the many bird species will die out. Finally, with the burning of forests, atmospheric circulation is altered throughout the world, surface albedo is changed, and the net carbon dioxide increment escalated.[17]

OCEANS

The critical point for fisheries came in 1971. The catch has fallen by 1 percent every year since that time. The south African sardine, the Peruvian anchoveta, and the Norwegian herring, to cite but some examples, have been so over-fished as to seriously threaten their breeding base populations. Black Sea Mackerel have disappeared, halibut catch has plummeted by 90 percent. The boils of filter-feeding fish (clupeoids) are greatly on the decline, and with them, many large marine animals which must feed on them. There is virtually no enforcement of catch quotas, nor are there any interspecies models for management being used in open waters. What happens, for example, if *Homo sapiens* begins large-scale harvesting of krill? Other upper-trophic-level organisms, including whales, will be adversely affected. Is the loss of diversity a tragedy? The current adverse relationship between our species and the harp seal, the capelin, pollack, fur seal, cod, and humpback whale suggests that it is.

Dissolved oxygen – which fish must extract from the water they take in – is being steadily consumed by bacteria feeding on feces, by chlorination, hydrocarbons, acids, alkalis, detergents, wastes, and chemicals across the spectrum – many first passing through and destroying crucial estuaries en route. Seven of the world's ten largest metropolitan regions border wetlands. By the year 2000, less than 40 percent of the original wetlands in the United States will remain, primarily as a result of urban outreach, drainage procedures, and channel cutting.

The landbased outfalls, the runoff, the atmospheric emissions of synthetic organic chemicals and heavy metals, the 700,000 tons of spent nuclear fuel that has already been deposited on the ocean bottom is all unregulated and escalating. The global inputs from human sources are ten times their natural deposition. As of 1982 in the United States, for example, approximately 1200 major oil spills have occurred. In 1967, Arvid Pardo, the Maltese ambassador to the United Nations, first proposed a Common Heritage of Mankind Law of the Sea. But this has turned into a nasty forum between the northern and southern countries vying to gain multinational sanctions over the control of nickel, copper, cobalt, and ferromanganese ocean-bottom nodules – worth an estimated $500 billion.

Human conceptions of optimality are biased by what turns a profit, perhaps more so in the oceans than anywhere else as a result of the long-standing illusion of the ocean's inexhaustible extent and absorption abilities. Scientific analysis of energy conversion coefficients between the base photoplankton crop (at the bottom of the food pyramid), and the upward trophic productivity may be off by a factor of 100. Selective fishing requires a corresponding re-emphasis on the tropho-dynamic approach to guestimating organic biomass and probable world catch. Fisheries seldom operate in sync with marine population cycles. Until recently, technology masked declining populations. Now, replacement populations are annually declining and are shorter-lived. Most importantly, human beings must begin realizing – as Thor Heyerdahl and Jacques Cousteau for years have warned – that the oceans – source of all life on the planet – are fragile.

ANIMALS OTHER THAN OURSELVES

Every day at least three species become extinct at the hand of mankind. By the year 2000, 20 percent of all living plant and animal species will exist no more. This represents – in addition to the imponderable fact of extinction – the tragic loss of *billions* of animals now and in the future. Various remedies for this atrocity have been advanced. In Africa, conservation by economic incentives has been proposed, referring to wholesale cropping, harvesting of animals for protein and regulated hunting. But once these economies turn sour, then the very argument for conservation is lost. Grizzlies are poached by organized rings in the United States — as are rhinos, elephants, tigers, and gorillas elsewhere in the world — but illegal kills are not the sources of extinction. Instead, agricultural, migratory, industrial, and domestic growth have resulted in the attrition. National Park authorities at Yellowstone have shown that our very officials, charged with protecting wild species — in this case the Brown Bear — may be insufficiently apprised by science, or dangerously biased in favor of special interest groups, other than animals in their care. A river otter needs as many as 80 kilometers of stream to forage in. Northern Spotted Owls may require 243 hectares of old-growth forest, a grizzly 160 kilometers of open terrain. That land is simply not there for them anymore, by the legal and consumptive standards and aspirations of *Homo sapiens*.

Peter Verney and Paul and Anne Ehrlich have provided compelling glimpses into the history and current abuses of the animal kingdom by man.[18] Such aggression has always persisted and may define the essential nature of conscience, penance, and doubtful self-regard that first bore down upon our just-upright progenitors. Some chemical imbalance goads us to zealously murder animals beyond our real or imagined dietary and social needs; to sport with, study ruthlessly, and capitalize upon the untold agonies of other forms of life.

According to Verney, the Roman *venationes* — killing orgies — reached their height during the Emperor Trajan's celebration of his Dacian conquest. Eleven thousand large animals were massacred over a period of four months — lions, rhinos, camels, elephants. The first English whaling expedition around the islands of Spitsbergen took place in 1610. By 1750, what had seemed inexhaustible was no more. In 1930, 29,000 blue whales were slaughtered. By 1978, the International Whaling Commmission quota was set at approximately 4000, with only insolent Japan still adamantine about feeding whale meat to its citizens.

Slaughtering beavers began in 1604 along the St. Lawrence River. Verney records dealer receipts in London for the year 1895: nearly 10 million skins. By 1912, the fur seal was virtually extinct. The African "champagne safari" and Indian *shikar* came fervently into vogue in 1850. As late as 1964, at least 50,000 leopard skins were being shipped annually from East Africa, and the Indian tiger was nearly extinct. Birds have perhaps been least understood, least loved. *Plumes fantasies*, the passion for feathers, has taken an extraordinary toll on every beautiful avian in the world. In 1844 the last Greak Auk was hunted down for an Icelandic collector. The quetzal, the Bird of Paradise, egrets (particularly the great white and snowy) have been brutally marketed. In one London auction, 24,000 clusters of egret feathers were sold. In Italy, 240 million exotic birds a year have been killed. The eulogies for animals are not heard by many, though poster sales of baby harp seals prior to the kill are high on the merchandising list. Fifty-four kilograms of beef are consumed per capita annually in the United States. Stray animals picked up in city pounds — the legions are uncountable — are held from 2 to 10 days then put to sleep. Scientists kill, maim, torture and *retorture* upward of 400 million animals a year globally.[19]

NOTES

[1] *Science*, 28 August 1981

[2] Colin Clark, *Population Growth and Land Use* (New York: St. Martin's, 1977).

[3] J. H. Fremlin, "How Many People Can the World Support?" in *Population, Evolution, and Birth Control*, ed. Garret Hardin (San Francisco: W. H. Freeman, 1969).

[4] *Environmental Quality Annual*, Council on Environmental Quality (Wash D.C., 1983); and Russell W. Peterson, "Laissez-Faire Landscape." *New York Times Magazine*, 31 October 1982.

[5] Jacques-Yves Cousteau, *The Cousteau Almanac — An Inventory of Life on Our Water Planet* (New York: Doubleday, 1981); and Luna Leopold, *Water: A Primer* (San Francisco: W.H. Freeman, 1974).

[6] Roger Stobaugh, Daniel S. Yergin, eds., *Energy Futures: Report of the Energy Project at the Harvard Business School* (New York: Random House, 1979).

[7] Kirkpatrick Sale, *Human Scale*, (New York: Coward McCann & Geoghegan, 1980) pp. 325.

[8] Jeremy Rifkin (with Ted Howard), Entropy (New York; Viking, 1980).

[9] Russell W. Peterson, "Laissez-Faire Landscape." *New York Times Magazine*, pp. 27-56, October 31, 1982.

[10] George Will, "A Word for the Wilderness," *Newsweek*, August 16, 1982.

[11] Conservation Foundation, *State of the Environment 1982 — A Report From the Conservation Foundation* (Washington, D.C.: The Conservation Foundation, 1982).

[12] R. Neil Sampson, *Farmland or Wasteland: A Time to Choose* (Emmaus: Rodale Press, 1981).

[13] *State of the Environment 1982 — A Report from The Conservation Foundation* (Washington, D.C.: 1982).

[14] Erik Eckholm, *Down to Earth: Environment and Human Needs* (New York: W.W. Norton & Company, 1982).

[15] Lester R. Brown, *Building A Sustainable Society* (New York: W.W. Norton, 1981).

[16] David Dickson, *The Politics of Alternative Technology* (New York: Universe Books, 1979).

[17] Paul and Anne Ehrlich, *Extinction: The Causes and Consequences of the Disappearance of Species* (New York: Random House, 1981).

[18] Peter Verney, *Animals in Peril: Man's War Against Wildlife* (Provo, Utah: Brigham Young University Press, 1979); and Ehrlich, ibid.

[19] P. Verney, ibid., and Tom Regan, *All That Dwell Therein: Essays on Animal Rights and Environmental Ethics* (Berkeley: University of California Press, 1982).

Chapter 8

Tropical Sorrows: Bangladesh and Brazil

Our species is the only one to lay claims to the natural world. We long ago drew up the battle lines. Today, that struggle is most dangerously pronounced in the tropics, where soil is in short supply and human food at a premium.

BANGLADESH AND FOOD AID POLICY

As early as 1812, a humanitarian Congress voted to provide $50,000 worth of food for the victims of a Venezuelan earthquake. But the modern experience of such aid has stemmed from domestic pressures and foreign-policy priorities: what to do with American surplus, how to use aid to compliment price subsidies at home, and how to effect political changes abroad under the mask of benevolence. *Program food aid* (by concessional credit arrangements), and *project food aid* (by outright grants) have thus evolved in skewed competition following public law (PL 480), the Agricultural Trade Development and Assistance Act of 1954. This act originated during a period of high governmental price supports, all begging the issue of surplus disposal by way of trade.

But these dual measures have enabled several presidents to press home political interests with low visibility: an agricultural trump card principally understood to be an aid vehicle, the Food For Peace initiative implemented in 1960 at the end of the Eisenhower era. President Kennedy appointed George McGovern to be its coordinator under the prevailing Kennedy doctrine of unconcealed foreign policy food manipulations.

By 1964, according to Mitchel Wallerstein, during the Shastri government emergency in India, food aid was fully tied to economic levers, as outlined in a memorandum to President Lyndon Johnson from Secretary of Agriculture Orville Freeman.[1] The document, in principle, is a good one. In practice, it was vulnerable to personality and political subterfuge: "Negotiation of new PL 480 agreements represents an opportunity to induce the recipient nations to take steps which they would not take on their own in improving their food situation." Following frantic appeals in 1964 from nerve-wracked Indian leaders, Johnson lost patience with India's problems, abolished long-term food aid guarantees, and introduced self-help measures and sporadic short-run food aid extension. This "short-tether" policy failed to achieve its goals. Prime Minister Gandhi moved closer to Moscow's position, and, in 1966, Johnson cut off all aid to India, pending a U.S. Department of Agriculture assessment of India's self-help efforts, an analysis which took into account such domestic catalysts as high-yield seeds, credit, tax incentives, fertilizer availability, and management procedures. A positive report forced Johnson to concede grain shipments, though they were entirely unsatisfactory. As it happened, in 1972, famine in India resulted in nearly 1 million deaths. The famine was part of a global food shortage — the worst since World War II. But the Nixon administration responded by shipping nearly one-half of its available surplus — not to the starving multitudes in Africa, Latin America and Central Asia — but to South Vietnam and Cambodia. United States' grain leverage probably persuaded the Soviets — who received much of our remaining shipments — to coerce North Vietnam into negotiating a cease-fire. At the same time, all food aid was withdrawn on political grounds from Chile. Special "reverse aid" concessions were made to South Korea, specifically to induce her export restraints on synthetics (which were competing with similar products in the southern United States, where many of Nixon's friends and constituents stood to make big profits).

By a heated moral gesture, Congress passed the Foreign Assistance Act of 1974 to counteract and reduce such politically motivated food concessions. The separation of food and state made beneficiaries of many countries. Foremost among them was Bangladesh, which has received 5.5 million metric tons of grains and cooking oil since its independence in 1971, or $2 billion worth ($11 billion total foreign aid from all countries).

Three million people were massacred in 1971 during the Indo-Pakistani war for Bangladeshi independence. Sheikh Mujibur Raman returned to his country from England victorious, abolished parliament, made himself dictator, banned the press, and proceeded to deal with in-country refuges by encamping them. Nine million had fled to Calcutta during the fighting, and many were now returning home. I visited the country in the winter of 1975, following a series of disastrous floods and of a famine that took at least 330,000 lives. The country was still in ruins since its war of independence. I wandered the streets, visited refugee camps, interviewed dozens of government officials, underground rebels, foreign aid experts, other journalists. At the time, the Anjuman-e-Mahifidul Islam organization was collecting 400 unclaimed bodies a week off the streets in Dacca. Five minutes from the Intercontinental Hotel, where crows fed off the fresh shanks of red meat being carted in for visiting diplomats, I saw my first corpses, strewn across sidewalks, dogs festering around the erupted carnage, passers-by doing their best to ignore the stench. Malnutrition, as well as famine, engulfed an estimated 60 percent of the countryside. Urban squatters were treated like criminals, rounded up and trucked off to camps such as Tongi and Dattapara. I saw poor and elderly being beaten. At Tongi, two biscuits and two hundred grams of powdered milk were being dispensed daily to the tens-of-thousands of dying victims. Fifty-thousand children were going blind each year from Vitamin-A deficiency. A shortage of spare parts in nearly every commercial and industrial sector meant that no amount of aid could keep Bangladesh afloat. Whatever foodstuffs did manage to survive the rat infestations at the docks were then dispersed to ranking officials who could sell off their cuts at a profit.

Meanwhile, 10,000 babies were born every day, for a population increase of forty-seven per thousand. In 1975, the density per square kilometer was 1360 people. Today it is nearly 2000. In 15 years, the population of Bangladesh should reach an impossible 180 million. The country is no larger than Illinois.

I remember mists — mucilaginous, thick with mosquito — hanging over the mustard fields, the cane bush, the mango, banyan, bamboo, and jack trees. I made out a young woman, carrying a child in her arms, eratically drifting along a muddy tributary, weakened. I approached her and in her harried face saw a racial tapestry, Austro-Asian, of prehistoric Dravidian, Aryan, Mongolian, Muslim, Arab, Turk, Persian, and Afghan. She was an incalculable victim of history and ecology. But she was no beggar; rather, a proud and truly handsome girl, qualities that I perceived in all the Bengalis.

The girl's parents have always had to answer to the zamindar who collected their taxes and crops by way of a tenurial feudal land system. Today, America spends $20 million a year on family planning to prevent the parents from having any more children. But from the parents' perspective, they are doing all right. Food production — at a 2.1 percent increase per year—has left a record amount of grain in their village storage. Rice is coming in three times a year. The lepers on the edge of town have all been offered treatment for their condition. No one has to pay any taxes. Government ministers are rejoicing in Dacca over their aid deals and this is taken to mean a brighter future for everybody. If the local residents' supply and demand of energy and food balances were known, in gigajoules and tons, they might even have cause for confident early retirement out on the farms. Their rice, jute, vegetables, straw, and mustard all came in well this year. They had the use of draft power and plentiful organic fertilizer, whose end-use suffered little competition for fuel (13 percent). Their own post-harvest assets amounted to nearly 100 kilograms, and that of their village, 96 metric tons.

Or such was a scenario prepared on the village of Dhanishwar by Harvard scientist John Briscoe in 1979.[2] Briscoe pointed out that every bit of residue was used in the village, making the village something of an ecologist's dream. But then Dhanishwar was very much a dream village. Most do not resemble it.

In fact, 90 percent of the Dacca government's development budget is aid-inspired. For all the land's blatant fertility, there is simply not enough to go around, nor is there anywhere near the amounts of necessary—emergency—fertilizer. Cattle dung is more often used for fuel. Muslim culture discourages the cultivation of many vegetables, and the government finds that it does not have the cash to finance projects that are foreign-sponsored. Most donors from abroad are now providing project aid, reminiscent of self-help scenarios prior to 1974; projects which will force or enable in-country manufacturers to cultivate sales. But grass-roots capitalization is not forthcoming from abroad, only austerity loans. Bangladesh needs hard cash and hard commodities, namely, fertilizer and petroleum.[3]

The country's vital statistics are not the absolute worst. At least a dozen other nations (including both Yemens, Qatar, Togo, and the Comoro Islands) all face equally desperate situations. What makes the ecological crisis in Bangladesh so hideous is the size and density of its population and its subsequent vulnerability—ever renewed—to famine. The country has a population of 90 million. The average age of a resident is 16½, making for an explosive coming generation. Life expectancy is 43. Per capita gross annual income is about $100, even during the very best of years. The questions we must ask are: Why is there poverty in Bangladesh, what can be done to ameliorate it, and is the situation as hopeless as most experts claim?

Some 75 percent of the people are in fact greatly below the internationally accepted poverty line. In a country that is 91 percent rural, and agricultural, the immediate crisis rests with the fact that over half the country also is landless. The alluvial plains themselves are abundantly fertile, easily capable of feeding the entire current population perhaps into the future; but not without a disciplined agricultural approach to land equity and productivity. For the most part, Bangladesh is not intensively farmed. Why not? There are three explanations. First, Moslem law of inheritance enforces the passing down of land to each child. This diminishes the parcels to below the size propitious to family sustainability. As the size of holdings decrease from generation to generation, owners are eventually whittled down into the cadre of the landless poor. The wealthiest 16.7 percent of the rural population owns more than 66 percent of the land, and 60 percent of the rural poor have barely a hectare to each person. Importation of new technologies invariably goes directly to those larger land holders who can afford the equipment and who have enough land to sensibly apply it. As they become richer, they will buy out the poor farmers nearby, thus fueling an inflationary landless class and perpetuating a feudal system by which those same disinherited families begin sharecropping on a now widespread basis: the sharecropper must cover any and all costs of production, but share produce fifty-fifty with the owner, who is often an absentee landlord able to circumvent the 3.25-hectare landholding limit by registering excess acreage in the names of children yet unborn. A sharecropper cannot cost-effectively eke produce from a tenth of a hectare. His profit would amount only to $20 for a year. Nor can he normally afford the up-front expense of sharecropping larger plots. He is out of business.

In the meantime, he and his wife are busy making babies, knowing that one in four will die before the age of five. Mortality rates among the landless are twice that of families on 1.2 or more hectares of land. The landless father may take a job in which his labor is rewarded with 25¢ a day, plus a meal for himself. He can buy 1 to 2 kilograms of rice with that money. With a large family, 1 to 2 kilograms is not enough to cover the 500-gram minimum nutritional requirements. Nor is the Bangladesh government likely to help. Although 57 percent of the government's total budget (again, 90 percent of its developmental budget) comes from foreign aid, very little reaches the grass roots. The U.S. Agency for International Development (AID) hopes for 25 percent grass-roots effect. Most food aid sent to Bangladesh disappears into cash accounts of the upper class and military. In November 1970 the delta was swept by a tidal wave, driven by cyclones. Three hundred thousand people died because of the overcrowding. The floods made for bumper rice crops. A breakthrough surplus meant less aid from Western countries; which meant in turn that the government had to borrow money from its own banks to finance additional grain storage bins. Extending its debt this way forced the World Bank to withhold a $1 billion loan. Four years later, the government — desperate for income — sold its main export, jute, to Cuba for a mere $3 million. But this violated U.S. sanctions against Cuba. As a result, with near certain famine imminent in Bangladesh, U.S. Secretary of Agriculture Earl Butz nevertheless withdrew all aid to the country, forcing it into the position of buyer. Bangladesh simply did not have enough money. Under force of famine, it stopped further sales to Cuba to quality for U.S. aid under PL 480. By the time negotiations were completed, it was November and the body count had already been taken on the streets of Dacca.

BRAZIL: THE POLITICS OF EXTINCTION

The harpy eagle glides elegantly above a primordial trellis of liana. Giant otters patrol their riverine kingdoms, while anacondas sleep away the dripping hours in the original branches of creation. Immense turtles throw themselves in and out of sun striae, while every imaginable fish darts luminescent through the hundreds of hidden tributaries. There are more species of fish in the Amazon River than in the entire Atlantic Ocean. White-lipped peccaries forage along the seeping varzea forest floor in search of fruit, the acai, tucuma, and orange caja. Many of the trees disappear 50 meters into the clouds above. There are some 550 million hectares of Amazon rain forest, 3.5 million square kilometers, nearly half of the Earth's water moisture, easily a million plant and animal species (only a small percentage have been identified). In but 2 hectares of Amazon forest, 173 floral species have been discovered on a base of 900 metric tons of living biomass. The Amazon occupies nearly half of Brazil, though only 4 percent of its population. It is this wilderness to which mankind is genetically best adapted. For several thousand years we have fashioned our homes according to its evolutionary laws – soothing shade, moderate sunny exposure, birds and green plants, and humid air. Even in ancient Bahrein such components made for Eden.

Archaeological relics suggest habitation at the mouth of the Amazon 5000 years ago. Today, the Indians that survive do so in accordance with many of their traditional, explicitly harmonious tactics. They hunt with blowguns and poison, fell boars and tapirs, set deadfall traps for armadillos and a wide range of rodents. They catch birds and pay obeisance to a staggering complexion of beliefs and deities. Emilio Moran cites the Mundurucu and Yukpa tribes that impose rigid constraints on the amount of game taken, fearing the loss of soul and life for anything approaching excess.[4] The Tukano "make love" to their animals when hunting, believing in a mutual pool – shared by all animals – of energy. Given the immensity of insect biomass in the jungle, many tribes consume dozens of insect species. Indian agriculture is characterized by swidden or shifting cultivation (Milpa), a system remarkable for its ability to purge the land of parasites and pathogenic organisms detrimental to crops. Allowing fields to remain fallow for extended periods, thus insuring fertile soil, is a strategem commonly employed by the aborigines, thus insuring the best pact available to them. Maintaining lengthy fallow periods requires a lot of land. This the Indians are now losing.

Prior to the permanent Portuguese invasion, commencing in 1616, the Amazon Indians were apparently superbly fitted to their environment, with a population of nearly 7 million. But the Portuguese were intent upon extracting valuable resources from Indian lands – manatee, river turtles, andiroba oil, cloves, jaguar skins. The settlers produced sugar, fought to enslave the Indians, imported African slaves (between 9 and 11 million Dahomeyans and Yorubas, 25 percent of whom died aboard the slave ships), and generally plundered everything in their path.

Charles Goodyear had discovered a process for vulcanizing rubber in 1839. At the time, the Amazon produced the only rubber known in the world. The nineteenth-century *seringalistas,* appropriators of the *seringal* or rubber-producing regions, enjoyed some relief from their labor shortages with the forced migration of drought victims from the northeast and with a Brazilian population that doubled each 33 years between 1872 and 1965. By 1877, the new rubber market had attracted a second huge wave of Westerners to the northeastern Amazon. The result: ephemeral profits, widespread famine, drought, and massacres. By 1910, 1.1 million migrants lived in the Amazon, averaging a $323 per capita income, all from rubber. Today, there are 3.6 million Amazonian creoles and 125,000 Indians. The creole per capita income today is $171. These latter *caboclo* descendents of Portuguese conquests, dependent entirely on the beneficence of the *seringalista* class, more often than not experienced the terror and deprivation of manhunts, beatings, and slavery – with the result that the rubber boom in no way contributed to the social or ecological prosperity of Brazil, let alone the Amazon. It was strictly an extractive, short-term strategy, eventually made obsolete by synthetics and British rubber innovations elsewhere in Asia.

Re-colonization schemes are historically popular in Brazil. They are fueled by promises of rich agriculture, of forest products, endless land, and alleged mineral sources. Covetous Western nations have been willing to invest in these promises. But the result has been huge deficits, much suffering, and a powerful chaos stubbornly sustained with utopian gusto. Always the "New Brazilian Society" was to be at hand.

In 1940, President Getulio Vargas proclaimed the untapped treasure of Amazonia to be Brazil's future. Castelo Branco's military regime of 1964 instigated Operation Amazonia and SUDAM (the development agency) under the ministry of interior. A duty-free port in Manaus and the Banco da Amazonia were created to stimulate private capitalization. In 1964 the Belém–Brasilia highway was inaugurated, connecting the far-north to the south. Six companies contracted labor and supplies to cut a 70-meter-wide swath on an east–west axis, reaching Peru and the Andean Perimeter Highway. Bulldozers and earth scrapers tearing into the jungle replaced the language of birds and the archaeval wind through moisture-laden canopies. A euphoria had set in: in 1971 the U.N. Food and Agricultual Organization suggested that the Earth could multiply its population by a factor of *eight* if only the Amazon were settled. The World Bank loaned Brazil $629 million to accomplish that task. Though the scheme began with the decision to implant small homesteads, by 1973 a policy shift occurred, placing priority upon the *glebas*, large ranches. The Federal Colonization Agency (INCRA) offered all colonists 100 virgin hectares of land along the highway, with generous financing for clearing, planting and building of households. Food subsidies – $35 a month for the first 6 months – the promise of electricity, and other amenities, were held up to the would-be frontiersmen as bright beacons. INCRA allotted titles of up to 600,000 hectares to the glebas.

Meanwhile, meager land parcels for biological and human preserve were hotly debated in Brazilian parliament. By law, the colonists are to leave half of their homesteads free of all impact. This is to maintain some semblance of biological integrity in the Amazon. But this ordinance has not been policed, and colonists, particularly the large land-holding ranchers, have transgressed not only fallow requirements but cleared entire properties. From 1968 through 1978, 70,700 square kilometers of the forest were mowed and burned to provide grazing for 336 ranches running 6 million head of cattle.* Every minute, 10 to 20 hectares of Amazon forest are lost. A perimeter road will eventually encircle 12,967 kilometers of the Amazon frontier, making for yet greater land access and land give-aways.

Cattle ranchers have disrupted the soil hydrology and hydrochemistry; have exterminated hundreds of thousands of species. Reliance by planners on macroscale maps (1:500,000) has deceived everybody. Such macroscale imaging fails to account for the subtle soil differentials (oxisol, alfisol, ultisol) and thus undermines the government's ability to recommend subtle agricultural or husbandry strategem. The Brazilian Institute for the Defense of Forest (DBF) has faced strong resistance from the government, whose position has been one of autonomous military capitalism, a calculated shifting of all ecological issues from the U.N. Environmental Program to the more lenient Food and Agricultural

**Latin American grass-fed beef can be raised at one-fourth the cost of raising Colorado beef, though the savings are hardly passed on to hamburger consumers. More than 25 percent of Latin American forests (including the Amazon) have been destroyed in the past two decades to produce beef for the U.S. market, but per capita Latin American beef consumption has fallen considerably.*

Organization forum. Brazil has a hungry nation to feed and 12 million automobiles to fuel. Its ethanol program has not been systematic, the sugar cane crops diminished as a result of ants and the loss of important ash from the bagasse. Ranchers have lost money, farmers have lost money. Only 20 percent of the originally intended 100,000 families have settled in the Amazon. High volume-to-value ratios have rendered most crops profitless. No great mineral deposits have been found, and the distances along the east-west highway from Estreito to the Tocantins river to the Peruvian border render transport of crucial materials utterly cost-ineffective. Of the millions of metric tons of wood dragged out for sale, the majority is wasted. Meanwhile, the precious nutrients lodged in the upper five centimeters of the Amazon soil – are destroyed.

But as of 1979, 400 huge developments were being encouraged for taxation yields and "to give men without land a land with men," as President Medici had first put it. Volkswagen, Georgia Pacific, U.S. Steel, Alcoa, King Ranch – these are some of the "men without land" invited to the Amazon. With such large parcels of land being given away for complete exploitation, there is absolutely no way, even with the Projeto Radar da Amazônia (RADAM) surveys, to oversee clearance. At the present time, 100 tons of topsoil are lost per hectare each year. The forests that are destroyed will take 1000 years to regenerate. With their destruction goes the refugia, remarkable island or mountain summits, deep in the jungle, sometimes no more than a hectare in size and containing the living continuum of a species diversity entirely preserved from the Pleistocene: color morphs, strange and beautiful speciations, butterfly wings of a color not seen since the time of dinosaurs.

All of this is being steadily destroyed. And there is human tragedy. At least eighty-seven Indian tribes have become extinct in this century. We have seen an overall Amazonian aborigine population decrease over the past 500 years from an estimated original 6.8 million to a present 125,000. Anthropologist Emilio Moran cites disturbing research indicating, for example, the decimation of the Parakana and Nambiquara groups within very short periods of time from outside invaders, ranchers, and roadworkers. The National Indian Foundation (FUNAI), solely charged with protecting the indigines, has actually been implicated in the rites of extermination. FUNAI's repacification techniques have been likened to military campaigns. Reservations have been partially destroyed, tribes beaten back, women raped. Under Brazilian law the Indians have no rights, are unable to sue. Stephen Bunker reports that seventeen huge tribal *regions* are on the verge of complete, irrepairable destruction.[5] As the highways intensify their own slash-and-burn techniques throughout the Amazon basin – and more and more agrovilas, agropoli and ruropoli villages (48–1000 family sized towns) spring up – the forests are razed, the inroads flooded; the Malaria vector

(Anaopheles darling) is introduced, intense soil erosion occurs; Fusarium fungus destroys Indian-cultivated plants, and river fish species die out.[6]

Only now has the revelation of global atmospherics impressed biologists. The Amazon is the critical link in the Earth's carbon dioxide clearing house. Amazonian Indians and forest species possess the richest repository of native wisdom and potential medical, and technological plant products of any other region on the planet. But the metaphor of human disruption – since the sixteenth century – is fully at work in Brazil. The country is a vortex of ecological imperialism and shortsightedness. A generation of embattled conscience has arisen in the Latin American writers who have responded to the political and moral crises with an anguished outpouring: Nobel Prize winner García Márquez *(The Autumn of the Patriarch),* Vargas Llosa *(The War of the End of the World,* set in Brazil's northeast and semidocumenting one of many massacres), Ignâcio de Loyola Brandâo *(You Will Not See Any Country),* Heberto Padilla *(Heroes Are Grazing in My Garden),* and Manuel Puig *(Eternal Curse to Whoever Reads These Pages).*

In the meantime, Brazil's economy has plummeted by 50 percent, and inflation has risen at an annual rate of more than 100 percent. Sao Paulo, whose *turgurios* slums are every bit as fatal as those in Bangladesh, has the highest projected growth percentage of any city in the world. In 1985 the city will contain 16.8 million people. Brazil's national debt continues to be among the worst of any country, at a current $80 billion.

But even the great illusion is beginning to show itself. The wealthy can get burned. The American Daniel Ludwig's forest pulp Jari Project was sold in 1982 to a Brazilian consortium of twenty-seven entrepreneurs for one-fifth of Ludwig's original billion dollar investment. A decade of ecological mismanagement had taken its toll.

NOTES

[1] Mitchel B. Wallerstein, *Food for War – Food for Peace* (Cambridge: MIT Press, 1980).

[2] John Briscoe, "Energy Use and Social Structure in a Bangladesh Village," *Population and Development Review 5 (1979)*: 615–641.

[3] Ann Crittenden, "Foreign Aid Under Fire," *New York Times Magazine,* 6 June 1982; and A. Crittenden, "Bangladesh Hunger Linked to Feudal System," *New York Times,* Nov. 24, 1981; and James W. Hatton, "Bangladesh Pushes Global Aid Plan," *L.A. Times,* Part I, p. 20, Oct. 11, 1981.

[4] Emilio F. Moran, *Developing the Amazon* (Bloomington: Indiana University Press, 1981); see also B.J. Meggers, *Amazonia: Man and Culture in a Counterfeit Paradise* (Chicago: Aldine/Atherton, 1971).

[5] Stephen G. Bunker, "Forces of Destruction in Amazonia," *Environment* 22 (1980).

[6] Moran, *Developing the Amazon*; and N. Smith, "Colonization Lessons from a Tropical Rain Forest," *Science* 214 (1981): 755–761; see also J.H. Nigel Smith,

Rainforest Corridors; The Transamazon Colonization Scheme (Berkeley: University of California Press, 1982); and Smith, "Colonization Lessons From a Tropical Forest," *Science*, pp. 755–761, vol. 214, no. 13, Nov. 13, 1981.

Chapter 9

Shock Lung

We cannot last but a minute or two without oxygen. Yet we do not hesitate to foul that narrow layer of atmosphere surrounding the planet and upon which we so earnestly depend. Los Angeles presents a countenance ready made for science fiction, disaster scenarios, and doomsday statistics. There may be some black humor to it all, but not for those stricken with cancer. In fact, though it is difficult to verify cause-and-effect statistics, it appears likely that more than 100,000 people die every year in Los Angeles as a direct or indirect result of air pollution.

The 1976 Seveso disaster, which resulted from a high-pressure TCP (trichlorophenol [Dioxin]) explosion at Hoffman-La Roche's Givaudan Icmesa plant in Seveso, Italy, could have been avoided. Givaudan's executives, interviewed by writer John Fuller, revealed their complete familiarity with the potential risks of the plant.[1] Insolently eschewing ethics from their net returns, they courted tragedy where wiser industrialists might have halted.

Tragedy they still have. An 80-kilometer radius was lightly covered with white ash. The ash was Dioxon, a chemical so potent and with so myriad a disease and malfunction-inducing spectrum that those who were contaminated are still waiting for cancer to appear, for their babies to be stillborn or born without limbs, for major central-nervous system afflictions to begin showing up. The pesticide DDT has a half-life of 15 to 40 years. Dioxin, a derivative of TCP, may last for centuries once it is leached into the soil or mixed with food chains.

Seveso was evacuated while animals fell from the sky in mid-flight. Birds in migration or with long feeding patterns could easily have taken parts-per-million of Dioxin into Milan, into fields all over Europe. Experts who rushed to the scene realized that there was no possibility of charting the toxic-flow perimeters. And their own professional distress set off an anguished series of discretions. Meanwhile, the residents of Seveso were relocated to distant hotels, where they waited in vain for epidemiological consultations. After 6 months, children began suddenly developing lesions. Controverting centuries of Roman Catholic dogma, medical authorities pressed for abortions. Then a first victim died. Lawsuits flooded Hoffman-La Roche. Chemical researchers in America recommended that the entire Icmesa plant be surrounded by walls, incinerated, lowered deep into the ground, and covered with concrete.

Spills of this magnitude are not rare. Life Science Products in Hopewell, Virginia, subjected more than seventy of its employees to Kepone poisoning. Brain damage and sterility are two immediate prospects for many of those workers. Under contract to Allied Chemicals, the Hopewell firm dumped nearly 6 million liters of the ant and roach killer toxic sludge into the James River. The effects are not yet fully monitored. Life Science Products was fined $17 million, a meaningless sum. Kepone, like Dioxin, is so resistant that it costs $8800 a kilogram to destroy.

The U.S. Air Force spent 7 years trying to figure out how to dispose of "Orange" (2,4,5-T), a gaseous toxin used to permanently destroy at least 129,500 square kilometers of forest in Vietnam. Finally, the voluminous weaponry was transferred to the Dutch vessel *Vulcanus* and carried into the Pacific, to be incinerated at 1444° C, 1600 kilometers from the nearest point of land. The extent of marine repercussions is unknown.

The polychlorinated biphenyl (PCB) invasion of Michigan – in which fire retardant with a PCB base was confused with magnesium oxide and put in mixtures of cattle feed in 1973 – is still being analyzed for its inevitable ill effects. The toxic substance entered the meat, milk, and eggs consumed by innumerable Michigan residents. Throughout the United States, 9 million kilograms of PCB have been produced. Infiltration is impossible to curtail. In New York City, scientists told a Senate subcommittee that it is no longer possible to find uncontaminated milk anywhere in America. In Japan, PCBs entered the rice oil of Kanemi and a documented 1000 people succumbed to chloracne skin lesions and other disorders.

If we stopped spraying pesticides today – such as the toxic nerve powder, sevin, which the Oregon State Fish and Wildlife Commission approved of as an agent against ghost shrimp in the tidal flats (which is in fact killing everything *but* the ghost shrimp) – it would take at least 40 years to find out where we are at environmentally. But in developing countries, there is a four-to-six-fold increase in pesticide use. A "Hazardous Substances Export Policy" has occasionally surfaced in Congress but so far is unlegislated. Western corporations continue to profiteer upon their global dissemination of dangerous chemicals.

At present the EPA is investigating some twenty-seven chemicals for possible sanction: They include fluorocarbons, zirconium (a possibly carcinogenic aerosol compound used to inhibit perspiration), vinyl chloride, nitrofurans, diethylstilbesterol (DES), the "morning after" birth control pill, kepone, and others. So far, only *four* hazardous chemicals have stringent regulations attached to them. This is extraordinary. The primary issue is the short-term logic and arrogance of power that prompts the major chemical producers to continue manufacturing chemical substances suspected and even known to be lethal. Their lobbies rival the Mafia in persuasiveness.[2]

Making light of ecologists' doomsday prognostications has become something of a vogue across America. New Yorkers and Los Angelinos boast of their abilities to withstand ozone, PAN irritants, and carbon monoxide commuter intoxication. Evidence reveals our physiological capacity to habituate to certain levels of pollution. Similarly, grasshoppers (stupid ones, dead ones) are known to occasionally become oblivious to newly introduced predators. The planet generates its own pollutants, to be sure. Bacteria can and always have metabolized organic and inorganic toxins. Some eukaryote cells have even shown versatility in the presence of high radiation levels. The very oxygen consumed by most organisms today was once a deadly pollutant. But the pollution load engendered by Homo sapiens is of an unprecedented dimension.

LOS ANGELES

At the first grey stroke of dawn, the garbage trucks converge. In this apartment complex, races mingle, the far-away stench of exotic feces burning in a landfill rises like philosophical pipe smoke. Immigrant dreams stream through broken-down neighborhoods at gunpoint. Millions are out and about on the roads – lawyers, real estate speculators, arms manufacturers, engineers, computer technicians – all burning in columns of bright day. When I breathe, my tongue and lips and eyes react, telling me to stop. The mind loses something critical to its mass, forgets, has no fluency, commands one dumb-out after another. A warm, thick, meaningless caress, puce-dirt sprawl, festering loll, maze of fast lanes and cheap grace, all strewn beneath the north side of Wilshire Blvd., that region of Renaissance faggot mansions, electrical gates, inflated oases of dubious culture. Wild coyotes straggle through the highland floral gardens of Los Angeles, beseeching their coyote God in the sky: Lord, Lord why did you do this to me; of all places in the universe, why Los Angeles? The other day I read that one of them picked up a little girl and dragged her toward its den. With broken neck she fell, quite dead. The coyote left her, took some nervous chews, and went off hungry for another night. I lived in Los Angeles for 3 years and eventually found some good words for the place. Now I have forgotten them, having moved back out into the *real world*.

The El Segundo Blue butterfly is making its final stand in a pitifully dwarfed acre, surrounded by the world according to Standard Oil, which have given the little fellow the benefit of doubt, put up fences as a symbol of something sacred, tenacious, but how enduring? One-hundred and sixty kilometers up the coast, near Santa Barbara, the California Condor, reduced to a few dozen, gasps against the onslaught of pollutants (ozone transports) inundating its air space. The butterfly has been with us for hundreds of millions of years, ever since the first angiosperms flowered during the Mesozoic era. As for the Condor, it too has been around since nearly the beginning, a distant relative of the Pleistocene Teratornis, a bird that may have been the largest animal ever to get off the Earth's surface, at least until Icarus.[3]

Acidic rains infiltrate much of the eastern United States and Scandinavia with a pH frequently exceeding 1000 times its normal acidity, but Los Angeles has a more baffling crisis to contend with. Yosarian, the deserving hero of *Catch-22*, lay in a hospital bed systematically reviewing the glittery range of calamities to which the mortal flesh is heir. Why do his antics and pillowy contemplations strike us as somehow charming, even heartwarming? Los Angeles exacts its own vengeance, and with a depressing absence of such literary edification.

Within the 4000 square miles of Los Angeles is abundant industry typical of most urban conglomerates. Let's draw up our own Yosarian index of these enterprises: They include iron and steel mills; foundries; nonferrous smelters; petroleum refineries; asphalt blowers; cement manufacturers; pulp mills; acid manufacturers; coke producers; glass furnaces; glass fiber producers; cotton gins; soap and detergent producers; gypsum processors; coal cleaners; pulverized coal; fuel oil; natural gas combustion; solid waste disposal; multiple-chamber commercial incineration; paint and varnish manufacturers, alkyl resin producers; polymerization plants; open-hearth furnaces; blast furnaces; aluminum smelters; chlorination; crucible reverberatory and sweating furnaces; brass, bronze, lead, magnesium and zinc smelting; calcine kilns; pot furnaces; calcium carbide plants; coke dryers; rock, gravel, and sand production; and uncontrolled smelt tanks. And what does all this zealotry mean for the American way of life? How about the following, for a starter: iron oxide dust, smoke, oil and grease, metal fumes, catalyst dust, ash, sulfuric acid mist, liquid aerosols, alkali and product dusts, chemical dusts, mists, aggregate dusts, coal and coke dusts, coal tars, raw materials dust, alkaline oxides, resin aerosols, charr, oil aerosols, ash dehydrated coffee dusts, cotton fiber, carbon black, and detergent dusts. Remember to add up all of Los Angeles' 112,750,000 kilograms of refuse every year (and growing). Stationary pollutant sources combine into liquids, solids, and gasses. They infiltrate our food and every biological cycle, reach absorption levels in the lung, the skin, in the wells and soil nutrients, in all other animals – even in the farthest reaches of the troposphere, where the ozone reactions occur.[4]

Along with the stationary forms of pollution (characterized by smoke stacks and their trails of white, blue-grey or ashen black effluent, depending on the inside temperature of combustion), are the more crucial, transportational pollutants – those coming from our cars. They actually account for 60 percent of the pollution problem in Los Angeles.[5]

Four hundred and fifty billion dollars of U.S. manufacturing value puts out nearly 2.2 billion kilograms per year of air pollution, or one kilogram per $16. Some of the pollution is chemically neutralized according to a multivariety of built-in environmental safeguards within the atmosphere. But most of it is not.

Without doubt, Los Angeles is the city with the worst air quality in the world. Its pollution index values range from the unhealthful, very unhealthful, to hazardous over 300 days every year. Mexico City, Calcutta, New York, Denver, and Athens are approaching Los Angeles in pollution severity. What does this mean? What are the origins of the pollution index, who stipulates its criteria, and based on what?

Pollution has a natural background: natural carbon dioxides; even natural carbon monixide accompanying such eruptions as Mt. Saint Helens. There is natural gas, ozone, even hydrocarbons. But it is the big city that oversteps production of such gasses, usurping for its own the frail balance that hovers in the seven or so miles of available oxygen around the planet. Los Angeles has ten times more dust and carbon dioxide, twenty-five times more carbon monoxide, and many hundreds of times more sulfur dioxide than the surrounding southern California perimeters – which in turn contain more pollutants than any wilderness area. Even the wilderness refuge is now infected by global pollutants. The climber who ascends Mt. Everest without using oxygen is sticking his face into a jetstream larded with a 30-year's build-up of radiation.[6]

There is 20 percent less radiation on Los Angeles' horizontal surface, 10 percent more clouds, 100 percent more fog in winter, 10 percent more rain, a higher temperature, an 8 percent drop in relative humidity, and 30 percent less wind than outside the city. Autoclimatic, the City of Angels additionally emits 635,040 kilograms per day of carbon monoxide, 36,288 kilograms of particulate matter, 27,216 kilograms of sulfur dioxide, 635,040 kilograms of reactive hydrocarbons, 703,080 kilograms of nitric oxides. Since 1977, stationary pollution sources have emitted 59,875 kilograms more every day as compared with the previous 5-year period. Los Angeles is not alone in this legalized form of self-aggression. Over 7000 communities in America suffer varying pollution syndromes. Los Angeles has approximately 10 million people spread out over more turf than any city in history. There are nearly 8 million vehicles in the metropolitan region. Add to this a meterological condition called the "Blocking Omega High Syndrome" – a continuous low inversion pattern over the coast – and you have the recipe for disaster.[7]

Over 100 people die every day in the United States from cancer, 90 percent of which is environmentally induced. In late 1982, for example, researchers at the University of California – Berkeley established a link between breast cancer and air pollution. The lung inhales air and with it comes various combinations of chemically reactive pollutant agents. Natural pollutants such as fog, pollen, and bacteria are small enough (10 microns) to be filtered out by the nose. But smoke, carbon black, metal fumes, and viruses are in the 0.01-micron range, thus possessing a far greater capacity for eluding the body's cleansing mechanisms.[8]

Natural atmospheric gasses will remain in the environment: argon forever; nitrogen for 10 million years; oxygen for 1-to-1000 years; carbon monoxide for 1 month to 100 years; sulfur dioxide, oil, natural gas, and coal effluent for 1 minute to 1 month; fluorine compounds for 1 minute to 1 week; oxides of nitrogen and hydrocarbons as long as 2 months. These chemicals accumulate, in other words. Yes, a natural background level exists for some of them. Any backpacker can tell you about it, though this courts a terminological ambiguity. Pollution is defined as that state of uncleanliness. Obviously, nature is never unclean and rather requires some other agent to defile it, to mix things up in sufficient magnitude as to induce the chaos of impurity. This begins to sound suspect. Why "obviously"? Perhaps pollution is a word like *taboo*, better left to the vagaries of anthropology, depending on each individual society. Pollution may be applied to a woman's period of menstruation, in some societies. But in Western cultures it refers to the specific products of industry, chemicals not normally associated with the natural condition – not to the degree of accumulation, anyway. When you hike through the woods of Vermont or British Columbia, you'll find minerologic corrosion, pollen, fungi, spores, stellate hairs, insect parts, moth and butterfly scales, diatoms, foraminifera, radiolana, spanish moss, fibrous material – particles of biomass small enough to enter the lungs. Superimposed on this ancient legacy of organic material are the city pollutants, chemicals that have drifted into the woods, to the seashore, like soot and rubber dust and ammonium chloride. The difference is simple: in nature, if you're allergic, your eyes will hurt, you'll sneeze, and perhaps start coughing. In the city, you're likely to die, if not today, then in half a lifetime.

Since 1950 the prevailing visibilities in Los Angeles have been less than 3 miles on the average, save for those welcome windy days when clarity rushes off the ocean brushing the surrounding mountains desert mauve, their true hue. But in summer this rarely occurs. The formation of smog relates to the condensation of gas-phase species onto existing particulate nuclei. In other words, human pollution forms around natural background materials, causing a reaction termed *aerosol*. The growth of these aerosol particles occurs through several different types of gas reactions on the surface of the nuclei. In the end, there is wash-out, gravitational settling according to a mathematical formula known as Stokes' Law, which takes into account size, weight, and air currents. For those really minute particulates, Cunningham's Correction, added to Stokes' Law, predicts the settling behavior.[9] Hanging over this aerosolization of life in Los Angeles is an even greater shadow: the democratic share-and-share-alike spoilation greedily underwritten by the chemical oligarchy, a $100 billion industry that produces 200,000 unregulated chemicals each year. Ten million known chemical structures now exist. The National Institute for Occupational Safety and Health, in its *Registry of Toxic Effects of Chemical Substances,* published by the U.S. Department of Health, Education and Welfare, lists 82,908 chemical substances under suspicion. These include 2000 carcinogens, 300 mutagens, 1711 teratogens, and 787 toxic substances. A toxic effect is defined as any noxious effect on the body, whether reversible or irreversible. No qualifying limitation about duration or concentration of exposure is stipulated in the *Registry*'s criteria. These chemicals are tested on dogs, rats, cats, rabbits, monkeys and birds. Some 50 million are tormented to death in the United States every year so that we in turn can find out to what extent we are killing ourselves. We have learned that, at the very least, 100,000 anthropogenic substances are toxic, thousands cancer inducing. The classification of the chemicals is accorded subtle abbreviations. For example, MMI tells us whether or not there is a mucous membrane effect — irritation, hyperplasia, changes in ciliary activity. The acronym CUM relates to cumulative effects, defined as any substance retained in quantities greater than is excreted or exhaled. The list is extensive, though far from complete. There are literally millions of chemical *reactions* about which we know nothing at all. Anyone can walk into a library, thumb through the *Registry* and examine the myriad LC 50s appropriate to Los Angeles. LC 50 refers to a 50-percent death count of all those animals exposed for *x* duration to the stipulated lethal concentration of a substance. It's all charming vernacular, modern man's *Book of the Dead*.

We need love, food, water, sleep. But we need oxygen far more critically. We are what we breathe, and we breathe it 15 to 18 times a minute, with 0.5-liters per breath. We should not be surprised, then, given the build-up of chemicals that so many environmental disasters have occurred. In 1930, sixty people died of air pollution in a river valley in Belgium; in 1946 an international smog situation affected lives in dozens of cities; in October 1948, twenty people succumbed during 4.5 days of extensive pollution in Donora, Pennsylvania. Autopsies showed edema, hemorrhaging, purulent bronchitis. No single substances existed in sufficiently high concentrations to be considered the overriding contributor to mortality. In London in 1952, *4000* people died of air pollution in 5 days. No specific cause was implicated. Londoners didn't know what was happening. Four thousand people can disappear in their respective homes, but it certainly won't seem like a disaster. In 1953, 400 people died of air pollution in New York. Again, in December of 1962 there was a worldwide outbreak, felt from Cincinnati to Osaka. There have been countless other *incidents* — everyday explosions and derailments that result in evacuations. Today, those incidents are continuous, changing the air. Our government did not establish national air standards (the very notion rather horrifying) before 1971. Two centuries before, one Percivall Pott first reported on the incidence of cancer among chimney sweeps. In 1905 Dr. Des Voeux had volunteered the term *smoke-fog death* while speaking to the British Smoke Abatement League.[10] About that time, writers were commending the automobile for the new horizons it opened up. Indeed, some people love Los Angeles' freeways. As to one enamored, the city can come alive at night, its seething neon a transcendental ebb and flow. The incandescent tails of traffic light join in an extraordinary ritual of motion, telling the story of our civilization much as the horse-and-buggy explained earlier times. Los Angeles is the one city conceived after the fact of automobiles. This too has about it a fresh sense not unakin to reverence.

The California Department of Health tells us that in 1975, 170,000 people died of respiratory ailments within the state. Of that number, 125,000 lived in Los Angeles. Because Los Angeles has a population roughly half that of California, only 85,000 deaths should have occurred, proportionately. Breaking the deaths down, we find that the afflictions consisted of respiratory cancers, asthma, emphysema, bronchitis, obstructive lung disease, and chronic interstitial pneumonia. But here's the catch. We have no accurate estimates of the extent of respiratory distress or chronic lung ailments that go unreported: The epidemiologist must simply rely on mortality statistics, health surveys, and quantified data generated from various social service organizations. These statistics have their use, but as far as comparing regions, we have no way to reliably relate them to air pollution. Death reports may not even mention respiratory ailments present at the time of death, ailments that you can be sure have contributed to the underlying illness, as in the case of massive coronaries. Health surveys normally treat one specialized segment of a population, and social service agencies produce statistics for administrative purpose, not research analysis. This, compounded by the absence of any law mandating the reporting of chest conditions other than tuberculosis, makes the epidemiologist's task nearly impossible.

Pollution enters the nasopharyngeal compartment and goes either to the blood, to the gastrointestinal (GI) tract, or to the tracheobronchial area. From there it either goes back to the bloodstream or to the GI tract, or down to the alveolar, pulmonary area, and then on to the lymph system, slowly. There is, in the nasopharyngeal region, in addition to your nostril hairs, a mucociliary clearance mechanism that conducts a rapid uptake of material into the blood. But particles too small get by and settle in the alveoli sacs of the lungs where they produce dead-air regions, culminating in emphysema, carcinoma, and other diseases. In analyzing all of this, pulmonary specialists speak of deposition fractions, relating the particle size to Brownian movement. When a certain submicron particle is said to have a deposition fraction of 0.05, it means that the chances of its being cleared are slim. Most people will not know to voice complaints when they feel ill from air pollution: Symptoms are subtle and neither does anyone realize it is pollution in the first place. Seven thousand people were taken ill at the Donora disaster in 1948. The other several thousand residents – the remaining half of town – were apparently unaware of what was happening. It was only in 1976 that the National Cancer Institute started a clearing house for the classification of environmental carcinogens. The Institute estimates that 350,000 carcinogens are currently floating about. This data well exceeds that computerized in the *Registry of Toxic Effects*. These substances exclude what chemists call "synergetic reactivity," the ability of chemicals to interrelate in new and unpredictable ways. This reactivity bodes more harm for our chemical future than all of the known health problems combined.

No disease specific to air pollution, no well-defined microorganism can be traced. The chemicals are elusive, variegate, in complex and highly unstable mixtures, forever shifting in the air and water and food paths according to whatever new smokestack, new gene-deforming, fetus-eroding deodorant or hair dye enters the market place; wafted by wind, absorbed by plants, drifting out toward China. Long-term exposure studies are still lacking. Epidemiologic research is like anthropology – there is no limit to the time of analysis nor to the many connections complicating cause-and-efect relationships and specificity. Children, who breathe as much as eight times the amount of air as their parents, will suffer all that much more – as will mouth breathers over nose breathers. Joggers who workout during rush hour are courting the onset of disease. And still, no threshold level for adverse effects in a large community can be established. We said that particles less than 0.05 microns will pass through the nasal cavity and throat and enter the lungs. Sixty-eight percent of lead particles in vehicle exhaust, to seize upon one of the pollutants, is smaller than 0.3 microns. Ten percent of all U.S. lead consumption occurs in California; 5 percent in Los Angeles.[11]

In 1955, California introduced the Pollution Alert system. Twenty years later, government administrators were convinced that the automobile and smokestack retrofits (i.e., the catalytic converters) were solving the problem, though even then there was widespread concern that the converters were actually adding to the problem. In fact, the El Monte South Coast Air District, which collects hourly contaminant data from seventy-five sources throughout the Los Angeles basin, indicated at that time a steady surge in air-pollution alert days. The catalytic converter created acid mist. Smokestacks, retrofitted with electrostatic precipitators, have a cleansing efficiency calculated by weight, not by pollution particulate size. These cleaning devices are said to have a removal capability of 99 percent. That might sound impressive until we realize that the remaining 1 percent actually relates to the enormous mass of 1-micron-sized pollutants that nose-dive into your lungs, uncurtailed. The macrophage cell – our body's primary mechanism for removing alien submicron particles in the lung – is largely incapacitated by carbon. What this means for the lung's future is unclear. In any event, the EPA's legal apparatus has not faced the medical ramifications of its own insufficient best available technology mandates. The utilities would face bond-issue cliff-hangers if they had to comply with regulations insisting on 100 percent nonpollution. This would be helpful. With higher prices passed on to the consumer, increasing pressure on the free marketplace to devise alternate end-use technologies would come into play. The EPA, working with industry, has explored several curtailment plans to avoid increased costs all around. These include "bubbles" for consolidating the emissions of several stacks into one, and then muzzling it; levying pollution fees; issuing pollution credits that could be bought, traded, and sold. In Los Angeles, a "new source review" rule in effect since 1983 will attempt simply to halt any new pollution sources from operating.[12]

But the market allocates social value. The brief history of the National Environmental Policy Act (NEPA), demanding environmental impact statements is the case in point: Both Tellico Dam and the Alaskan pipeline were exempted from NEPA because of fiscal priority and national security. The market can not respond quickly. In a strained economy, in an oppressed environment, self-seeking people do not benefit the common good. They overshoot it. The EPA, which is understaffed and grossly under-budgeted, has sent only a trickle of industrial transgression cases to the Justice Department. With pressure from General Motors and U.S. Steel, and with administration laxity, pollution standards and laws will continue to deteriorate, especially as coal — with its barrage of effluent — makes ready to become America's number one energy source.

When a fireman succumbs to heat and smoke inundation, he is likely to experience a deadly reaction known as "shock lung"; his entire respiratory tract suffers exreme trauma. The same can happen to victims of pollution. Workers suffering from degrees of shock lung lose nearly $40 billion in wages each year from unearned sick-leave.

Each breath contains all of chemical history. You are this moment passing molecules to your lungs that Aristotle himself lived by. What will the future generations breathe? Imagine the amount of automobile in the very blood your heart struggles to make good. Los Angeles' freeway traffic — the engines themselves — consume 34,012 metric tons of air every day. Human respiration requires more than 54 million metric tons per year throughout the United States — the same amount of pollution that industry is producing. We add all the figures up, and speculate into some future August afternoon. Tourism in Los Angeles is at an all-time high. Millions and millions of additional drivers are on the road, people from elsewhere whose lungs have not adjusted to Los Angeles' atmosphere. Industry is spewing out its wastes, and the electrical utilities are heating up to provide rampant air conditioning. A heat spell exacerbates the whole nightmare. Not so much as a breeze lifts the ugly pall. Such a situation is already with us. In 1968 Senator Edmund Muskie introduced a bill to establish a pollution disaster fund. The House rejected it. The bill defined a "pollution disaster area" as any region whose natural environment had been so altered as to render it unfit for permanent residency. It was decided that too many cities already qualified.

EVACUATING LOS ANGELES

The Summer Olympics, 1984, Los Angeles. Could 10 million people get out of Los Angeles? No way. Here's the scenario. The Office of Emergency Services for Los Angeles employs disaster specialists — catastrophe connoisseurs earnestly in tune with the vagaries of escape artistry. Their prognoses for the city (applicable to any city) are no comfort. Twenty percent of Los Angeles' population will resist evacuation and head for bomb shelters, of which 6200 have been identified in the county. Most are downtown. The Department of Defense issues a shelter construction manual stipulating that there be slightly more than 3 meters of space per person, 2 meters for head room, and a ventilation rate sufficient to maintain daily average temperatures of 27.7°C. It requires an air supply of 8.5 cubic centimeters per minute per occupant. But the shelters won't work. The intake valves are clogged. There is no sewage disposal or provisions for food or water. Many shelters are infested with rats. When people start asphyxiating, the construction manual recommends placing eight layers of cotton handkerchief over the mouth and nose. And that's the last we'll ever hear of them.

What about everyone else? Many would head in their cars to points throughout southern California. Two and one-half million people will go for peripheral regions; 7.2 million will move central, through the San Fernando Valley; the rest of the escape roupes are unpredictable.

Bottlenecks outside of the city will force outgoing vehicles into incoming lanes. Relocation will be halted by the carrying capacity of those junctions. En-route refueling will require 40 million liters of gas to service 700,000 vehicles going further than 402 kilometers. A typical service station pumps 5670 liters a day and services 100 autos. Urban stations generally have three 37,800-liter tanks (one for each grade) and are resupplied once a week from the distributor. But during relocation, some very different measures will be necessary: three refuelings per minute would have to occur, and that is impossible. It translates, again, into standstill, eruptions of violence.

Buses, trains, aircraft, and boats could – under ideal, orderly circumstances – remove a major percentage of the population (nearly half) within one week. But that would require people getting to the bus stations, depots, and airports. If all movement stops on the freeways, people are unlikely to move by foot, nor could they make it. Furthermore, all scheduled aircraft in California are subject to federal government usage during any emergency, translating into priorities for military aircraft. A pollution disaster would make Los Angeles' nuclear plants vulnerable to sabotage.

As for food distribution in the outer evacuation circles, Los Angeles possesses insufficient inventories at the wholesale, retail, and consumer levels to support even a short-term relocation. The U.S. Department of Agriculture estimates that rural populations have on hand nine days per person of food. At 2000 calories a day, outside of the city, folks might last the week, given unlikely equal calorie distribution. But the city itself will run out of food. And there are terrific water constraints. Contra-flow (unidirectional) evacuation efforts will make it impossible to provide responsive aid to disabled vehicles and victims along the exodus corridors. And with the contra-flow, Los Angeles' 15-percent work force – government, industry, food processing, and distribution – has no access back into a crippled city. With so many on the road continuously, surface carbon monoxide levels will reach staggering highs. Drivers are likely to pass out at their wheels, their vehicles colliding in a bumper-car hallucination. As it is, the Los Angeles Rapid Transit District's 250 maps of the freeway system predict traffic volumes showing that 2.3 accidents per million vehicle miles (98 percent freeway usage as defined by normal rush-hour configurations) will result in a standstill. Evacuation conditions *guarantee* such standstill.

As for medical evacuation, even more futile. While most of California maintains six ambulances per 10,000 people, Los Angeles has only one in 20,000. The city has 354 surgical beds per 100,000 people, with a usual 59-percent occupancy rate. This leaves 10,000 beds available. Who could get to the hospitals; who would want to? Problems, inadequacies, indecisiveness compound chaos.

In the mid-1970s, the Defense Civil Preparedness Agency contracted the JHK & Associates in San Francisco to make a confidential analysis of the feasibility of evacuating Los Angeles and San Francisco. The study **determined, however unlikely, that Los Angeles would require a minimum of 8 days for evacuation. By that time, during any environmental disaster, the dead would already be dead.**

Such a crisis could be manipulated. During the summer of 1976, a conference of the Committee on Disarmament met in Geneva to draft a treaty that would ban the military use of environmental modification techniques. Smog[10] — some heretofore unknown chemical structure capable of aggressive synergetic reactivity with traditional city-bound emissions, of molecular reproduction at an exponential rate — could easily be loaded in an ordinary-looking diesel truck, placed on a Los Angeles freeway at rush hour, run by remote control, and made to jackknife, its contents releasing into the environment. No driver, no license plates, no way to counteract the instant diffusion of the chemical reactions. Such science fiction would not be so sobering if the major cities had not already prepared the groundwork.

NOTES

[1] John Fuller, *The Poison That Fell From The Sky* (New York: Random House, 1977).

[2] Michael Brown, *Laying Waste: The Poisoning of America by Toxic Chemicals* (New York: Pantheon, 1979).

[3] Paul and Anne Ehrlich, *Extinction: The Causes And Consequences Of The Disappearance of Species* (New York: Random House, 1981), pp. 215–218.

[4] *Atmosphere-Biosphere Interactions: Toward a Better Understanding of the Ecological Consequences of Fossil Fuel Combustion* (Washington D.C., National Academy Press, 1981); and R.H. Perry, and C.H. Chilton, eds., *Chemical Engineers' Handbook*, 5th ed. (New York: McGraw Hill, 1973); and L.A. Piruzyan, A.G. Malenkov, G.M. Barenboym, "Chemistry and the Biosphere," trans. by N. Preroda, *Environment*, vol. 22, Dec. 1980, pp. 25–30.

[5] E.F. Obert, *Internal Combustion Engines and Air Pollution* (New York: Intext Educational Publishers, 1973).

[6] A.C. Stern, ed., *Air Pollution*, 2nd ed. (New York: Academic Press, 1968).

[7] W.P. Lowry, and R.W. Boubel, *Meteorological Concepts in Air Sanitation* (Corvallis: Oregon State University, 1967); and O.G. Sutton, *Micrometeorol-*

ogy: A Study of Physical Processes in the Lowest Layers of the Earth's Atmosphere (New York: McGraw-Hill, 1953).

[8] *Air Pollution Primer*, National Tuberculosis and Respiratory Disease Association (New York: 1969); and *To Breathe Clean Air*, National Commission on Air Quality (Washington D.C.: 1981).

[9] Cecil F. Warner, and Kenneth Wark, *Air Pollution—Its Origin and Control* (New York: Harper & Row, 1981).

[10] John Arbuthnot, *An Essay Concerning the Effects of Air on Human Bodies* (London: J. & R. Tonson, and S. Draper, 1751).

[11] A.C. Hexter, and J.R. Goldsmith, "Carbon Monoxide: Association of Community Air Pollution with Mortality," *Science*, vol. 172, 1971, p. 265; and H. Schimmel, T.J. Murawski, and N. Gutfled, "Relations of Pollution to Mortality, New York City, 1963-1972," Paper No. 74-220, Annual Meeting, Air Pollution Control Association, 1974.

[12] *Trends in the Quality of the Nation's Air—A Report to the People*, E.P.A. (Washington D.C.: Government Printing Office, 1980); and Tom Alexander, "A Simpler Path To a Cleaner Environment," *Fortune Magazine*, May 4, 1981, pp. 234-254; and J. Clarence Davies and Barbara S. Davies, *The Politics of Pollution*, 2nd ed. (Indianapolis: Pegasus Books, 1975).

Chapter 10

Alternatives

Human beings have always sought paradise. Some found it in their back yards, and others traversed whole continents in pursuit of a fantasy. Work and technology have escalated the logic of paradise – from the earliest writings of the Church Fathers, the beginnings of monasticism, to the creation of pyramids, ziggurats, and Chinese heavenly capitals. ***Today, technology offers our species the chance to reacquire that balanced place in evolution.***

> *We have a rule for health. First thing in the morning, even if you don't have to crap, you should at least go out to the field, squat, take a pull on a cigarette, fart and come back.*
>
> *When the grains are threshed and stored outside at night, a ring of fire ashes should be drawn around the pile at sunset to ward away evil spirits.*
>
> *In this age man has begun to eat more meat and dairy products, and supposedly, man will eventually begin to eat human flesh as well. Then, after all the human meat has been eaten and there are no more people left, the* Satyayuga, *the Age of Truth, will begin again.*
>
> *Buffalo dung just doesn't have the spiritual power that sacred cow dung has.*

The best way to get rid of head lice is to snatch some big buffalo lice from the water buffalo and put them in your hair. They eat the hair lice right away . . . When you get old and your hair turns white, your lice turn white, too.

The marten's piss smells so bad it make the bees' eyes water, and they swarm.

After cooking the millet to make mash for alcohol, you have to spread it out on a bamboo mat to cool. At that time when it is cooling never let someone into the house who has just arrived from a steep uphill climb — their smell will spoil the mash for sure.

The expense of buying plastic bangles will surpass the cost of a gold bracelet.

Change always comes. We own land and we say that it is 'our' land, but two and three generations later, who is it that owns our land? In a hundred years none of us will be here. Even the trails of my youth have changed or been moved, and I have seen streams and rivers change their courses.

Nepali Aama: Portrait of a Nepalese Hill Woman*

THE DIALOGUE

Our technology represents millennia of astute money, exploited labor, and inventiveness. In 1844, the Textile Factory Act in England first legally employed the word *factory*, referring to mills that had adopted steam engines. The Act was the first indisputable signpost of modern industrialist psychology. Men like Eli Whitney, Robert Owen, and Charles Babbage worked in a contagion of possibility — the sickness of contraptions — to multiply manufacturing. No innate law of accumulation dictates human nature anymore than does restraint and the learned suspicion of all excess. The anthropological record spotlights hundreds of strikingly harmonious instances of full employment, self-sufficient ecological collectives — be they Gé-speaking Suya in the Amazon, Bimin-Kuskusmin of Papua New Guinea, Gran Cordillera Igorot in the Philippines, or Karakoram (Pakistani) Hushi. But the Western tendency has been toward quantification of goods. Motion and time studies at the turn of the nineteenth century culminated in Henry Ford's Highland Park factory, with its unit operations. Mechanical refrigeration, digital computers, cybernetics ("the science of control" first coined by Andrè Ampère in 1834 and revitalized a century later by Norbert Wiener), and

---

**trans. B. Coburn (Santa Barbara: Ross-Erickson, 1982).*

the adoption of the internal combustion engine to agriculture marked the advent of industrial metaphor, the process of leverage beyond the visible and easily comprehensible: exponential magic.

For 2000 years, Western philosophers and merchants, poor poets and friars have argued the dangers and the blessings of technology. Invariably, the critics and upholders find themselves equally dependent on the basic contrivances of their age, from sewing needles to matches. Romanticism in every era has attempted to expiate the sins of mechanical bondage, to liberate the human spirit from machines that have, they argue, escaped human control, made work tedious and demoralizing, and forced on human consumptive habits inordinate drives and gadgetry.

Spread radiantly across the rural farmsteads of van Gogh – in the grime and doubt and dim despair of Samuel Beckett's parables and T. S. Elliot's "The Wasteland" – we are reminded that technology creates an elite class and so disenfranchises the masses, crippling man by severing him from the natural world of his upbringing. Finally, as echoed in the universe of a St. Exupery and Cervantes, such technology abets the very dissolution of man's soul. All of this is quite verifiable.

But a compelling counter argument has been neatly encapsulated by technology's leading apostle Samuel Florman, who has carefully examined the indictment of five recent humanists, all of whom have castigated mankind's dependency on machinery.[1] The writers Florman discusses are Jacques Ellul, author of the provocative *Technological Society* first published in France in 1954; Lewis Mumford, once a leading historian of technology, turned critic (Mumford's book, *In the Name of Sanity*, is one of the most salient pleas for the rehumanization of spirit); René Dubos' *So Human an Animal*, which was awarded a Pulitzer Prize, Dubos himself vanguarding humanist biology, medicine, and ecology; Charles Reich and his soft, but celebratory *Greening of America*; and finally, Theodore Roszak's essay, *Where the Wasteland Ends*.

Florman responds to these critics with the following suggestions: that man is by nature a technological animal and that machinery is deterministic. The Roman architect Vitruvius had likened human technology to nature's own organizing talents. Mankind simply borrowed all of its good ideas from the natural prototypes.[2] Florman insists that job boredom is low on the order of American grievances, that anti-technologists have idealized peasant life, forgetting how uncomfortable and brutalizing it can be, and how technology has come about precisely to immunize us from its unnecessary indignities. Regarding consumers, we are informed that engineers devote more time creating hi-fi equipment and concert halls than in fashioning motorcycles. But when the market demands loud motorcycles, it will get them. As far as a technological elite, well, hasn't technology always given the masses more freedom? Against the damning insistence that machines are cutting us off from our roots, the engineer is quick to point out that wealthy people could have comfortable abodes in the wilderness but prefer penthouses in the city. Poorer people could stay on their farms, but obviously get tired of the lonely monotony and hard, unprofitable drudgery. Millions of people, it is true, survive well and good without the woods and semitropical turf of Pleistocene East Africa. Likewise, domesticated cats seem to make out all right. If a farm is considered natural, what about a bridge? And haven't the great mystics endeavored to reveal paradise in the miniature havens of a domesticated, urban backyard? If prehistoric cultures adapted to circumstance, have not modern peoples, from tropics to arctics, done the same, as was pointed out graphically in the 503 pictures from sixty-eight countries of Edward Steichen's wonderful book *The Family of Man*?

The humbled answer to all of these incisive commentaries is a resounding affirmative. Human beings adapt, marvelously. It is our glory *and* our disease. We have seen the mind adapt to nature, as Florman attests, adapt so well as to create its own surplus in the neocortex, extra grey matter just in case times get rough. But a 2000-cubic centimeter brain may have nothing to do with 2000 cubic centimeters of brain power. We may be *headed* into the technological jungle where electronic bird calls can lull the most damnable, fixate temper to primeval sleep – but is this the answer? Engineers are dedicated craftsmen; mass man is as dedicated to his comforts. Critics stand to win little ground in such sizeable company, that is unless their proofs and intimations cross over into the disturbed territory of self-doubt, global inequity, overkill, and consumptive fatigue. Ethics and alternatives are never popular. The eye is strained to accept filth and the Earth's own exhaustion. There is no disputing the comforts and personality of technology. Its role in human culture has always been revolutionary, all-transforming. Only in rare instances has a leader, a people of any previously mechanical penchant, suddenly reversed its outlook, and never for more than several generations. The issue is not whether loopholes exist in the defense of technology. They do: free-enterprise zones, for example, do not inspire edifying working conditions, nor do the modern day Coke Towns, assembly lines, and underpayed workers in every industry. Industry itself, as we have so far identified, is prone to producing ecologically damaging excess. It is predicated on the faith in absolute profits or absolute state controls. The real issue embraces the extent of our wisdom in singling out those emphatic liabilities whose produce, impact, and long-range rationale put into jeopardy the consumer and his environment. What *Homo sapiens* is, and what it *does* begs evolutionary contradiction. We are a species that dreams, wherein lies great vulnerability.

POVERTY AND TECHNOLOGY

Poverty foreshortens the future, said the anthropologist Oscar Lewis. Great wealth does so as well. On either side of the distortion, fiscal incentives and daily bread are equally urgent priorities. Technology weighs in between hope and futility, extracting alter ego, giving heat, illuminating, making concrete all of those verbs by which we perceive ourselves to be in the world: our comfort, our purpose. There is, of course tremendous margin for disruption, whether we know it or not. We can adapt to nearly anything while our sperm, mettle, and blood, even sanity, remain rather homeostatic. Without water, true enough, we'll fall over backward, grovel dramatically, squeak, and then expire. If we let our blood temperature drop to around 92 degrees, our train of thought, urine, and steady hand go all to hell, and we die. Without enough food, a different sequence of paralysis ensues. For all of our agility, blitheful wits and rainbow-swooning upness, we can be (and often are) farting, banal brutes with more chances of success than we're worth.

Technology is our smooth-over, the currency for exchange of differences, our guardian spirit. It is ready at the push of a dial to forgive us, illuminate our darkness, quell our doubts over the talk show or telephone.

A tool is a great gift. A gift implies exchange of feelings between people. The monolith, the central grid of electricity, the far-off government, risks cutting off those supposed beneficiaries. Whereas the module — the dispersed seed, the local forum — ensures participation. Where there is involvement, there is comprehension, deeper feeling arousal. In the face of malnutrition, decisive technology has the responsibility to *feel*. But in a world as paradoxically cut-off from its constituents — nation from nation — as is our world, technology often adopts a narrower calling: selfish steel, protectionist power. In warding off confusion, the individual obtains his technology the easiest way he can and complies with elitist, even totalitarian energy policies. All he wants is to continue on his journey. With enough conviction, he might set out to prove a point. He might reconstruct his own energy sources, dig his own well, put in his own generator, secure his eccentric variances from a county building inspector. With his friends, he may influence local legislation to prevent the state utilities from building an unnecessary generating plant. He may encourage bond issues and volunteer his time for ecocycling. He may go on to broader spheres, board an aircraft bound for Africa laden with antibiotics.

Governments have increasingly modeled themselves after this person, a tendency in the last century nothing short of remarkable. Government burden-sharing has been construed as a gift, the gift of coherence, especially in America, whose prosperity is explained by more than mere wilderness windfalls, Indian genocide, slavery, monopolies, corruption, and unassailable disparities between the rich and the poor. America has an intoxicating goodness. Dozens of countries have it. But American goodness has something else. Every American *feels* this to be so.

STEERING THE MARKETPLACE

United States energy policy in the 1980s can be perceived as a conservative affront to decency as well as an ultraliberal regard for efficiency and conservation. The turnscrew tightens around the level of government subsidies: When it is loose, free enterprise is on its own, expected to streamline and maximize the economy. When it is tight, the whole system is strained by government spending, high deficits, lopsided, compensatory interest rates, increasing bankruptcies, foreclosures, and the hesitancy of business to commit to new jobs or investments. The measure of a stable economy has two windows for remaining sunny – military budget and the integrity of America's energy collective. Depressed housing starts and high unemployment are the results of military paranoia, and the very least we could expect. Energy per se, and not the ball bearing, is also an effect of military paranoia. Unwilling to cut the Pentagon's budget the federal deficit has slowly pulled the blinds down on energy, creating a monster, a black box, where clean air and sunshine would have worked more cheaply and intelligently. Save for demonstrative investments in nuclear energy and synthetic fuels – two energy strategies of specific relevancy to the military – the trend toward energy independence in the United States has not enjoyed federal interest. Rather, the private sector has been exhorted to employ all manner of ingenuity in devising its idealist schemes. The bottom-line liberal complaint is not that private enterprise shouldn't free America from energy dependency but that government, in lieu of such military addictions and social welfare antipathies, must make concerted efforts fiscally to assist those business-sector innovations to help make them competitive and timely. The French government has, since De Gaulle's nationally subsidized railroad locomotive project, proven the wisdom of such collaboration. Furthermore, continued support of the nuclear nexus over alternate energy sources denies the very Sagebrush Revolution which Washington's "new federalism" has aspired to. The Nuclear Regulatory Commission's (NRC) 5-rem radiation standard for normal background radiation in the United States is twice as high and twice as dangerous as standards in other countries (e.g. Sweden). Additional uranium tailings mean additional mortality in America, but the U.S. government has formulated its own morality, its private trade-offs. Catering to the militarization of nuclear fuels violates Section 407 of the Non-Proliferation Act, which commits the president to protect us from radioactive, chemical, or thermal contaminants arising from peaceful nuclear endeavors. For every transgression against common sense, there is usually a law somewhere on the books ready to reject it. Concerned citizens need only probe the statutes.

The direction of capital flow through a system can be modulated to protect human rights, while protecting the environment. Planned economies are often perceived by Americans to be depressing, freedomless. But every egalitarian tendency comes recommended by our own evolution, which relied from birth on the act of sharing. Technology is an ingenious way of sharing among a large number of people. That means that technology must be made available to people if there is to be a fruitful balance. Science that is inaccessible – a nuclear weapon, for example – benefits no one. The accessibility of science means grass-roots technology, technology that works humanely. Science is upbeat and requires a good deal of money. Business is upbeat and can take certain risks, but must have some assurances down the line. Together, investment in new science technologies demands the government's help. I'd like to examine six very recent breakthroughs in science that serve to evidence the compatability of technology and the market place.

1. Gabor Somorjai, head of a chemical research team at the Lawrence Berkeley Laboratory, has developed a pragmatic, cost-effective method of extracting hydrogen from water using sunlight and rust. Two discs of iron oxide, containing magnesium and silicon, are immersed in water of sodium sulfate. The water molecules are charged, separated and catalyzed. Sunlight liberates electrons from the hydrogen ions. This is exceedingly clever – we have plenty of rust and sodium. Hydrogen can, presumably, be burned as cleanly as natural gas, though in endless supply.

2. A professor at Ben-Gurion University in Israel, Yirmiyahu Branover, has created a power generator able to convert solar and geothermal energy into electricity without any attendant pollution and at half the price of conventional electrical systems. It is a very small machine, requiring virtually no maintenance.

3. Whittaker Corporation's AgriSystems Division in Somis California invented a new hydroponic lettuce-growing method. It will increase conventional lettuce yields 100 times, reduce labor hours by 75 percent and improve the leaf's quality.

4. Morris Wayman, a chemical engineer at the University of Toronto, is now able to convert carbon dioxide into fuel oil by exploiting the partnership of two plant bacteria that enjoy making whoopee together – namely, concentrations of fuel.

5. The Cotton Technological Research Laboratory in Bombay formulated a novel method for producing biogas from willow dust, a waste product of cotton textile mills.

6. The Department of Energy (DOE) has come up with efficient fluorescent light ballasts that greatly reduce the energy loss in transforming the electricity into light. The government spent $1 million on research. Expected yearly savings throughout the United States are $3 billion, or nearly 1 percent of America's annual energy cost.

Since 1973, conservation efficiency has increased U.S. energy savings by 15 percent. But as Lee Schipper, a researcher at Lawrence Laboratory says, "Little of this research is taking place in private companies, which don't stand to gain much individually from these efforts. Few utility companies have the resources or expertise to conduct these tests. Nor do individual builders have the time or money to perform experiments in thousands of houses; none can wait two or three winters to see what pays off."[3] Those who have been able to wait, to sit out negative cash flows, have been the trust giants – George Hearst, Simon Guggenheim, William Clark, John Rockefeller – men who built their own system, controlled the train routes of product distribution; owned the refineries and the smelters; controlled the very price of the market place as well as the geography of resources. The system determined the type of energy, not vice versa. For hydrogen to replace oil and natural gas, a new market incentive would have to be effected, either by government subsidy or legal mandate. Such incentives for energy convertability are available, in theory, beyond mere conservation efforts. Many of the "new" alternate energy developments harken back to old times; all have the approval of various legislative actions to ease them into the market place; all are versatile technologies, domestic sources of energy with fewer environmental degradations. Each could satisfy the current law stipulating that the utilities must, when feasible, pick up electricity generated by small producers. Feasibility refers usually to peak-hour power and to the proven reliability of a wholesaler. Legislative incentives include tax breaks from the Farmer's Home Administration, alternate fuel, research and application loans and guarantees from the Energy Security Act of 1980, the U.S. Department of Commerce Economic Development Administration, the Public Works Development Grants, the Department of Energy Appropriate Technology Small Grants program, the Geothermal Loan Guaranty program of the Geothermal Energy Act of 1980, the Small Business Energy Loan program, the Energy Tax Act of 1978, and the Windfall Profit Tax, which extends to 1985.

The laws governing environmental impact can claim no consistent application from state to state, but the laws nonetheless are on the books. They can only help to legitimize alternate approaches against the escalating cost of refurbishing outmoded and dirty traditional fossil fuel sources and medically disastrous nuclear facilities. The array of legal codes at work in the United States pertinent to environment and technology, for all of their shortcomings, provide an extraordinary glimpse into the agenda of self-help, which biologists like Gregory Bateson have perceived to be the central selective mechanism for leaping over logical and illogical types in nature.

What are the legal codes endeavoring to protect Americans from runaway technology? I mention only the *major* categories and regulations. These include: the *National Ambient Air Quality Standards*

(NAAQS) for seven major air pollutants – lead, sulfur dioxide, ozone, hydrocarbons, nitrogen dioxide, carbon monoxide, and particulates; the 247 air quality regions in the United States are responsible for preventing significant deterioration and implementing "attainment" status, though the stays of execution are always ambiguous and varied. The *Clean Water Act of 1977* sets standards for all discharge of wastewater; the *Safe Drinking Water Act of 1973* protects groundwater. The *Comprehensive Environmental Response, Compensation and Liability Act* (the "superfund") has sought to remedy hazardous wastes disposal. Other environmental impacts come under the purview of the *Endangered Species Act*, the *Fish and Wildlife Coordination Act*, the *Wilderness Act*, the *Wild and Scenic Rivers Act*, and the *Federal Land Policy and Management Act*.[4] It is an imperfect, grossly understaffed paper dragon of police work whose faults have been expertly analyzed in a thousand other forums and publications. But the laws *are* in existence. The laws and the realizations that stimulated them are all recent. The very notion of alternate technology suggests the existence of a major shift in thinking about just how much energy, and what kind, humans really require.

Alternate technology addresses the prescriptive powers of nature, the congestion of self-made perils, and an unprecedented concern with mortality. Like the wooden shrines in Korea redolent of pilgrims' footsteps, or Ilya Repin's portrait of Tolstoy wielding a plough, the new approach to mechanization proposes to enlist the highest degrees of science in accordance with the oldest observances of restraint and modesty. Behind every Rockefeller was a bib and a bed-time. We see yellows and reds more easily than greens and mauves to find the fruit of our first arboreal chain of being; our hands are best suited to holding a modest portion, our fingers adapted to scooping out the sweet insides. We can only eat so many guavas before falling to sleep. And these basic rules of the Stone Age are at the bottom of all rugged individualism, matriarchy, and invention. It's a question of steering the market place, socializing the access to capital, and safeguarding the privatization of experience. This is no easy juggling act when so many billions of aspirants – hungry, burdened, angry, and hopeful – are involved.

THE XANADU SYSTEMS

Energy use in the United States dropped 3.5 percent in 1980, though the GNP stayed the same. This was the largest yearly savings in American history, says Amory Lovins.[5] Renewable energy afforded 3600 trillion btus in America from 1977 to 1980 (or 7 percent of our energy supply; 16 percent globally). By the year 2000 the renewables' share of the global energy budget will double, and with it will come more jobs. The battle is far from over, though many – from William Tucker to Amory Lovins (wildly opposing perspectives) – believe that the rudiments for transition to soft energy paths are firmly in place. Still, current global research on coal and breeder processes outmatches alternate energies research by at least twenty times.[6] The outward effects of alternate energy usage will involve a partial reorientation of Western *fat*. We will lose weight without immediately recognizing the changes. But slowly, the face and body in the mirror will make us happier, more carefree, better lovers all around. Less land will be paved over, slums reconditioned, durability stressed in every part of production. Only those technologies that extract quality in volume will be permitted. Nonproducers will be the very ones to receive preferential price breaks, and these boons to developing nations will come on no platters from the OPECs and Trilateral Commissions. They will come, rather, by the logic of investment in sheer population numbers, numbers that work by ancient preference in traditional light industry. Gallium aluminum arsenide solar cells, methane, water, sun, collectives – this is the coming definition of the calorie and its role in the public sector. It is the coming of age of Xanadu science, the *technology of paradise* that human beings have in their power. Xanadu, recall, was the Mongol's summer palace outside of Cambaluc (Peking), a resort with all the luscious amenities to quell the restless heart of Temuchin.

WIND*

The DOE has spent $200 million in the past 7 years exploring an equation of primeval importance:

$$p = c \times A \times V^3,$$

Where p is the power, c is the power coefficient relating to the energy conversion efficiency of a turbine, A is area, and V is velocity. Offsetting oil and gas consumption, approximately 20 percent of U.S. industrial energy could be derived from wind turbines working at a full load under 20–30 mph wind velocities. Such winds have been mapped by Robert Sheahan and can be found along the Pacific Northwest, portions of the Great Plains and Colorado–Wyoming border, the Texas Gulf, Lakes Erie and Ontario, and in unusual locales such as the channel under the Golden Gate Bridge, whose southwest corner will be used for an installation that will light the bridge by wind power. Sixty companies are involved in wind-machine manufacture. In the 1920s, the Savonius

*The most definitive and inspiring discussions of alternate energy are found in Richard Sheahan's *Alternative Energy Sources: A Strategy Planning Guide* (Rockville, Maryland: Aspen Systems, 1981), Lester Brown's *Building a Sustainable Society* (New York: W. W. Norton, 1981), and Brown's *State of the World in 1984* (New York: W. W. Norton, 1984). I have relied heavily on these sources in the section to follow, and am indebted to the authors.

rotor, a vertical-axis windmill employing the Magnus (lee-side differential pressure) effect, was developed in Finland. This is still the most common form of the technology.

HYDROPOWER

The United States generates 71,000 megawatts of electricity per year with water power. Fifty thousand unproductive generating sites still exist in America (dams not being used to generate electricity). Put into operation, these sites would harness 300,000 more megawatts. The Public Utility Regulatory Policies Act of 1978 provided the incentives for this pure energy source ($c \times$ flow $\times$ head $= p$, where c is the power coefficient related to the energy conversion efficiency of the hydraulic turbines [head] and generators). Many of the late nineteenth-century American hydropower plants were retired with the advent of cheap fossil fuels. In France, 90,000 sites could be harnessed; in the Himalayas the potential is staggering. A major portion of Chinese energy needs could come from strategically placed dams on the Yellow and Yangtze rivers, as well as minihydros proliferating for exclusively local uses. The chief problem with hydropower, of course, is the environmental effect on otherwise free-flowing wilderness conduits and the animals depending on them. Pumped hydrostorage facilities – the dropping and raising of water – is one alternative and already composes 2 percent of America's electricity.

SOLAR ENERGY

The solar resources are unlimited, and the price of various installations is continually coming down. Photovoltaic cells cost $300 per peak watt in 1973. Today, that cost is less than $1. By 1990 it is expected to be 10¢. Japan expects to heat 30 percent of its commercial buildings with solar energy by 1995, and Israel more so. Photovoltaic cells alone could contribute 25 quads, or nearly 25 percent of total U.S. energy needs. Already, one Papago Indian reservation (Schuchuli, Arizona) is completely electrified by such cells. In Canada, the "Saskatchewan Model" home has relied on so perfect an insulating plan as to eliminate the need of a furnace and of any electricity for heat. In Africa, solar ovens and dryers can transform the time and energy required to prepare meals. Soybeans normally require 8 to 10 cooking hours, which adds to the severe wood shortage and consequent desertification. With solar fermentation they can be prepared in two, nonimpactful hours. Solar ovens can parboil otherwise long-cooking grains and beans.[7] Solar ponds, flat plates, heliostats, parabolic trough and dish concentrations (passive and active systems) relying on tax incentives and market solubility will transform the look and style of human consumption.

SOLID WASTE ENERGY RECOVERY

By 1978, more than 200 plants in Europe and fifty in Japan were converting urban garbage into energy. The United States only had six. Most vegetal residues can be converted into fuel-grade alcohols. Water hyacinths grow, to this effect, at remarkable rates every day in warm, sewage-enriched urban pools cultivated to yield alcohol or methane. Other methane conversion techniques accrue from animal feedstock deposits and are particularly cost effective in lots of more than 1000 head. World Bank Third Window loans could subsidize research and expenditures for smaller methane production units in the Third World, revolutionizing energy usage by providing fuel without sacrificing soil fertility (which results when dung is turned directly into fire). With forty-nine cattle per hectare in India (thirty-seven per hundred persons), the Asian subcontinent offers the perfect testing ground. China has pioneered the technique in Sichuan Province since 1974, employing at least 7 million methane digesters nationwide. Westerners produce between 1.8 and 2.25 kilograms of solid waste per capita each day. Eighty percent is convertible at 9920 to 11,022 btus per combustible kilogram. The United States has an estimated resource base of 750 million metric tons of waste products per year.

WOOD

Noncommercial forest inventories (the 30 percent of residue from commercial logging) and abundant miscellaneous forest incidentals (e.g. railroad ties, wood chips, etc.), could satisfy 3 to 5 percent of America's total energy demand. In 1980, 2.6 percent actually came from wood, as opposed to 1.2 percent from nuclear generation. In the Third World, reforestation programs are the essential complement to any wood consumption. China's Green Wall will eventually encompass 81 million hectares, 4022 kilometers across Mongolia. The Third World poor rely on wood to meet approximately 80 percent of their fuel needs. To sustain this, we need a ten-fold reforestation increase in many of the worst-afflicted countries. Faster-growing species, better cooking stoves, explicit watershed conservation programs, silvicultural gene conservation and germ plasm banks, more efficient harvesting and wood processing, and the transformation of forest to cropland only where arable soils are assured are priorities of the coming decades. More efficient Third World cooking stoves could cut the consumption of firewood by as much as 50 percent. In New England, 150 factories are now using wood rather than oil for heating. Militating against dense urban use of wood, however, is the resulting air pollution and release of toxins.

PEAT

Biomass accumulates throughout the world at approximately 1.36 x 10^{14} metric tons each year, at a 1500 to 2500 quad convertibility factor — a figure equal to 15 American consumption rates, five times what the entire human population currently requires. One particularly attractive source of biomass energy is peat — the humus that falls into water where atmospheric oxygen is excluded. Biological decomposition promotes a carbon retention at the initial coal-forming stage. Peat has the potential energy equivalency of 150 percent of all total domestic energy reserves of gas, oil, and shale oil. Seventy-six power plants in the Soviet Union are peat based — 2 percent of Finland's total energy firing. Hemic, semidecomposed peat has the best energy value. It is low in sulfur and more volatile than coal. With 45 percent dewatering, hemic peat — as currently exploited at the First Colony Farms and Minnesota Gas Company — possesses an 8800 to 13,000-btu-per-kilogram firing content. Hydraulic mining can flood a bog, driving the submerged fibrous mass through a slurry, then into filters and dewatering holds. The combustible product can be fired in utility boilers previously designed to burn coal. Sheahan provides a compelling financial prospectus for peat energy capitalization, based on successful adaptations in Ireland.[8]

ALCOHOL FUELS

Ethyl alcohol, a liquid fuel, burns with greater heat and cleanliness than fossil fuels. It can be produced — as it is in Brazil, Zimbabwe, Kenya, and South Africa — from innumerable grains, fruits, and legumes. Henry Ford provisioned the fuel's use with his Model T. One metric ton of corn yields 318 liters of alcohol at a cost of 29¢ per liter (1983 prices). Valuable ethanol byproducts, such as distillers grain, then become available. The conversion of food waste from the processing, packing, and transport sectors will mitigate an otherwise self-defeating food–fuel competition and stimulate price stability for energy and agriculture. But substituting for fossil fuels is no panacea: Automobiles — with all of their attendant waste — are still encouraged to continue in operation and manufacture. However one fuels it, the automobile is a monster of over-all consumption.

GEOTHERMAL ENERGY

As early as 1894 geothermal energy was heating homes in Boise, Idaho. The first electric powerplant using geothermal energy was installed near Larderella, Italy in 1904. The major region in the United States for such energy is the pressurized reservoir along the Texas–Louisiana Gulf coast, though numerous smaller locales offer convertible hot dry rock and direct geothermal energy (e.g. the geysers in northern

California). Elsewhere, the Pacific horseshoe ring of fire represents untold reserves of thin-Earth heat. Sixty-five percent of all home heating in Iceland is geothermal. Lester Brown estimates that California's electricity might be 25 percent geothermal by the year 2000. The Soviets hope to supply 5000 megawatts of geothermal energy from deep within the Avashinski volcano on the Kamchatka Peninsula.

RECYCLING AND GENERAL CONSERVATION

An entire technology-in-waiting stems from precious materials normally cast off. Millions of serviceable garments made of natural fibers and oil-based synthetics hit the waste bin without ever the chance to be renewed, as the edicts of fashion eat up invaluable materials across the land. Laws discouraging scrap recycling must be revised; major energy transnationals need to examine the nonenergy business sector that has begun to recycle metals. Steel scrap saves 47 percent of the production cost per year, in some cases. Returnable bottle legislation in a half-dozen states already spares the unnecessary production of a half-million metric tons of aluminum, 1.5 million metric tons of steel, 5.2 million metric tons of glass, and 46 million barrels of oil. Lester Brown determined that 62 percent of auto fuel consumption would be saved if presently understood energy-conscious measures were fully implemented.[9] Given the public sentiment for voluntarism America could easily pave the way for a whole new economy of preservation.* President Carter understood that little sacrifice would be exacted, and to some degree Americans entered into a conservation contract under the Carter administration in the wake of OPEC-induced exigencies. But that mood has been denied under the Reagan administration. And headlines proclaiming the endless stock of fossil fuels and precious metals have merely encouraged added greed and ignorance. Julian Simon's *The Ultimate Resource* (1981), and Princeton professor of physics Gerald O'Neil's tales of future space colonization – "Lagrange 5" self-replicating systems, capable of absorbing *80 trillion Homo sapiens* – have stoked heady delusions.

We should consider four important areas of savings: 1) *home energy* – many possibilities, including clock thermostats and vented dampers on the oil burner, an aquastat temperature control on the boiler; storm windows; a passive solar greenhouse; plugging air leaks; attic, duct, and wall insulation; 2) *cogeneration* – 60 percent of a utility powerplant's fuel value normally becomes waste heat and thermal pollution but can in

*Sixty percent of the respondents of public opinion polls indicated that they were willing to volunteer their labor for standing in line, cashing in returnable bottles, etc.

fact be reconverted into usable energy; 3) *compressed-air storage* – capable of reducing the energy consumption of air compressors responsible for over 50 percent of all operating conventional gas turbine electric generators; and 4) *more efficient batteries* – (nickel–zinc, over lead–acid), and the reduced production of high-entropy mineral commodities with nonreusable designs. Stamped, uncontaminated alloy parts would facilitate this latter need, as but one example.

SYNTHETICS

Synthetic fuels in the United States may produce approximately 1 million barrels of oil per day – 6 percent of U.S. consumption in 1990. Government support, at present, for coal gassification, shale oil, and tar sands research and development bodes for a return to the nineteenth century "town gas" production, with subsequent air and water impacts.

AGRICULTURE

Third World governmental efforts to share farming risks through crop insurance and improved land-holding systems will at once liberalize regimes and help check the fragmentation of traditional family holdings. Over half of the Earth's cropland is devoted to rice, corn, and wheat. Fewer than twenty other plants compose the majority of the human diet – millet, sorghum, peas, beans, peanuts, yams, sweet potatoes, cassava, sugarcane and beets, coconuts, and bananas.[10] The cultivation of other protein-rich foods holds compelling promise. They include the Central American amaranthus, of which over 800 species have been identified; the Andean quina (1500 series), which will grow in salty soils, and fast-growing seacoast eelgrass and kelp. Ecological farming will be mixed, polycultural, and rotated; human and animal wastes will be put to complete use; human labor will substitute for herbicides and pesticides. Many farming cooperatives already practice collective land husbandry, ecological sensitivity, the cross-fertilization of weather-and-disease-resistant species, an intimate knowledge of, and reverence for vegetational zones, microclimates, herbal and medicinal self-help, and total systemics. They include the Amish in Pennsylvania; horticultural communities in Egypt, Taiwan, China, Japan and Sri Lanka; the Israeli *kibbutz*, Soviet *kolhoz*, Mexican *ejidos*, Tanzanian *ujamaa*, and the Qollahuaya *ayllus* of Mt. Kaata in northern Bolivia. These cooperatives are immune to the ravages of most other *frantic farming*, as I term the two extremes – impoverished, and overindustrialized agriculture. Farming must out-produce and out-sustain the computerized, pyrotechnics of corporate farming. (American agricultural ethics and economics must turn once again to traditional family farming as manifest in the

160-acre organic livelihood and beautifully evoked by American Wendell Berry and Japanese Masanobu Fukuoka).

Commercial plant products for such diverse needs as sperm oil equivalents, anti-coagulants, and alkaloid-bearing flowering plants (e.g. guayule, jojoba, leucaena, periwinkle, Cinchona bark, Rauwolfia, Euphoriba lathyris, foxglove) offer yet another realm of valuable alternatives to higher-entropy synthetics. Only 2 percent of all angiosperm species have ever been tested. Nearly eighty major drugs are already more cheaply extracted from living plants than they can be synthesized, but stepped-up efforts to inventory the standing pharmacopoeia of floral life on the planet is critical.

The United States produces three times as much food per hectare as India. But this increased production takes nearly ten times the energy in technology and accessories (see Chapter 7). Each metric ton of nitrogenous fertilizer consumes 5 metric tons of coal and one metric ton of steel for its manufacture. The Chinese, with scarcely any technology have brought several new species into cultivation able to naturally fix nitrogen. By escalating our trophic pyramidal tastes (feeding cereal to pigs rather than to humans), we have degraded the equality of consumption by a factor of ten the world over. One cow eats the equivalent of ten people.

With our laser contour cropping, integrated pest management (IPM), and high-yielding varieties, we have managed to produce food inferior to rapidly disappearing native cereal strains. The story of Teosinte *Zea diploperennis* in Jalisco, Mexico is the best case in point. Found in 1977 by a student of botany at the University of Guadalajara, *Zea* is a perennial containing twenty chromosomes suited to cross-breeding with corn for a fungal resistance previously unenjoyed by corn monocultures. This one species will very likely augment corn production spectacularly. But it was on the verge of being bulldozed into extinction when it was discovered.[12]

Studies with the European elm bark beetle, spruce budworms, codling moths, and khapra beetles are encouraging for pheromone-baited trapping techniques and enhanced chemical communication profiles useful in eliminating harmful pesticidal practices. Such chemical interference has been subject to vigorous critique, but it appears to be a lesser evil. A more promising development involves plant genetics. Green Revolution fertilizer builds up salts in the soil, degrades irrigation schemes, often results in the exhaustion of fallow cycles, and is simply too costly for most Third World farmers. High-yielding strains have relied primarily on this fusion of chemicals with natural strata. Plant geneticists are now working to create plants that do their own fertilizing by improving the 1 percent rate of photosynthetic efficiency (the approximate energy conversion range throughout the plant kingdom), perfecting protein composition to meet the human nutritional requirements, and breeding resistance to diseases that otherwise account for a nearly 35-percent attrition rate among American crops. United States agriculture is particularly relevant here because it produces more than one-half of all the world's agricultural exports). Computer models for bioactive plant genes will eventually heighten the attractiveness for private industry to finance these ventures. But we must not forget that plant genetics caters to a concept that is not ultimately fruitful: the monocultural ethic—a high-volume profit reliance on biological monotony. Monoculture is vulnerable; plant cross-breeding and protein manipulation merely attempts to lessen the vulnerability.

KELP

Only the tropical rain forest rivals kelp's 60 centimeter per day of growth. One plant with 200 fronds can live several years. It stays within 30 meters of the sea surface to better absorb the sun. Giant kelp is found only off the Northern Pacific Coast, primarily in southern California. A kelp forest can rejuvenate itself in a single year. Like poseidonia sea grass, kelp forests foster huge and varied populations of marine life. Pumps for catching upwelling nutrient, and methane gas conversion of marine photosynthesizers are presently being studied by the U.S. Navy in San Diego. Anaerobic digestion offers the potential of 23 domestic quads of energy a year—1200 quads globally.

AQUACULTURE

Cattle require nearly 1000 days before marketing; shrimp need only 200 days of aquaculture. Salmon hatcheries can use the fish's homing sense, freeing the animal to feed on natural foods in open waters and thus eliminating artificial feed costs. Even at a rate of return of 2 to 3 percent, this is proving cost effective. Aquaculture accounts for 25 percent of all the fish consumed in China, 10 percent in Japan, but only 4 percent in the United States (65,000 metric tons). All of these percentages could be increased. The sea's tidal properties could also be exploited for hydraulic energy. One hundred sites have been identified already worldwide (e.g. Minas Basin, Bay of Fundy), but not without associated effects on water level. Even a 15-centimeter rise will wipe out numerous species.[13]

POPULATION

Nepal has eliminated tax deductions for children. In China, severe penalties are imposed on parents having more than two children. In Bangladesh, itinerant minstrels sing of birth control in outlying villages where women prefer to remain indoors rather than facing embarrassment before the family planning councilor. Fifty-five million couples now take birth control pills. Over 110 million women in China use the controversial IUD, and millions of couples have undergone voluntary sterilization to date throughout the world. Western nations have slowed their population growth to an increase of 0.6 percent a year – a negligible rate when compared with the 1.8 percent mean increase among less-developed countries (LDC). LDCs will account for an increase of 79 percent, or 5 billion people, in the year 2000. Less-developed countries have a built-in childbearing momentum in the coming generations, though China's declining fertility rates shift the overall fertility rate among women from 4.3 to 3.3 children per woman. Life expectancy in the year 2000 is projected at an average of 65.5 years. Uncontrolled urban settlement, extreme inner-city strife and malnutrition, increasing disparities between the north and south population profiles, has *no* easy alternative short of catastrophe or altruism and the immediate use of the new technologies we have been discussing. But as it stands, the world's population will increase by 55 percent between 1975 and the year 2000.[14]

NOTES

[1] Samuel Floorman, *The Existential Pleasures of Engineering* (New York: St. Martin's Press, 1976)

[2] Vitruvius, *De architectura*, X.1, 4.

[3] Lee Schipper, *L.A. Times.*

[4] Richard T. Sheahan, *Alternative Energy Sources: A Strategy Planning Guide* (Rockville, Maryland: Aspen Systems, 1981); see also the President's Commission for *A National Agenda For the Eighties* (New York: New American Library, 1981).

[5] Amory Lovins and Hunter Lovins, *Brittle Power: Energy Strategy for National Security* (Andover, MA: Brick House Publications, 1982).

[6] Mans Lönnroth; Peter Steen; and Thomas Johansson, *Energy in Transition: A Report on Energy Policy and Future Options* (Berkeley: University of California Press, 1980); See also Hanry M. Peskin, Paul R. Portney, U. Allen, V. K. Neese, eds., *Environmental Regulation and the U.S. Economy* (Baltimore: Johns Hopkins University Press, 1981).

[7] Irene Tinker, "Energy for Essential Household Activities," mimeograph (Center for International Development and Technology, June 1981); See also Robert Riddell, *Eco-Development – Economics, Ecology and Development, An Alternative to Growth Imperative Models* (New York: St. Martin's Press, 1981).

[8] Sheahan, *Alternative Energy Sources.*

[9] Lester Brown, *Building a Sustainable Society.* See also Brown's superb chapter, "Reconsidering the Automobile's Future," pp. 157-174, in *State of the World 1984: A Worldwatch Institute Report on Progress Toward a Sustainable*

Society, (New York: W. W. Norton & Co., 1984). See also Douglas Considine, ed., *Energy Technology Handbook* (New York: McGraw-Hill, 1977).

[10] Paul Ehrlich; Anne H. Ehrlich; and John P. Holdren, *Ecoscience – Population, Resources, Environment* (San Francisco: W. H. Freeman, 1977).

[11] Wendell Perry, *The Unsettling of America: Culture and Agriculture* (San Francisco: Sierra Club Publishers, 1977); and Masanobu Fukuoka, *The One-Straw Revolution* (Emmaus, PA: Rodale Press, 1978).

[12] U.S., State Department, Council on Environmental Quality, *Global 2000 Report to the President*, 3 vols. (Washington, DC: Government Printing Office, 1980): Vol. II.

[13] Wesley Mark, *The Oceans – Our Last Resource* (San Francisco: Sierra Club Publishers, 1981).

[4] *Global 2000 Report*, vol. I; See also C. Maxwell Stanley, *A Guide to Survival – Managing Global Problems* (Muscatine, Iowa: Stanley Foundation, 1981).

Chapter 11

Prometheus in Arcadia: Japan, Bhutan, and Switzerland

There have been fleeting moments in history when an entire people banded together to achieve harmony. One might cite twelfth-century Cordoba or eighth-century Fujiwara Japan. Today, three countries challenge the rest of the world's lassitude. In Japan, Bhutan, and Switzerland, fruitful rationality is at work. These societies share a genius for alternatives – alternatives to violence, alternatives to deprivation.

JAPANESE NATURAL FARMING

Masanobu Fukuoka's farm near Shikoku, overlooking Matsuyama Bay, comprises less than 1 hectare of rice field and 5 hectares of mandarin orange grove. With a planned sequence of efficient labor, traditional Japanese farmers grow a crop of rice and one winter grain every year in the same field without ever depleting the soil's nutrients. Fukuoka builds on this legacy by adopting a radical "do-nothing" approach, as he terms it. It's not exactly labor-free, but nearly so. Here is how it works. He sows the rice seedlings in the fall, along with white clover and a winter grain; rice straw covers all of the crop. The rice lies dormant beneath barley (sometimes rye) and clover, which spring up quickly. Meanwhile, the oranges are taken in. In the spring, the barley is gathered, spread for drying, processed by hand, and stored; the straw is scattered for mulch. Rains weaken the clover, and the rice begins to sprout through the mulch. The rice is harvested in October and hung for drying while autumn seeding begins. Fukuoka gathers between 635 and 775 liters (1100–1300 pounds) of rice per 0.2 hectares. His method yields approximately the same amount as that of ridge-furrow and chemical farming. The difference is that Fukuoka's fields gain added fertility every year. He has not touched a plough to them in two decades. His fields hold water better than technologized land, and his rice has as many as 300 grains per head — an exceptional richness.

A former microbiologist, the venerable Fukuoka lives with his students on a mountain cut irritatingly in half by a freeway. Still, the mountain provides the farmers not only their fruits and vegetables, which they grow, but wild herbs, riparian shellfish, and sea vegetables at the coast a few miles distant. They keep goats, bees, and chickens and live without electricity — only candles and kerosene lamps.[1]

Writes Fukuoka,

> *The world exists in such a way that if people will set aside their human will and be guided instead by nature there is no reason to expect to starve . . . . When a naive scientific knowledge becomes the basis of living, people come to live as if they are dependent on starch, fats, and protein, and plants on nitrogen, phosphorous, and potash. . . . .To believe that by research and invention humanity can create something better than nature is an illusion. I think that people are struggling for no other reason than to come to know what you might call the vast incomprehensibility of nature.*[2]

BHUTAN

The Bhutanese have always been self-sufficient, awesomely in tune with their environment, and rigorously Buddhist. Each family maintained its own manicured lands, reaped its own wealth, and, with predictable surplus, traded for luxury items with Tibetans. Today, the World Bank classifies Bhutan — with a GNP of under $100 million and a per capita income of $70 — as the poorest country in the world, along with West African Mali. For comparison, Bhutan's neighboring countries, Sikkim and India, produce a $90 and $130 per capita annual income, respectively. Nepal shows a $110 income per capita, Bangladesh $100, Brazil $532, El Salvador $390, and the Soviet Union $2300. The world per capita income average, ranging from Bhutan to the United Arab Emirates, is $1,350. Martinique, South Africa, Gabon, Yogoslavia, and Reunion all fall within that mean per capita range. But what does this really say? I have seen dismal poverty in Kuwait, where the yearly income is supposed to be nearly $12,000. Wealth and poverty statistics can not unequivocably describe yearly incomes in Hong Kong: less than $1600.[3] All these figures are in 1975 dollars, requiring inflationary adjustments that widen the gaps between rich and poor. But what does such poverty mean? In Bhutan, the term is nearly meaningless.

The country has always been the *envy* of those few visitors able to see her. Only the Swiss and Japanese – with whom Bhutan shares a similar ethos – have been accorded the same cultural stature. In all three countries, money was traditionally less useful than natural resources. Today, Bhutan maintains that lofty insight.

Samuel Davis was the first foreign artist to paint in the Himalayas and record dispassionately the matter-of-fact details of life in eighteenth-century Bhutan. His drawings, done while Davis was attached to an early British mission, offer no falsely construed romance, yet shape a tranquil picture indistinguishable from the Bhutan of the 1980s. During the Enlightenment, the German philosopher Johann Herder actually hailed Tibetans as the most nobly civilized of all human beings (though he had only heard about them). Davis' *Diary*, first read aloud in 1830 to the Royal Asiatic Society in London, ignored many of the yardsticks of the good life that we today do not so take for granted:

> *But, after all, these advantages and this happiness [Davis is referring to "equality and general prosperity"] are of a negative quality, and not such as would tempt the more enlightened part of mankind to change conditions with the inhabitants of Boutan. They are for ever excluded by the nature of the country from making any considerable progress in arts, manufactures, and commerce, and therefore not likely to acquire any very eminent degree of science, taste and elegance. They might, it is true, become better soldiers if they were more suitably armed, but such an improvement might only induce them to disturb the peace and invade the property of others, without contributing any needful security to their own, which is already, from the unchangeable ruggedness of the country, as unassailable as they can desire.*[4]

Davis the artist is more generous. He brings a loving resolution to the eye, one undeniably saturated with calm, diffused paradise glow. It embodies all the perfection of a painting by Claude Lorrain set out in the very integrity and pastoralism of the country that came to be known after one of its numerous mountain passes, *Shangri-la; la* meaning *pass* in Tibetan.

George Bogle, who preceded Davis into Bhutan by 10 years, commented that the Bhutanese were sword bearers and wore clothes according to their individual taste without regard to matters of social rank or birth. As early as the mid-seventeenth century, rumors about Bhutan had spread throughout Asia and Europe. One inveterate French Orientalist, J. Tavernier, collected news about the country from his six fact-gathering missions to the Far East. Wrote Tavernier, "They know nothing of war." The first Westerners to reach Bhutan in 1627, the Portuguese Jesuits Cacella and Cabral, were impressed with what they found:

> *The King takes pride to be loved for his meekness. He is much esteemed for his abstinence . . . eating only milk and fruit. And he retreated for three whole years just before we arrived. During this retreat he stayed in a small house built on the very steep side of one mountain . . . He received his food—without speaking to anybody—by means of two ropes which were descending from his meditation hut to the houses below . . . He said that he was passing his time praying . . . and painting . . . He showed us the best . . . one small image of the face of God (sculptured in white sandlewood) small but beautiful.*[5]

From the Sanskrit *Bhot ant*, meaning "end of Tibet," or *Bhu'uttan*, meaning "highland," ancient Buddhist writers referred to the country as *Mon Yul*, paradise of the south, and *Druk Yul*, land of the Thunder Dragon, where the Drukpa Kagyupa sect of Tibetan Buddhism came to rest. Tibetan chroniclers also referred to the southern empire as the Lotus Gardens of the Gods. British emissaries to the country continually remarked on its inaccessibility, "shut out on every side from the rest of the world."[6] But thirteen European expeditions penetrated the country. The Sino-Indian border war of October 1962 signalled to the Bhutanese hereditary patriarch the onslaught of the modern world. In fact, Bhutan had fought earlier battles with Tibet during the seventeenth century and with Great Britain over East India Company interests in the Assamese tea regions. British forces actually swept through southern Bhutanese strongholds and forced the Treaty of Sinchula on the Shabdung religious authority in 1864. But Bhutan never suffered from any attempt by outsiders at colonization, and this fact—in conjunction with her indigenous spiritual and ecological convictions—created something of a Himalayan Eden, the very one traditionally revered.

From 1000 to 7600 meters the altitudinal variation of Bhutan's 47,000 square kilometers casts a spell over its people. Surrounded on the South by steamy tropics, and on three sides by prehistoric glaciers, Bhutan abounds in hundreds of extraordinary plant and animal species: blue sheep, snow leopard, giant panda, musk deer, rufousnecked hornbills, Himalayan bears, Ward's torgon. Blue pine, cypress, rhododendron, and oak trees thicken the precipitous slopes with scent and allure. In the rich alluvial valleys, rice, wheat, maize, potato, and barley grow plentifully. Further south, the tropics and tea plantations cultivate oranges, pineapples, and bananas. Tigers, golden langur, elephants, and rhino inhabit the very shadows of unexplored massifs. The earliest Indo-Mongoloid animists, Sharchops and Ngalops, trace their hieratic migration to the same religious wave that settled Japan in the sixth century A.D. Contemporary *dzongkha* is written in classical *Ucän* script, spoken in the *Cheokay,* traditional manner – a high, effeminate Mandarin-like eloquence. English is taught in the schools.

Architecture, social life, diet, ice skating, Finnish saunas, rose-scented stamps, all intellectual endeavor, is mediated by Tibetan Buddhism, which is fundamental to the country's politics. Two out of eight members of the Lodoi Chopdah (Royal Council) are religious Lamas. Nearly 50 percent of all young men enter monasteries. Bhutanese Kagyupa Red Hat Buddhism dates to the reveries of Lama Marpa and the Tibetan poet-Saint Milarepa who passed down an intoxicating series of austerity measures – aesthetic, humanistic, ecological contemplation – very much ingrained in Bhutanese history and contemporary preference.

In 1907 Ugyen Wangchuck merged civil and monastic duties previously separated. The country joined the Colombo Plan, entered the Universal Postal Union in 1969, and gained membership to the United Nations in 1971; her status: nonaligned. When Nehru and then U.S. Ambassador to India John Galbraith visited Bhutan, it took them 10 days on the back of a yak to reach the capital.

Bhutan has a pronounced shortage of labor and – again like Switzerland – imports it. Nepalese help in her production of wool, textiles, dyes, silk, tobacco, and herbs. Ninety percent of the workforce engages in agriculture and animal husbandry. Bhutan's density of population is thirty-three per square kilometer, one-fourth that of Switzerland. Agricultural research stations, sophisticated irrigation, fertilization, seed selection, and rotation techniques have been long used. The Bhutanese have discovered such minerals as limestone, gypsum, marble, zinc, and copper, but no full-scale mining projects are planned. Bhutan has specifically embarked only on those low-impact, low-capital industries that would maintain its traditional environment. Small hydroelectric units have been launched at Thimphu and Paro, the summer and winter capitals. Bhutan's 1550 kilometers of road (single lane) connect a country the size of Switzerland. Twenty thousand students attend 124 schools, all tuition and expenses assumed by the government. For the 0.9 million inhabitants, illiteracy is still very high. Illness is not. In the mid-1960s, a British medical expedition led by Michael Ward discovered, to its delight, virtually no serious medical problems in Bhutan. One woman complained of a dreadful head ailment experienced years before – the diagnosis: headache. In a village to the northwest, the expedition members arrived during a blizzard. Lamas greeted them, apologized for the weather, then played their shenai flutes and bone horns and drums.

Incantations were recieved and, what do you know, the weather immediately cleared up.*

In Bhutan, poaching can be punishable by a 10-year prison sentence. But such violations are virtually nonexistant. The Bhutanese constitution anticipated E. F. Schumacher's "small is beautiful" catch-all. It is the Bhutanese national mantra. To this end, the country's department of tourism has restricted most entry. I led the second American group into Bhutan in the Winter of 1975. That year, there were thirty-six visitors in all from abroad. Now, approximately 100 come per month. The King intends to keep that number as the upper limit, never to suffer the ravages of Nepal at the hands of meddling anthropologists, mountaineers, insensitive tourists, and conniving art speculators and thieves. The Bhutanese are also candid about their preference for no sexual contact with non-Bhutanese.

The King, Jigme Singye Wangchuck, who became the youngest monarch in the world at age 19, has completely turned his monarchy over to the conscientious democracy of his people, as represented in the Tshogdu (assembly) and Lhengye Shungtsog (his cabinet). This inherently homogeneous body of equal rights dates to ancient times in Bhutan, more ancient than Magna Carta, than the Allthing Icelandic democracy of the Thingvellir natural amphitheatre (900 A.D.). In male and female relations, divorce is common and easily consummated, even with some nonchalance. Premarital sex is normal. Polyandry is still practiced, as elsewhere in the Himalayas, with women taking several husbands. Tibetan Tantrism — yogic, sexual, sacred marriage Buddhism — has its hold on Bhutanese art and sensuality. The frescoes of Paro Dzong number among them certain mandalas from medieval times, depicting the movement of consciousness between the soul and all of nature. These guiding threads have become holy formulas, rites of passage meant to transcend the ambiguities of human contradiction. The cosmography of Buddhism has endowed mountains, caves, waterfalls, and whole regions with religious provocation. Like Japan, all of Bhutan is a sacred place of pilgrimage.

The young king's coronation in 1974 was noted in the world press for the innocent arrow that lodged in a visiting dignitary's leg during an

---

*Much the same thing occurred during the Sikkimese royal wedding in the 1960s. Lamas in the court of Queen Hope and the late Chogyul elected for the precise timing of the 20-minute ceremony many months in advance. The wedding was scheduled for the summer, at the height of incessant monsoons. Diplomats from all over the world braved torrential storms and gutted mud roads to attend. Minutes before the ceremony began, the storm abated for the first time all summer and the sun shone all afternoon. By evening, the storm resumed for another month.

archery competition. "Don't worry," laughed the king, "it happens to me all the time." In fact, he prefers basketball and plays it every day with passersby in a field near the 100-room capital building. Taschichho Dzong (*dzong* referring to the medieval-style monastery-fortress architecture for which Bhutan was always famous). Not only is the king young, but nearly every other official in the country as well. Only the old persuasive Buddhists, who exert tremendous power over the leaders, stay modestly hidden. When I visited the country, I wore around my waist nearly $30,000 in payment to the Bhutanese for their gracious services. I was responsible for twenty people, each charged $150 per day to cover everything from chauffered Landrovers to steaming cauldrons of Mongolian cherry plum spare ribs served in the fancily adorned, hand-wrought royal visitor's chambers, where we all bedded down beneath thick Tibetan quilts done in patterns of snow leopard and Himalayan dragon, rock music piped in overhead for our presumed enjoyment. No one wanted to broach the subject of payment. I finally forced the cash upon some unsuspecting teenager who shyly wrote me out a receipt for the money.

Of the 300 Bhutanese students who have gone abroad to college, not one has remained away from his homeland for more than a few years. Admittedly, the West holds some allure. Karma Tenzing, an official in the Ministry of Trade, and graduate of Adelphi College in New York, still orders religiously his argyle socks from the Bloomingdale catalogue.[7]

SWITZERLAND AND THE POLITICS OF RESTRAINT

On the night of June 27, 1787, around midnight, the historian Edward Gibbon wrote the last line of his six-volume *History of the Decline and Fall of the Roman Empire*, to which he had devoted 12 years of leisurely contemplation. He peered out from his terrace in Lausanne to Mont Blanc, highest point in Europe, a gleaming epicenter for symbol-making ardor, the land of Frankenstein, of political justice; that transcendental realm that Shelley knew to be above the petty turmoil of human wars and strife. Monk Lewis, Lord Byron, Samuel Coleridge, and William Wordsworth, were among those roused to action by the sight of the mountain's advancing glaciers. The peak was first climbed in 1786, followed by a British ascent just 2 months after Gibbon's work was completed. The climber, Colonel Beaufoy, published a narrative of his ascent in the British *Annals of Philosophy* years later, indicating the early significance accorded such daring. There was Gibbon, staring at the ghostly white massif, smelling the luscious acacias, and falling entranced to the reflection of a full moon on the lake below his estate.

Under 5 feet tall — and by all accounts the butt of mockery — Gibbon was, like Rousseau, a solid republican. He was affected, a lover of opulence. Like his other nearby contemporary, Voltaire, he believed resolutely in the ethos of gentlemanly decorum. Rome's agony, wrote the historian, was all Christianity's fault. Citing Livy, Gibbon reasoned that the Roman conquests through half the world were but perturbations of self-defense. Both Gibbon and Voltaire sneered at monarchy, finding relief among the Swiss and Genevoise. Between them, Gibbon and Voltaire produced more than 100 volumes, maintained world-roving correspondences. Both lived long and interesting lives, securely cached beneath the inperturbable mountains. Voltaire was the more vital, unbiased and brilliant of the two, a passionate defender of human rights. Gibbon emerged as the most widely read and imitated of any historian since Thucydides. The little polymath had fallen in love with Switzerland, moving to the lake for the first time in 1753. Throughout his work there is the Platonic ode: democracy is a "charming form of government, full of variety and disorder, and dispensing a sort of equality to equals and unequals alike."

What brings us to Gibbon's terrace? It was here, in 1780, writing at the conclusion of his third volume that he speculated on the future of mankind, taking as his model the curious fate of his beloved Rome. And it is characteristic of Switzerland—ever since the union of Uri, Schwys, and Unterwalden in 1291—that she should become a window on the world, the New Jerusalem. As for the lake, it is lucent, Olympian blue, manifesting seiches and sudden changes in height, unnerving winds from four directions, and then delicious calm—a meteorological reflection on world politics.

> *The balance of power will continue to fluctuate. . . . but these partial events cannot essentially injure our general state of happiness, the system of arts, and laws and manners. . . . If a savage conqueror should issue from the deserts of Tartary, he must repeatedly vanquish the robust peasants of Russia, the numerous armies of Germany, the gallant nobles of France, and the intrepid freemen of Britain; who, perhaps, might confederate for their common defence. Should the victorious Barbarians carry slavery and desolation as far as the Atlantic Ocean, ten thousand vessels would transport beyond their pursuit the remains of civilized society; and Europe would revive and flourish in the American world which is already filled with her colonies and institutions.*[8]

Gibbon went on blithefully to discuss the new era of military might, availing itself of gunpowder, mathematics, and chemistry, and easily able to defend mankind from any latent upsurge of savagery. With mankind's continued and assured refinement, the naked, abject spirit from which we all had come would be finally and forever clothed. "We cannot determine to what height the human species may aspire in its advances towards perfection; but it may safely be presumed that no people, unless the face of nature is changed, will relapse into their original barbarism."[9] The *Discourse on the Origin of Inequality,* written 26 years later by Geneva's disinherited son Jean-Jacques Rousseau, took issue with Gibbon's perception of human nobility. The *Discourse* fully illumined the apotheosis of "natural man" – of man prior to all metallurgy and agriculture and the encrippling society of men. Gibbon and Rousseau were symbolizing past and future from heavenly vantages. Where was the veracity, the engaged radicalism?

Rousseau flattered the whole body of Genevan ministers, remarking on their zeal for justice, gentleness, and even temper; their wisdom, moderation, and conjugal union "exerted only for the glory of the State and the happiness of the public."[10] In fact, Rousseau was not altogether tickled over the treatment he had received at the hands of the Protestants in Geneva. It was they who first imprisoned Rousseau's father, a watchmaker, and then showed little sympathy for the son's later works. Only by endorsing the Protestant creed was Rousseau able to claim his citizenship. But by then he had gone off to his hermitage near Montmorency. When Voltaire received a copy of *Discourse* he wrote to Rousseau, "Never has so much cleverness been used in trying to turn us into beasts; it makes one feel like walking on all fours, when one reads your work. However, it being more than sixty years since I lost the habit, I feel unfortunately that it is impossible for me to get it back, and I leave that natural gait to those who are worthier of it than you or I."[11] Rousseau crafted a startling analysis of modern man's debauchery by first uncovering its roots at the inception of government and avarice. By prolonged reflection on the original paradise – with its welcome vocables and food; its female; its tranquility, lack of gods, lack of violence, everywhere a blushing fertility – he established a pious beginning whose future could only be marred. Incremental erosion of Eden, the trespass of time; and with it, socialization, fire, weapons, propositions wielded to assist unnecessarily in the generation of thought, permanent shelters, domestication of animals, and so forth – this inexorability, said Rousseau, was the loss of wilderness, the component of humankind's sin. We were cast into a wilderness only to remind us of the true wilderness we had forsaken. This was the original doppelgänger, Eden and its dualistic other. "The fermentation raised by these new leavens at length produced combinations fatal to happiness and innocence." Need transformed into subjection; the concept of right altered to keep mankind in "perpetual labor, servitude, and misery."[12] Only in the mountains of Switzerland could the noble savage still raise his head and remain a free spirit. There were such people, usually chamois hunters. When the British tourists first came trekking into Argentiere, Grenoble, and the Bernese Oberland following the post-Napoleonic opening of the Continent, many were to remark upon the "goiter'd idiots" hunting up on the mountain.[13]

Rousseau died two years before Gibbon was to make light of his thesis, as Voltaire had. But Rousseau did not disappear. His style and purpose, the gravity of romance that masqueraded as ethnology, was trumpetted during the French Revolution. As for Gibbon, he died before seeing his beloved Switzerland temporarily subjugated to the will of a dictator. Eventually freed of her annexation, Switzerland would resume her middle-Europe eminence, above vicissitude, grand mediator — that rare bird in an age of dogged bureaucracies.

Many Swiss today would hesitate if asked the name of their president. Like the Federal Assembly over which he presides, the Swiss leader is not exactly in the limelight. Anonymity, frugality, unpretentious referendums, little need of lobbying, rigidly fixed budgets — these are the characteristics of what may be the most stable and edifying of all Western governments today. It is certainly the richest country per capita in the world, neck-in-neck with ambiguous Kuwait. But Switzerland worked for her wealth, and this makes a civilization. Her welfare paternalism has outlawed the kind of poverty discernable in Kuwait, where money scavenges upon resources.

A country half the size of Maine, we can fly over Switzerland in 20 minutes. There are twenty-six sovereign cantons, three major ethnic groups, two religions, and four language nationalities. For more than a century, Switzerland has insisted on peace for itself, and tried to lend some of it to the rest of the world. Her mountains and remarkably manicured copses and glens, all manner of meadow and vale, lake and mossy ledge and grazing heifer suggest the European paradise. But Switzerland is actually the second most industrialized nation on the Continent, next to Belgium. The rudeness of industry is most keenly felt, perhaps, when trains come screaming up otherwise inviolate valleys. The Alps have served politically to ensure a fortress within a fortress, veritable policy — *réduit national* — the defense citadel. America relies on missiles; the Swiss congregate in their rarified ski chalets, hemmed in by towering walls of granite and glacier. The Swiss are not without weapons.

While Italian inflation was rising at 17.1 percent, the Swiss franc edged upward a mere 2.5 percent. By controlling the bank reserves and the currency (the strict definition of monetarism), the Swiss have managed, without privation, to sustain a steady-state economy. The Swiss *did* voluntarily accept a substantial drop in real gross domestic product in response to OPEC inflation. But their recession did not affect unemployment, which continues to remain constant, somewhere below 0.25 percent. This is because the Swiss are not set on long-term growth, and because of the *gastarbeiter* element — foreign workers, of whom there are 8 million in Western Europe: 11 percent in France, 10 percent in

Germany, 8 percent in Britain, and 20 percent in Switzerland, not including additional seasonal and border-crossing workers. During the mid-1970s recession, about 25 percent of the imported laborers were compelled to leave. But they're now back again, basking in the beautiful country and enjoying a dependable wage in a prestigious currency.

Poverty is nearly nonexistent in Switzerland, save for some mountainous farming communities surviving off tourism and government agricultural subsidies. In many cases mountain real estate speculation has driven land and building costs well beyond the means of alpine residents, some of whom at least have managed to cash in on the bonanza – a syndrome not unfamiliar to most American communities. The Swiss banking laws of secrecy were established during the Holocaust to protect whatever German Jewish wealth could be gotten out of the country. Tax evasion in Switzerland is a civil, not a criminal offense. The statutes are the same in Holland.

"They be borne only of war," wrote Sir Thomas More, referring to this people 500 miles east of Utopian England. The Swiss were able to sustain all heat and cold and labor. For this reason the Utopians employed them as international mercenaries. The Swiss have hardly fought since More's day. Yet their military preparedness is comprehensive. Everywhere are nicely hidden tank traps. Each new home contains a fairly substantial wine cellar and shelter. Every male has certain military obligations from the age of twenty to sixty. The total cost of defense to the Swiss GNP is about 2 percent (nearly 30 percent in the United States). To understand the Swiss military requires at least a rudimentary discussion of Swiss politics and its unique role in international arbitration.

One can cite the usage of political abstaining as early as 1393 when the town of Zurich concluded a treaty with the Emperor engaging itself to *unpartysches Verhalten*, impartiality in the event of war. The term *neutralitet* was used after Switzerland's defeat by Francis I in the Battle of Merignan in 1515, which successfully checked all future expansionism of the cantons. Switzerland accepted her limitation and elevated it to a principle she has ever since upheld. The Treaty of Westphalia recognized, in 1648, the Swiss Federation. By 1674, the Diet of the Confederation officially declared its neutrality, a foreign policy resulting from failures in the protracted conflict between France and the Hapsburgs in upper Italy during the first half of the sixteenth century, as well as religious schism at home and the realization that the Alps could very well make good so radical a proclamation. Many of the cantons had for some time been neutral by treaty, pledged to *stillesitzen*, or sitting still. The rise of a European balance of power further solidified the effective role of a neutral Switzerland.

In 1798 the French army defeated the Swiss, and the country was transformed temporarily into a French-controlled republic. But in 1802

Napoleon's Act of Mediation re-established the obstinate Swiss Confederation, adding six more cantons to the original thirteen. In 1815, Geneva became a Swiss state, along with Valais and Neuchatel. The Declaration at Vienna on the 20th of March, 1815 made for formal acknowledgment by all the signatory Powers of the Vienna Congress under Prince Metternich of the "perpetual Neutrality of Switzerland."[14] A report on the Committee for Swiss Affairs of the Vienna Congress makes it clear that Switzerland's neutrality is granted only on the express condition of her "ability to make it respected."[15] Switzerland's religious wars were ended for good in 1847; 128 lives had been lost. The following year a new federal constitution modeled after that of the United States was drawn up in the spirit of 1291, a concatenation of free men in autonomous cantons. The most recent canton is Jura, number 26. Two members of each canton belong to the federal bicameral parliament, the *Ständerat*, and there are population-proportioned members of a *Nationalrat*. *Dörfligeist* and *kantönli-geist* — small-village and little canton spirit — remain the cherished, decentralized parochialism at the core of all local, cantonal, and federal elections and referendums.

Sunday is the voting day, occurring usually once a month. The annual outdoor legislative assembly of all qualified voters is called the *Landsgemeinde* — open and oral democracy from the Middle Ages, still practiced in Appenzell, Unterwalden, and Glarus. At the demand of 50,000 citizens, federal legislation must be submitted to the electorate for consideration and immediate acceptance or rejection. Any private citizen who musters 100,000 signatures, can propose constitutional amendments that must be considered by parliament and submitted to referendum. There is no other system like that in the world.

Of equal importance is neutrality, first stipulated in the Vienna Congress, and twice mentioned in the Federal State Constitution of 1848. "The maintenance of independence and neutrality" establishes an important link between national defense and pacifism, complete disarmament viewed as incompatible with permanent neutrality. The Swiss joined the League of Nations but with its collapse determined to align with no other international bodies that might diminish their neutral convictions. As a result, Switzerland is party neither to the United Nations nor the Common Market and has had to militarize itself more than most nations in Europe, proportionately. The Swiss constitution forbids the maintenance of a standing federal army, limiting the number of permanent troops to 300 per canton (for a total of 7800). But prowess in mobilization is formidable. With the outbreak of World War I, under the command of General Henri Guisan, 850,000 auxillary forces and local guards — all armed — were put on the ready. Switzerland did not fight in the Great War, though 7000 Swiss died in the service of France. Four thousand Swiss also fought in the American Civil War, mostly on the side of the Confederacy. Bernard Shaw satirized these mercenaries in his play *Arms and the Man,* better known in its operatic form as the *Chocolate Soldier,* referring to a demoralized front trenchman, Bluntschli, who preferred pieces of chocolate to a weapon, though ultimately he suffered a change of heart and became an expert survivalist. The practice of volunteering in foreign armies was outlawed in 1927. The League of Nations was to confirm the compatibility of Swiss neutrality with international collective security, though the Swiss had to agree to economic sanctions under Article 16 of the Covenant, thus making for a change from "integral" to "differential" neutrality. The Swiss had a terrible problem on their hands during the Abyssinian crisis. They finally renounced economic sanctions against Italy and dropped out of the League. Since that time, the Swiss have maintained full, neutral integrity.

In 1938, Federal Councilor Motta said that "we reject the doctrine which attempts to confuse the neutrality of the state with the neutrality of the individual; on principle, only the state is neutral, and will steadfastly remain so." This important distinction absolved Swiss Germans of all obligations of allegiance during World War II and was again a dramatic revalidation of personal and cantonal liberty, marking a philosophic pathway for all governments – *the exercise of restraint as the sum total of individual preferences.* Germany avoided challenging the Swiss neutrality. Had she done so, the Swiss would have blown up the St. Gotthard tunnel, effectively stopping all transport of coal along the Gotthard line between Germany and Italy. Switzerland was not invited to join the United Nations and still is not a member. She will join only if her neutrality is recognized as integral, meaning relief from the normal obligations, such as placing her armed forces at the disposal of the U.N. Security Council and granting full rights of military transit.

There have been other geographical neutral zones in history, areas of demilitarization, free cities, cities under League tutelage, areas of inviolability, nonfortification, disarmed islands, free-passage treaties, and natural barrier zones (e.g. the Alps). But of all the thin, often negligible arrangements of nonbelligerency, Switzerland's credo is a model for emulation. The country forces new mindsets upon all those who go there to think. Innumerable world organizations have chosen this cultural hub as their home. Every year, Geneva and New York host between them 6000 world congregations, producing over 1 million pages of crisis documentation. Geneva, the favored city, is on a vantage point between humanity's critical mass of apathy and its critical mass of hope. Looking toward the resplendent mountains, Gibbon's and Shelley's mountains, all seems imbued with vision and color and peace.

Voltaire hailed Geneva in his epic "The Civil War in Geneva" as "proud, noble, wealthy, deep, and sly."[16] An episcopal sea for hundreds of years, it passed from Roman, to Burgundian, to Frankish, to upper

Burgundian hands before being annexed in 1032 to the Holy Roman Empire, later to the Duke of Savoy, then to France as the Département du Léman until liberated by Austrian troops in 1814, at which time it was admitted to Switzerland. Franz Liszt eloped from Paris to Geneva with the Comtesse d'Agoult. Once a traditional asylum for persecuted Protestants, by the late nineteenth century it had become a haven for more radical types — Mazzini, Kropotkin, Bakunin, Lenin, Mussolini, and Kamenev.

The city's character of arbitration and global service is perhaps best exemplified by the strange, unhappy life of Henri Dunant, born in Geneva in 1828. He wrote a book eccentrically upholding the Augustan legitimacy of Emperor Napoleon III (*The Empire of Charlemagne Reestablished*), which he connived to present to Napoleon on the eve of the battle of Solferino. June 24, 1859, found him in one of the bloodiest skirmishes in history. Thirty-thousand casualties in 10 hours; fifty-five per minute. And there was Dunant, standing in his freshly pressed white clothes, carrying his gift of a book. A trance set in and he began numbly cleaning the wounds of those soldiers who managed to survive. He organized townspeople to help him, and was so moved by the collective energies that he wrote a new book, *Un souvenir de Solferino*, in which he carefully pleaded for some neutral organization to care for wounded throughout the world. *Souvenir* was published in 1862. The Queen of Prussia's husband Gustave Moynier, member of several existing international charitable societies, liked the book and initiated a conference in Geneva in 1863, adopting the symbol of the Swiss flag with inverted colors — a red cross — for its own. Delegates from sixteen countries unofficially attended, but the following year the Swiss government took on regular sponsorship of the conventions. Dunant was squeezed out of the organization, and thereafter fell into painful obscurity after failing to establish a Jewish state in Palestine. He retreated to the Alps of Appenzel in 1887 and lived the life of a hermit for the next 20 years. A Swiss-German newspaperman found the "hermit of Heiden" living in destitution in the mountains. The story broke on front pages all over Europe. Dunant was a star and received the Nobel Peace Prize, along with Frédéric Passy (founder of the Interparliamentary Union for Arbitration and Peace) in 1901.

A short drive along Lake Geneva to the northwest, in the direction of Gibbon's house, takes us to Gland, Switzerland, where yet another international body works towards peace and amelioration. The International Union for the Conservation of Nature and Natural Resources (I.U.C.N.) produced in 1980 an ecological masterplan called the "World Conservation Strategy" of profound importance. The ecological approach to peace takes as its premier assumption the tension that must result from environmental constraints. The Strategy has its own background in earlier efforts at achieving such aims: the first National Governors' Conference convened by President Theodore Roosevelt in 1908 to examine the United States' stewardship of natural resources; Hoover's Research Committee on Social Trends; Franklin Roosevelt's National Planning Board; the Paley Commission under Truman, and Paley's subsequent Resources for the Future Group; the Princeton conference, "Man's Role in Changing the Face of the Earth" in 1955; the publication of Rachel Carson's groundbreaking book, *Silent Spring;* Aurelio Peccei's Club of Rome, with its WORLD 2 and WORLD 3 models (the Limits to Growth group at M.I.T.); the U.S. National Environmental Policy Act of 1969, and Executive Order 12114 calling for a global commons and global ecological impact statements; and finally The *Global 2000 Report to the President,* initiated under President Carter, and Willy Brandt's Commission on the *North/South Report.* I have skipped numerous laws and conferences but these particular actions have led systematically to the Swiss *World Conservation Strategy,* to Switzerland's own identification with global heartache.

The Strategy's aims are these: to maintain essential ecological processes and life support systems; to preserve genetic diversity, using species and ecosystems only in truly sustainable ways; to halt the present widespread abuse of land (abuse lauded as being rationally motivated for profit as well as bare survival) that is hastening desertification, the erosion of precious genetic resources, the destruction of forests (particularly in the tropics), and the complete obliteration of watersheds, with consequent pollution of lowlands, oceans, wetlands, and the accompanying impingement and annihilation in unimaginable number of plant and animal species and the crucial dynamics between them. With the destruction of soil and the international urban outreach has come the simultaneous decimation of indigenous tribal groups, and these too have been adopted by Swiss philanthropy – the South Australian Aborigines, the Amuesha People of Central Peru, the Cuiva Indians of Colombia, the Inuit and other Indians in Canada, the Canelos Quichua of Equador, Araucanians in Chile, 40 million non-Han minority Chinese. There are hundreds of other threatened peoples.

But the Swiss' shouldering of the world's conscience is not a free and easy burden. There is a reverse side to altruism. It is repressive, and abhorrent, like a latent demon in the glacier. The mountains of Switzerland have held back for centuries a secret sorrow, of life jettisoned, held precariously aloft, lonely, beyond the place of comfort or familiarity or the warmth of a human smile. The Swiss novelist and philosopher Max Frisch relates contemporary Swiss mountain life in a story so haunting as to cast some doubt on the meaning of Arcadia. *Man In The Holocene* is the story not merely of Switzerland's mission, but of all places and people whose fate hangs dangerously over escarpments; who have on principle dispensed with guardrails.[17] Half-dozen types of thunder and lightning hasten maleffects in the sympathetic nervous system, ailments threatening to topple civilization; vacuities, grievances, intermittent boredoms, avalanche and debris, Neanderthal vestiges of self – the simple life, the quiet life.

Since the Stone Age, change has occurred on a particular stretch of mountain somewhere near the Swiss border with Italy. The protagonist Geiser – the fitful, Leonardo-like raconteur whose odyssey into madness takes on the landscape's own character in Frisch's tale – presents Swiss personality in the guise of glaciology, dinosaur logic, the natural history of salimanders, and the Quaternary. In reporting on these similitudes, he conveys the sedentary reality of Switzerland with its near perfect bureaucracy and hazy memory, Permian semantics, Carboniferous day-in-day-out routine. There are legends, existentialist antics up on cliff-sides, unedible cats, mudslides, helicopter services, an old woman, infrequent callers, the dependable train, bees and woodpeckers, unmowed pensioners' lawns, beds of streams that will never change. A tavern is dead, an umbrella is dripping.

Geiser's life reeks of a claustrophobia that has no moment of sunlight. The cold valley is drenched in timeless drizzle. He stares huddled and bare against this painful collision of wintry long hours. Gone is the romance of Switzerland. Only the fossils of human foible and pertinacity will live on to the compassionate end; only the very mountains will emerge to whisper a remnant rhapsody of *Homo sapiens* in Switzerland, tenacious, good-hearted, and troubled – reflecting the lousy world around them which they have set out to heal-by-example.

NOTES

[1] Fukuoka, *One-Straw Revolution*.
[2] ibid.
[3] Population Reference Bureau, *World Population Data Sheet* (Washington, DC: Population Reference Bureau, 1975).
[4] Michael Aris, *Views of Medieval Bhutan: The Diary and Drawings of Samuel Davis, 1783* (Washington DC: Smithsonian Institution, 1982).
[5] Blanche C. Olschak, *Ancient Bhutan: A Study on Early Buddhism in the Himalayas* (Zurich: Swiss Foundation for Alpine Research, 1979).
[6] Captain B. Pemberton (1838), quoted by Aris, *Views of Medieval Bhutan*.
[7] Michael Kaufman, "Saving the Himalayas' Fragile Eden," *Asia*, January 1982; see also G. N. Mehra, *Bhutan – Land of the Peaceful Dragon*, (New Delhi: Vikas, 1974).
[8] Edward Gibbon, *The Decline and Fall of the Roman Empire*, 7 vols., ed. J. B. Bury (London, 1896–1900).
[9] ibid.
[10] Jean-Jacques Rousseau, "Discours sur l'origine de l'inégalité," in *Oeuvres complètes de Jean-Jacques Rousseau*, ed. Pléiade Bernard Gagnebin and Marcel

Raymond (Paris, 1959 –).

[11] ALS, Voltaire to Rousseau (quoted in L. G. Crocker, *Rousseau's Social Contract: An Interpretive Essay* [Cleveland: Case Western Reserve University Press, 1968]).

[12] Gibbon, *Decline and Fall.*

[13] Gavin de Beer, *Travellers in Switzerland* (Oxford: Oxford University Press, 1949); see also Herbert Kubly, *Native's Return* (New York: Stein & Day, 1981).

[14] Walter Hofer, *Neutrality as the Principle of Swiss Foreign Policy*, trans. Mary Hottinger (Zurich: Spiegel, 1957); see also Carol Schmid, *Conflict and Consensus in Switzerland* (Berkeley: University of California Press, 1981); and Christopher Herold, *The Swiss Without Halos* (New York: Columbia University Press, 1948).

[15] ibid.

[16] Voltaire, *Oeuvres complètes de Voltaire*, ed. Adrien J.Q. Beuchot (Paris, 1877–1885).

[17] Max Frisch, *Man in the Holocene* (Harcourt, Brace, Jovanivich, New York: 1980).

Chapter 12

A Visible Hand

*Since the time of Adam Smith's (*Wealth of Nations *(1776), laissez faire capitalism has presumed many things about human nature that we now know to be in error. Greed has no inherent restraints. Utilitarianism suggests an antidote to greed, but does not prevent the built-in dangers of a multinational corporation. Those dangers include the unchecked destruction of the environment and the exploitation of labor in the Third World. There is a temperate middle path, an approach to economics that derives its wisdom from the smooth workings of any ecosystem.*

THE LAKES OF LORE

I occasionally bushwack through the Coast Range of British Columbia to a high, glacial cirque in the Tantalus Mountains. An endless array of cascades and hanging ice walls has scooped out and filled the basin with a lake known as Lovely Water. It is the most sparkling, vital body of water that I know; Bremen blue to its unsullied depths, occasional diatoms and surface-suspended green algae catching the summer sun, insects and bass competing up high, trout skulking the hypolimnion bottoms, where there is nutrient-free darkness. If I wade out across the shoreline granite, I can peer down 30 meters into this water. Such tarns were common in Pleistocene North America and Europe. They tasted fantastic, were abundantly healthy, steady-state.

A shallower New Jersey mesotrophic lake, in proximity to humans, has an entirely different story to relate, one thickening by the year with sedge and overgrown weeds. Organic decay of every kind is exhausting the oxygen supply and wiping out the cold-water fish. Eventually, the lake will become eutrophic, a bog, and finally a forest.

Tantalus was the mythologic Greek King whose crimes were so excessive that for punishment the gods stood him knee deep in a freshwater lake. Every time he tried to drink of it, the waters receded. He was no longer in paradise, (Hesperdes) but in Hades. The obtuse economics of a lake's natural history is analogous to human affairs and the intercourse of money and goods. It might seem that nature has no winners or losers. The trout dies, swept up into a Taoist whirlpool of evolution, sweetly merged over millennia with muck, pickerelweed, pine trees where lake once lay. This may all be of little comfort to the fine fellow whose once luminescent entrails have slithered, lodged lonesomely in a restless process, mute molecules, summer vacationers, toilets and leaky cesspools, winterized roads and drainage culverts. The early oligotrophic, sparsely fed waters have been filled with the sewage of fishermen – nitrogen and phosphorus – excreta, gas leaks, all of the big city's effluents. The algae have covered the lake, the carp are now belly-up, and mosquitoes are waltzing on their insides.

The EPA studied 800 lakes in America between 1971 and 1977, lakes receiving human use. Sixty-eight percent of them were *dying out* – or *changing*, depending on your preference. Nature indeed fosters transformation. Ecclesiastes long ago sanctioned such wilderness dynamics. Only human beings catered to the notion of a *problématique*. The point is this: While we are alive and can recognize pleasant or adverse conditions, how are we going to live in the most integral fashion? Aristotle took a noble stab at the query and said, with the first biologist's resolve, that mankind's ultimate state of happiness (eudemonia) was that he should become himself; that whatever the circumstances of change (evolution), our inner nature was destined, must fulfill its contract with nature. Economics is a word first applied by whatever Greek kept his or her house in order. Extending the Aristotelian connotation, economics is the art of maintaining one's own nature in a world of circumstance. We fall from the womb into a reciprocal address, our genotype governed – in whatever concentric circle of cause-and-effect one wishes to begin the story-telling of human behavior – by environment. A lake has its own biological and spiritual budget to balance, as we do. Whether it remains the same – like Lake Lovely Water – or becomes a Lake Erie, or the malodorous quagmires of outer Hoboken, New Jersey, there will be perpetuated a balance. A eutrophic lake has responded predictably, and the changeover will be welcomed by dazzling new species – from the mats of blue-green algae, to dragonflies and snapping turtles, digger wasps and water moccasins. Creatures that sting and bite prefer a habitat of dense nettle country. What about inflation – when money gets harder and harder to come by, when the oxygen is squeezed out of the economic system? We said that the trout die first, the carp last. Who's who in human society? And where, if any, is there a balance to be grasped? Is there some polemical, difficult maxim lying deep in financial transactions, a latent power akin to natural selection that economists and consumers simply aren't talking about; some easy adjustment that would allow us to have our lakes, our swamps, and forests too?[1]

LAND

The linchpin of world trade and domestic security is not oil, nor "plastics," but land and with it, agriculture. Land is always the last word, the true politic, our final comfort. Americans have, with good reason, held confidently to their assumptions about soil rich beyond measure. The U.S. produces more food than any other zone of good Earth. But since 1972 those beliefs have been eroded and continue to fight for their fertility. As American harvests decline, we have less grain to trade, prices are higher in support of farm subsidies at home, and the number of the world's hungry increases, generating ever more tension between nations, widespread misery, and death. Haili Selassie's 30-year regnancy was expunged in the wake of Ethiopia's famine. Economically speaking, famine means oxygenless money, spiralling inflation.[2]

Everyday, more farms are foreclosed and liquidated. The average American farmer has nearly 202.5 hectares of privately owned land. Until recently, all ownership was private. Farmers have been traditionally anticollective, antibusiness. But America's agricultural system, employing one-fifth of the nation's labor force (most of it processing and distribution, *not* growing) and generating 10 percent of the GNP, is being quickly subsumed to the oligarchic 1 percent elite cadre of high-profit operators and computers. Huge farms with corporate monocultures are seeing the end of the small independent farmer. Within 20 years, more than half of the existing farmers will be forced into mid-career retirement. America's farm land will be owned by those very few multinationals that can regulate all prices and determine who lives and who dies in nearly every other country. Growers have not experienced so many debt loads and bankruptcies since the Depression.[3]

The New Deal policies emerged erratically, in no singular body of law, against a context of 25-percent unemployment. Local governments were paralyzed. Today, 11 million workers are unemployed, many more underemployed. There are 438 bankruptcies a week – an 11-percent increase over the 1970s – a 45-percent increase in business failures, and 31 percent mortgage delinquencies. If we had such figures in the written ledgers, say, of Herculaneum, it wouldn't mean much to us today, especially given the more cosmic callings of Mt. Vesuvius above the town. In many respects, the overall wealth of American capitalism similarly absolves us from fear of the daily economic tribulations. But not farmers, who are hit the hardest. And when the farming estate crumbles, the rest of the world's economy is not far behind.[4]

The wealthiest farmers knowingly deplete the quality of their soils to compete in a failing market place. The Carter administration had hoped to revise domestic agricultural management, lending firmer support for the small farmer, the ethnic, organic, American way of life. But the Reagan administration, full of unworked-out premises, has done just the opposite, reducing federal outlay and forcing rural bankrupcies through the withdrawal of promised loans. Government support prices now are well below the actual cost to farmers of production.

The notion of the free market place, as first set down by Adam Smith (his "invisible hand" of distribution and price stabilization, of supply and demand during times of scarcity), rested on assumptions of inexhaustible resources and limited demand, a scenario utterly incompatible with today's overpopulated, resource-deficient, interdependent world. It is said that a line with finite ends contains an infinity of dots; that the invisible hand will always alight upon those dots, that fabled jug of endless milk, to feed those in need. With decisive flexibility and youthful energy, college students have never squeezed more than eighteen of themselves into a Volkswagen Bug. And this finitude extends to every other human activity on the planet. The invisible hand has been replaced by a simple-to-read law, whose denial inexorably leads to the *tragedy of the commons,* the problem of the public goods. A moral, self-restraining husbandman cannot survive a market place that eats, steals, devises, and applauds as if there were always a tomorrow — not without government encouragement in any case. And then what?

The scenario of *tragedy* is predicated on an ecological model of incremental apocalypse. Termites eat and eat, greatly undermining a structure, any structure, like the great old wizened Russian Olive tree that collapsed in my family's front yard after 50 years on the very morning we were selling the house. Algal bloom, over-grazing, air pollution, these economies all operate according to the same principles of balance and overshoot. Lacking a market for wilderness, for safe drinking water, what individual crusader could be counted on to speak out for the Earth, let alone a minnow? People trained and incited by everything their nation has invented to seek maximal benefits and quick comforts, are not sufficiently educated or poetic enough to restrain themselves. Only the government — presuming that it follows a higher rationale(?) — can intervene to protect us from ourselves. This is old stuff, like the redwood tree in northern California that had been given a placard, "Chop me down before I kill again!" in honor of a president who had, weeks before coming to visit the redwood groves, said that trees produced carbon dioxide; carbon dioxide was a killer, and therefore, the more trees we chopped for lumber, the safer the world would be.

Jimmy Carter had proclaimed his administration's goal of reaching 20 percent of America's energy needs through the stimulation of renewable energy alternatives, but the Reagan group perceived that less than 10 percent of America's renewables could, or should be, tapped. Supply side economics has perpetually seen the free market forces as crucial to American productivity. This is fine, but not in absence of a plan. Ineffectual to an unprecedented degree, the Reagan administration has hoped that the American experience would shape the American experience. The adage, "No government is good government," worked among Panglossian societies where hierarchy, ritual, custom, and behavior were rigidly adhered to under the law of one's inner directives. There may have been a chief who took counsel, but the only government was that of individual responsibility. Of course, when the tribe grew to an awkward or fractious size, some portion of the clan routinely broke away, left for a whole new quadrant of jungle. In today's world, where the exclusive consolidation of land and wealth precludes most acting out of individual separation on the part of the little guy, a source of wise, impartial arbitration is mandatory – and that is govenrment. Without it, a whole new set of ecological constraints are quickly marshalled in the name of Fascism, the very Latin root meaning to bind with an axe that bundle of previously individualized rods.

NATIONAL ENERGY PLAN

The Reagan administration's energy plan called for the sharp increase in coal production to offset the admitted decline in oil imports and domestic production. Concurrently, that administration viewed nuclear energy with an eye toward raising its output by a factor of four before the year 2000. Synthetic fuels and techniques for recovering secondary and tertiary oil would be spurred intensively by the government. Approval for reprocessing spent nuclear fuel (and the sale of nuclear reactors abroad), the discontinuation of the Solar Energy Research Institute, and the simultaneous decision to help the nuclear industry by financial guarantees and enforced easing of licensing procedures – were all enacted. The "free market" has become its puppetry. In the name of the free market, any decision can be justified. Others use the phrase the *real world* to the same end.

From the very beginning of natural gas regulation, the government has asserted a powerful bias in favor of energy industries over the consumer – industries which, for half-a-century had proved their ability to buy their presidents.* In a similar manner, as recounted in Thomas Wolf's novel, *You Can't Go Home Again*, washing machine manufacturers succeeded in capturing the consumer's curiosity, addicting him to the idea of a machine he could ill-afford and did not particularly want.

More than half of all residences in the United States are rental. Abroad, that percentage is yet higher. But the tenant is not given the choice or responsibility to invest wisely in the free-market sector. He responds to price. But we're speaking of more than mere cost effectiveness. Cheap energy is one thing. Environmentally damaging cheap energy is quite another. The *visible hand* is called for. Government must itself oversee a free market that has demonstated its disregard of most medical and ecological reflexes. Studies at the Princeton Center for Energy and Environment and at the Harvard Business School confirm the extreme marketability of alternative, soft energy sources, sources more fully democratic. Increased tax credits for technology, investing in alternate technology, tax stimulation of rental properties that have begun insulation systems, the delimiting of all double and triple tax assessments, and the long-debated sliding-scale flat rate taxation are methods for revitalizing the government's immediate role in the transition. A *conscientious tax objector's status* for those refusing to see their hard-earned dollars fueling U.S. military expenditures is perhaps the most important legislation to be yet debated in the economic sector.

Labor needs a greater share of its productivity. This is neither Marxian nor strictly socialistic. Lane Kirkland of the AFL–CIO has suggested at least one major area where such shares can be easily dispersed: by effecting the corporate flow of investment capital through the leverage of accumulated welfare and pension funds. Right there, in middle-income industry, is a partial solution for redirecting from capital-intensive to labor intensive policies. The Chinese, Japanese, and West Germans have all done so successfully. A free market will not liberate the dispossessed, the low wage earner, or the unemployed. Nor will high deficits foster an economy capable of bringing the necessary benefits to the young and old, the infirm, the culturally deprived. One California Senator's ingenius plan to make social security payments optional for all those under forty-five, as a means of diffusing a $200 billion welfare deficit, is the first step toward legitimizing destitution among a wide sector. In this way, the government could keep up Pentagon expenditures while disavowing the rights of low-income Americans to expect anything from their government. The poor in America, like the poor anywhere, cannot be expected

*Regulation began in 1954 when the Supreme Court ruled that prices for natural gas were to be based not on the resource itself, but on the cost of handling it.

to sacrifice their meager annuities. Altruism must come from the wealthier echelons in greater shares than the ludicrous "trickle-down" theory, a proposition that has always conjured in my mind the image of the rich pissing on the poor. What we need is a new "trickle-up" theory. If the American upper class is worried about sloth and laziness, addicted welfare beneficiaries, and the chronic nonworker, then the rich, who have the time to investigate strategy, must implement some other scheme. It must be a scheme that can bring the minority back into the stream of productive life, rather than banishing them to desperation.[5]

A free market, by allocating supply and demand, theoretically ensures that the consumer will not squander scarce local resources. This is called microeconomics, and it is fraught with contradiction. But the free market works against itself by encouraging the very same exhaustion of natural resources on larger, macroeconomic levels in the guise of the monopolist. Four new indices of GNP have been devised to better calculate the true wealth and distribution of a nation's energy quality:[6]

1. MEW – *Measure of Economic Welfare,* which examines the "quality of life" in addition to the GNP
2. GNC – *Gross National Cost* (I & II). This treats the fraction of GNC produced by renewable resources and waste recycling, and that fraction of depletion stemming directly from nonrenewables.
3. NRU – *Natural Resource Units.* The government mints and distributes these units in recognition of a finite number of said units. The consumer is free to juggle his units, sell them, or spend them all at once. But the units are allocated in a manner that prevents their overall consumption from exceeding an ecologically sustainable maximum.
4. NNWI – The *Net National Welfare Index* was developed by the Organization of European Communities (OECD). It is similar to MEW and stipulates what shall condition the *quality* over *quantity* of existence.

These remedies stem from the limits-to-growth paradigm, based on a study done by Club of Rome in the early 1970s. Resources are indeed finite. There are not enough to go around equally for everyone on the planet. The market and rising prices will not likely induce responses to scarcity with sufficient dispatch, and might even intensify pressures of depletion. Programs for assessing technology thus go hand in hand with reformulating the GNP. One of the problems with the Club of Rome report was the over-use of the word "growth," a concept inbred in all people's feelings toward life.

National energy policies, government strength, labor revitalization, and a visible hand extended toward the common prayer of all people demands a set of macroeconomic considerations. We have mentioned tax credits. One other method of reemployment involves chartering a "nature corps," a workforce trained or retrained by public expenditure to accommodate the needs of degraded public works projects and failed private sector agriculture: energy farms in the Southwest; small-scale, decentralized, and innovative industries; and institutes that would no longer discredit income and environment. We would see a reduced workload as more labor came into its own, and an increased rate of production; a slower, more modest existence, with greater decentralization. Cities would decline in size and complexity, architecture would assume a new look; our money would become philanthropic money, more stable, less obscene. If the United States cut its per capita energy consumption in half, we would use as much energy as we enjoyed in 1940, when students at UCLA paid 25¢ in Westwood for a tunafish sandwich, chips, malt, and dill pickle – the level, granting inflation, of the French economy today. Though not entirely: with advancing technologies we could use our energy more efficiently. This would trim America's corpulence, (Americans are 1.3 billion pounds over-weight), renew our vigor and credibility and allow the 50-plus-percent of all scientists and engineers engaged in military work to consider alternate forms of employment as the Pope has advised. Between 24.3 and 30 percent of America's $1 trillion + GNP goes for defense. Such outlays are criminally berserk, as the majority of voters are coming to realize. Take your own monthly income after taxes and apply one-third of it to defending your household with military appropriations. What would your life look like? How easy would it be to get along as usual with your neighbor, providing that he had also accumulated such armaments at his front door? You'd never want to get drunk together, or look at his daughter. When my neighbor waters his lawn, he has a habit of generously assaulting homes on either side. I'd hate to see him with an army tank.

THE GREEN REVOLUTION

In 1941 President Camacho of Mexico knew that the only safeguard to his newly elected regime was quick industrialization. To this end, he brought down a team of agricultural scientists from the Rockefeller Foundation. What transpired was the discovery of a new, high-yield dwarf maize, bred uniquely in Mexico. The higher-yielding variety's (HYV) time had come, and it embodied the Green Revolution. Other Central American countries became involved in the mid-1950s. India came into the club in 1956, and then Nigeria. During the 1960s the International Rice Research Institute in the Philippines cross-bred for IR-8, a rice strain that doubled yields throughout the Far East. The Rockefeller and Ford Foundations, the Food and Agricultural Organization, Agency for International Development, and the World Bank and Canadian International Development Research Centre financed most of this effort. The result was more food, but the Green Revolution has hastened the destruction of farms and stressed inordinate reliance on cash crops and fertilizers, thus circumventing traditional techniques now seen to be more politically astute, labor conscious, and economically appropriate. In addition, the use of fertilizer in many areas has reached a striking rate of diminishing returns: For every seven units of fertilizer applied, only one unit of food is produced. And there is what Kenneth Dahlberg calls a "nutritional illiteracy" underlying the Third World's heavy dependency on selling export crops to pay back immense debts to the developed nations. Says Dahlberg, "The most direct link between malnutrition and the green revolution lies in the loss of pulses and other protein-rich crops as they are displaced by HYVs."[7]

Accompanying the Green Revolution was a spate of capital-intensive Third World projects, implemented with the spirit of the Marshall Plan and Western European collaboration. Sarah Voll has documented sixteen such schemes, many of them white elephants, in Gambia, Mali, Senegal, Tanganyika, Uganda, and Sudan. They were plagued by vitamin-deficient crops, erosion, inadequate feed for flocks, the inappropriate importation of technology from the West, heavy and unexpected transport expenditures and crop diseases, confused priorities, overcapitalization, untenable shifts from commercial to tenant farming prior to the showing of profit, inadequate feasibility studies, and the erroneous circumvention of local expertise, taste, and labor. Ultimately, these projects failed because of an "unwarranted faith in Western technology to overcome obstacles in unfamiliar surroundings."[8] Outgrowths of imperialistic mentality, cash-crop victimization, and domestic energy and technology programs that concentrated on extractive, high-profit reasoning, such were the failed preludes to the Green Revolution. They suggest the doomed geography of the invisible hand, calling more stridently than ever for a global commons, the working together of all nations.

NORTH-SOUTH

The high-volume fiascos of the Green Revolution cash crops (a few *have* paid off) are the very kinds of enterprises that qualify the undernourished nations for concessions of aid and loans from the West. In the summer of 1944, at Bretton Woods, New Hampshire, representatives of the United Nations met to discuss plans for an International Monetary Fund (IMF) and an International Bank for Reconstruction and Development. The primary goals of this highly charged get-together were to stabilize foreign exchange rates, facilitate international free trade, and to make currency more accessible to members. Among those present was

John Maynard Keynes, who proposed that the foreign exchange esperanto be something other than the multinational-based dollar. Lord Keynes also recommended that nations be allowed to solicit overdrafts from a collective bank, wherein the member nations would have equal drawing and voting rights, rather than having to submit to the requirements of external private investment, and the "exhorbitant privileges" (said Charles de Gaulle) of such money. Foreign investment capital has, since the end of World War II, been awarded all of its noblesse without accompanying oblige. Arguably, it earned its wealth at the expense of the Third World. But racial atrocities aside, other equally vindictive persuasions today implicate the one-sided policies of Western money.

The word *interdependent* has been much thrown around, driven into battle, vanguarded on the pillar of ecological tribute. Does it really matter to Americans whether millions of sub-Saharan nomads die out? What are the political ramifications, the destabilizing regional and global effects of local extinctions? The very questions of "discriminating altruism,"[9] of triage and sinking lifeboat theories (ivory towers from the past), immigration quotas, and sharing the pie confuse nationalism and the enthusiasm for indigenous privilege with *humanity*. If we are to take our species' purpose and evolutionary extant with any degree of seriousness, then the survival of all people matters. But much of the North's tactics of political expediency, stinginess, and outright tyranny have tested the staying power of an opposing viewpoint—one born of domestic crisis and the limitations of leadership. This latter course, at the heart of Ayn Rand's philosophy of Objectivism ("Reason, individualism and capitalism" of "rational selfishness"), compelled her hero John Galt of *Atlas Shrugged* (the title ironic in this context) to swear by his life and his love of it that he would "never live for the sake of another man" nor ask another man to live for his.[10] This is sanctimonious Darwinism, the New Federalism. Why soil one's own garden? I phrase the question this way to beg the issue; why soil it? Because we need the soil. Why be thy brother's keeper? Because we need our brother. Transgressing these codes, the West has sensed reprisal and so escalated its sacred cows: unshared capital, bigger, more heinous armaments.

Scrambling to obtain their pittances, Third World countries have engendered hospitable climates for Western "slave-labor camps." Average free-zone wages amount to 24¢ an hour, while free-zone governments accumulate huge external debts, now at $500 billion or 220 percent of their total export earnings. Spiralling stagflation in the West furthers the Third World collision course with bankruptcy and famine.[11]

The North comprises 25 percent of the world's population and 80 percent of its income. The South has a 25-percent mean infant mortality rate, 50 percent hopeless illiteracy, and estimates of between 20 and 50 percent malnutrition and famine. Hundreds of millions in the South are afflicted with river blindness, malaria, sleeping sickness, and schisotosomiasis. By 1990, the cereals gap between the haves and have-nots will be nearly 150 million metric tons. The Third World produces 3000 new refugees every day (not counting over one billion squatters, humans without a roof over their heads). Third World food demand is growing by at least 3 percent annually, far outpacing its ability to produce given the West's current approach to Third World self-reliance, austerity, and the "bitter medicine." Final consumer prices on Third World-produced exports mean a meager 25-percent share for those producers. The Western markets again dictate middlemen and tariff squanderings.

In October 1981, twenty-two leaders of rich and poor countries met at the plush Caribbean resort of Cancun, Mexico. Impoverished folk were seeking a "massive transfer of resources" from their rich diplomatic colleagues. The Group of 22 urged channeling more military spending into foreign aid. As it stands, most Western donor nations give approximately 0.35 percent of their GNP, Scandinavian nations exceeding this. The United States and Japan imparted but 0.27 and 0.23, respectively. And the Soviet Union managed a bare 0.04¢ on its GNP ruble. The Group of 22 wanted to double foreign aid, to fashion commodity cartels that would give them economic coercive powers over their own exports; and they wanted a larger share of drawing rights (SDRs), more clout in the International Monetary Fund, energy-intensive aid, better tariff arrangements, and an industrial shift from northern to southern countries, including an increase in their share of financed technology and research and development (they have but 3 percent of global R&D at present). The poor countries have called for a "planetary bargain." This means land reforms—getting farmers into the legal, financial and conceptual system of food production; the elimination of wage and price controls, the untying of aid, and preferential trade schemes enabling them to process their own exports rather than shipping raw goods at the expense of valuable dollar appreciations and scarce domestic labor. These Third World-processed imports would actually encourage productivity in the West at a time when more than 30 percent of production potential in London, Detroit, Los Angeles, Munich, Paris, and other cities is unused and unemployed.

At the time of Cancun, the United States had largely ignored the earlier U.N. Conference on New and Renewable Sources of Energy held in Nairobi, Kenya two months before. The outcome of that congregation of 5000 was largely to raise consciousness, much as UNEP's 1972 Stockholm Conference had done. But much more was needed for the 1980s than mere good-will. Three days before the African world conference, the Reagan administration did its homework and made some token suggestions. It outlined its intention of granting private enterprise in developing countries more American aid. A Bureau for Private Enterprise was founded by the administration that would make recommendations following a series of Third World reconnaissance missions to establish viable grounds for new industries there. The Dutch Philips' Gloeilampenfabrieke Corporation had built a plant in Eindhoven to test for technologies appropriate in LDCs. Japanese and West German companies had done the same. Of all the *Fortune* 500 overseas corporations, 10 percent are owned by Third World governments and businessmen. But these seeming successes are worn thin by far greater trials. Taiwan and South Korea have achieved a degree of prosperity. The Ivory Coast has as well. But all three economic over-achievers are deeply strained, in need of low-interest loans, of a visible hand to compensate for overpopulation, too little land, too little capital, and too small a trained industrial cadre – contrary attributes to those that offered nineteenth-century America the vision of eternal wealth, exploitable labor, and land presumed nearly valueless by law. For all of Mexico's Pemex Oil, the country is beleaguered by an $80 billion deficit, forcing it to sell more than 50 percent of that oil abroad, in strictest controversion of Mexico's long-time policy of resource nationalization, energy for the people. Every achievement of wealth and order produces thermodynamic disorder elsewhere in the system.[12]

The World Bank estimates that the developing world's energy needs can be greatly reduced by 1990, perhaps by as much as 30 percent. But is the World Bank providing loans for alternate energy policies? No. The OPEC nations are looking for different markets: changing from increasingly self-reliant Western nations to those LDCs that must rely upon Western subsidies to finance oil purchases and their own domestic oil R&D. Such transactions will define the character of the world's financial system if the multinational cannot be persuaded to explore untried technology in the huge labor fields of the South. The first major loan by the IMF to increase Third World oil exploration was given to India in late 1981. The amount was $5.68 billion, more than half the original IMF budget in 1946. The United States' foreign policy has fostered the same energy-intensive exploitation abroad as at home. Is this the answer? The World Bank puts 30 percent of its lending resources into agriculture, 80 percent of which goes directly into food production in developing countries. In the next 10 years some $600 billion will be necessary just to maintain present standards of world hunger. Shipments of food in the early 1980s accounted for 8 to 10 million metric tons; the total world trading volume was more than 250 million metric tons. The real breakdown during times of famine is not a lack of food, but of distribution and absence of hard cash to purchase it. In the thirty-one poorest nations, the average life expectancy is 45, versus 70 in Europe; the gram-total of protein intake is about 1/10th what it is in the West; per capita GNP is just over $200; only 20 percent can read and write. The Green Revolution has temporarily increased wheat yields throughout the world by 50 percent and rice by 30 percent, but there has not been the attendant egalitarian purchasing, again as a result of insufficient per capita income. As in the United States, the energy-intensive gains in Third World farming have rested almost entirely with large landholders, not with the poor family farmers. Further exacerbating famine conditions is the very nature of Western price subsidies: Japan, the Soviet Union, and the European Economic Community (EEC) keep domestic prices fixed so that global shrinkage does not exact inflation-countering responses in the Western countries. Supplies do not adjust, because Western farmers are not compelled to produce more. As a result, international prices soar and reserves plummet.

In India, agricultural production has increased in the past 20 years, but the protein intake by the poor has gone way down. So much for the Green Revolution. What has happened? First, overpopulation has easily outpaced food production. Second, despite energy-intensive efforts and huge investments from all over the world, market incentives have fostered a shift from nutritional to cash crops. There is no enforced policy, as yet, for prioritizing the cultivation or subsidization of nutritional crops. Indians consume less grain than is used to feed animals in the developed world. Between 40 and 50 percent of the country (nearly 700 million people) are beneath the absolute poverty line, with a per capita caloric intake 10 percent below world minimum standards.

Such paradoxes of progress were summarized by Willy Brandt's Commission on International Development Issues. Their report brought the world leaders together at Cancun.[13] The Brandt Report analyzed the most likely outcome of a $20 billion a year transfer to developing countries in the form of aid, loans, and investments. Econometric results showed that Western exports would rise some 3 percent every year thereafter. In addition, developing world imports to OECD countries would tend to moderate inflation. The new Norwegian government determined to lend no money to Third World projects likely to incur indigenous environmental damage. Interpreted to encompass famine, the prostitution of labor, and raw goods, the renunciation of cash crops, and local social scientific and agricultural know-how, such a stipulation enforces the stronger linkage of developed country investors, aid organizations, and the world's banking empire. The Brandt Report, along with the State Department and Council on Environmental Quality's *Global 200 Report* (1977–1980), powerfully bring the burden of ecology, economics, and peace to the attention of world leaders and citizens.

The Western powers need the Third World like never before. This sounds complacent, ungalvanizing, without sufficient proof. But the fact remains — however one perceives, computes, or argues the plethora of data – the Third World will possess 80 percent of the world's population in the year 2000.

Cancun did not achieve any earthshaking results. Conferences never do. There are clear alternatives to economic disequilibria and political inequity. But welfare and reform agendas are embroiled in a basic dilemma: the lack of consensus regarding what is good and evil about human destiny. The status quo feels no compunction to redress its blunders or those of others; and the oppressed can find no satisfactory, consistent affirmation of their plight from any administration the world over. The Third World has become the Wandering Jew. Revolutions – in Lebanon, Libya, Uganda, Iran, Angola, El Salvador, and South Africa (soon) – produce a change, but at the cost of untold human agonies. Against such outcomes – outcomes that could be avoided from the beginning, but are not – Western economies, laissez-faire voters, fix their home-life assumptions and narrowness, arming against the paranoia at large.

Liberalized policies of trade must occur in earnest in the coming years. America will be called to lead this movement, both under global pressure and mounting public opinion (particularly the majority of women's vote, which is strongly anti-Republican), declining soil fertility, steady unemployment, the disappearance of social welfare, and military spending deficits; crises that must completely revamp the free-market presumptions, intolerance, and shortsightedness. Every denizen of this planet must insist that his voted representatives conduct a campaign on behalf of such issues; and that the voter himself observe nature diligently, attuning his style of life to those observations; by supporting heart and soul the many environmental organizations; monitoring one's everyday purchases on the basis of the various multinational corporations from which those goods come; by becoming politically active in local forums, crying out, honoring the fundamental codes of American civil disobedience when necessary. It *is* necessary. The political process itself offers us the best technique for change.

A glacial lake is a beautiful gift and it comes about but once in a million years.

NOTES

[1] Roger Swain, "Troubled Waters of Our Lakes," *New York Times Magazine*, 20 September 1981, p. 59.

[2] Kenneth Dahlberg, *Beyond the Green Revolution: The Ecology and Politics of Global Agricultural Development* (New York: Plenum, 1969).

[3] Sarah Potts Voll, *A Plough in Field Arable: Western Agribusiness in Third World Culture* (Durham: University of New Hampshire Press, 1980).

[4] Garrett Hardin, "The Economy of 1981: A Bipartisan Look," presented at the Congressional Economic Conference, Washington, DC, 10 December 1980.

[5] Ayn Rand, *Atlas Shrugged* (New York: Random House, 1957).
[6] Willy Brandt, *North-South: A Programme for Survival; The Report of the Independent Commission on International Development Issues under the Chairmanship of Willy Brandt* (London: Pan Books, 1980).

PART 4

Twilight of Dead Reckoning

D.P. McGee. Courtesy of Simon Wiesenthal Foundation.

Chapter 13

Fire Mind

Fire is the ultra-living element. It is intimate and it is universal. It lives in our heart. It lives in the sky. It rises from the depths of the substance and offers itself with the warmth of love. Or it can go back down into the substance and stay there, latent and pent-up, like hate and vengeance. Among all phenomena, it is really the only one to which there can be so definitely attributed the opposing values of good and evil. It shines in Paradise. It burns in Hell. It is gentleness and torture. It is cookery and it is apocalypse. It is a pleasure for the good child sitting prudently by the hearth; yet it punishes any disobedience when the child wishes to play too close to its flames. It is well-being and it is respect. It is tutelary and a terrible divinity, both good and bad. It can contradict itself; thus it is one of the principles of universal explanation.

Gaston Bachelard, *The Psychoanalysis of Fire* (Boston: The Beacon Press, 1964).

We thus begin to understand the truly essential place occupied by cooking in native thought: not only does cooking mark the transition from nature to culture, but through it and by means of it, the human state can be defined with all its attributes, even those, that, like mortality, might seem to be the most unquestionably natural.

Claude-Lévi-Strauss, *The Raw and the Cooked, Introduction to the Science of Mythology* (Chicago: University of Chicago Press, 1969).

Primal man had the habit, when he came in contact with fire, of satisfying an infantile desire connected with it, by putting it out with a stream of his urine. The legends that we possess leave no doubt about the originally phallic view taken of tongues of flame as they shoot upwards. Putting out the fire by micturating – a theme to which modern giants, Gulliver in Lilliput and Rabelais' Gargantua, still hark back – was therefore a kind of sexual act with a male, an enjoyment of sexual potency in a homosexual competition.

Sigmund Freud, *Civilization and Its Discontents* (New York: W. W. Norton and Co., 1963).

He who is near Me is near the fire.

Jesus of Nazareth

But what is violence, as the words is used . . . it is simply fire – fire in all of its forms.

S.L.A. Marshall, *Men Against Fire*, 1947

Although fire has made human culture in its origins, it now threatens us with complete ruin. The neurological deception of flames has resulted in an addiction to firepower – whether it be in the form of the earliest matchlocks or thermonuclear reserves. There is at least one tribe left on Earth with no knowledge of fire, a people spectacularly adapted to the snowy highlands in which they live. This remote nomadic group may well represent the last remaining portion of wilderness within the human mind.

LUSTRE AND LOATHING

Floating around the planet every day and night are 44,000 lightning storms discharging 8 million blasts onto the Earth's surface, 100 every second. During July 1977, the largest fire in California's history raged from one site of lightning burst to a second one, across 100,000 hectares of Ventana Wilderness in Big Sur. I lived two miles directly below the break. The fire obeyed no predictions, neither admitting to helicopter infrared assessment (because of 9-kilometer-high swirling gasses) nor moving in ascertainable directions. It defied every kinetic theory of volatile substance, transcending the Maxwell–Boltzmann law of molecular energy and momentum. An entire territory of incandescent combustion went up from years of entangled dry rot, bore down into the delicate subterranean strata of micronutrients, razed all matter to ground zero, and for nearly a month raged in complete freedom. By the end, a poisoned, smoldering wasteland greeted the dazed inspectors. Miraculously, the Tassahara Zen Monastery – deep in the fire's path – remained uncharred.

An estimated 1 million animals were killed, as well as the entire living species of Juniper Sera evergreen. For weeks after the fire, the aquamarine waters of the Pacific were obfuscated by a rain of zinc-hued particulate matter. The surviving mountain lions, bears, red fox, and coyote were skittish in the aftermath. I recall the lonely wail of one lioness resounding somewhere up the mountain every night, beneath the copper-glowing moon.

The fire and the battle to entrap it; an aerial inferno, the collapse of entire ecosystems – fascinating, repulsive, lodged in some substratum of cognition where lustre and loathing converge. Within half-a-year, following extensive seeding, signs of green broke through the jet black remains, only to be gutted by unchecked floods. But with so much ash, the soil would eventually be replenished. Not long after Mount Saint Helen's eruption, young shoots of evergreen and a variety of alpine flowers were springing back to life beneath the very caldera.

In Siberia, similar conflagrations have burned for months across taiga. In 1825, a forest fire in Maine and New Brunswick burned more than 1 million hectares. The fire storm of Dresden, resulting from U.S. air strikes at the end of World War II, killed 300,000 people; 100,000 in Hamburg. The history of urban fires is the record of combustion's magic, of psychic numbing and contradictory allure. In a millionth of a second, the heat generated by a nuclear explosion exceeds that of the sun's surface. Human beings apparently crave fire.

In the woods, fire selects for new growth, cleanses, rejuvenates, kills. In the human imagination, the role of fire is indeed bewildering; it articulates the central technologies of all culture, arrests some passions, incites others.

The history of chemistry is closely associated with the fundamental experience of fire: Robert Boyle's "New Experiments touching the Relation betwixt Flame and Air," Robert Hooke's discovery of "nitre air," and Georg Stahl's elusive "phlogiston" (the very material of fire, basis of metallic calx), and the later true combustible, "inflammable air" of Henry Cavendish. In Sweden, Karl Scheele noted the diminished volume of air in which sulfur and phosphorus were burned, this leading to his theory of "fire air," a pure substance present in all matter, ready to be ignited. But it was Antoine Lavoisier (1743–1794), specialist in salts, gunpowder, gypsum, thunder; an economist, farmer, administrator, adjoint *chimiste* to the Academy of Science in Paris, who, in 1777, renamed it dephlogisticated air. Lavoisier called it *oxygen,* meaning acid producer, and based his work on the calcination of metals. It was oxygen, Lavoisier argued, that – when combined with a burning substance – made for combustion. But what first produced the burning entity? With his knowledge of oxygen, he and Pierre Laplace went on to analyze water, demonstrating the possibilities of quantitative organic analysis. For his efforts (which included establishing the metric system) Lavoisier was arrested with other former government administrators, tried in less than a day by a tribunal, and beheaded.

Matches, tipped with sulfur and phosphorus, came into being during the 1830s at a time shortly before the German missionary Johann Krapf (1810–1882) reached Mombasa, composed the first dictionary of Swahili, and reported the existance of a tribe that had no knowledge of fire, its members living as monkeys in a bamboo forest in southernmost Abyssinia.

ORIGINS AND ARCHETYPES

In 1826 a cliff at Dorset, England spontaneously ignited and burned continuously for 4 years. Such damp outcrops – veined with coal or oil shale – may contain nodules of pyrite which, when oxidized, quickly become combustible. Used as a hammerstone against flint, such nodules can – in a man's hand – produce incendiary sparks. Among Eskimos in Greenland the word for fire (*ik-nek*) refers to the striking of stones.[1] Spontaneous combustion of middens, the friction of branches, volcanic effluent, lightning, all of these happenings may have been the original sources of fire. We are predatory feeders, and thus warm food is natural – or so many argue. The eminent nineteenth-century British mythologer-anthropologist, Sir James George Frazer, in his compilation of fire myths, believed that nothing could be worse than having to heat food by the warmth of the sun, or in one's armpits; and that such disinclination spurred our ancestors to do something about it.[2]

How early were such ancestors? As late as 1956, knowledge of fire usage in Africa was assumed to be no older than 50,000 years, according to free carbon deposits discovered in South African Makapan breccia atop late Acheulian hand axes. In 1981 that period had been extended 1.5 million years. Our vision of human history must increasingly humble today's supposed ingenuity. For the odds against acquiring a healthy, controlled blaze in the wild (as anyone stripped bare, rendered useless, cold and hungry there would sure attest) are formidable.

The champions, priests, deities, and saviours of fire make up a Hall of Flame larded with psychoanalytics. Animals invariably play a crucial role in the mythic acquisition of fire. No beastiary at work, but everyday, domestic birds, fish, and mammals are often elevated to strategic intercourse with the gods. There is usually an adversary, some being in possession of fire who is hoarding it. Others always seem to find out and then steal it, employing animals. The fire merges with the sun, with lightness and goodness.

The variety of antics are worth relating, as Frazer has done most comprehensively. Hawks and fire-tail wrens brought fire to the Grampian Mountains of southwest Victoria, Australia. Other bringers of fire included a fish eagle of the Admiralty Islands; grouse, sage hens, hummingbirds, and owls in the Rocky Mountain Uinta Ute country; a kingfisher on the Andaman Islands. Cuttlefish figure in the Vancouver Island Nootka tribal myths. Their neighbors, the Tlatlasikoala, instead look to the deer for having originally stolen fire for them from their god Natlibikaq. The Mexican Cora Indians celebrate the opossum for having climbed the texcallame tree, where it obtained fire from the iguana's wife and mother-in-law.

Among the Bergdama of Southwest Africa, fire exists in every tree. The male is the borer of hardwood, while the female is the soft, flat board. Together, fire can be drilled. Among the Basongo Meno, fire came from the accidental boring of a hole in the end of a rib or raphia palm from which traditional fishing traps were fashioned. A Lengua Indian of the Paraguayan Chaco first discovered fire by watching a bird in its nest happily feeding on cooked snails, its supply of burning twigs nearby. The aforementioned Utes rubbed hard greasewood in a hole of soft sagebrush. Similar sexual rubbing was known and practiced by the Paom Pomo Indians of California. An old woman at Wagawaga, on Milne Bay in New Guinea is said to have introduced fire from her body to cook taro and yams. Another woman of the Lukwasisiga tribe on the Trobriand Islands gave birth to the sun. In gratitude for the moon, the sun gave the gift of its fire to the woman. A Guiana myth refers to an old woman who kept the first fire in her vagina. Other myths propose ants, mason wasps, coyotes, ravens, and bronze-winged doves as the first possessors of fire.[3]

The religious origins of fire take us a step closer to contradiction. Zoroastrian doctrine pits light against darkness, good against evil, Ahura Mazda against Angra Mainyush. According to the theologian Mani (216–77 B.C.), light will eventually win over. Thirty years later, it actually did, during the emperor Constantine's crossing of the Alps.

Constantine's panegyrists describe his vision of a flaming cross, seen earlier in a dream. With the routing of Maxentius' forces at Saxa Rubra, the Unconquered Sun — a Roman pagan deity, Sol — was incorporated in the Christian Church. The Hittites worshipped their own sun goddess in the city of Arinna. The Japanese look to another sun goddess, Amaterasu. The Ainus of Japan worshipped a male sun, as did the Mexicans, in the image of the fire god Xiuheuctli. In Egypt, Osiris (Ousi), son of Beb, was the god of Earth but engendered the worship of the sun and fire. He abolished cannibalism, instructed in agriculture, and produced bread and the vine. He married Isis, his younger sister, fell victim to his brother Set's jealousy, and at age twenty-eight was killed. Isis resurrected him, and Osiris came to be identified with the Nile's annual floods, with sunset and sunrise. He went to paradise, and all Egyptians were to follow him. It was Osiris who came to sort out the perpetual struggle between darkness and light, fertility and the unfruitful, in the youthful guise of tall, dark, slender hieroglyphic. He sang and was of gentle disposition. Patient, optimistic, capable of rousing musical solos, he was the essence of civilizing energy. His consort lived on as the fertile Egyptian plain, beneath the Delta, annually flooded by her mate; constantly embattled with her brother-in-law, white-skinned, red-haired Set, the ugly one, the warrior god Sutekh, to be revered by the barbarian Hyksos at their capital, Avaris. Rameses II would become known as the Beloved of Set, an aggressor. But by the Twenty-Second Dynasty, Set was fully equated with a vicious beast, and Osiris assumed the sacred animal heads of falcons and jackals, crocodiles and rams, the bulls of Apis. Archaeologists sifting through the rich loam still find crocodile cemetaries along the Nile. In warring Assyria, Sutekh is Gibil, fire god, vanquisher of enemies. In Greece, Osiris is Apollo, god of the mellow light.

Aristarchus of Samos (fl. ca. 280 B.C.) placed the Sun at the center of the cosmos, though this opinion was reversed a century later by Hipparchus of Nicaea. The mythical demigod Phaethon, not one to argue details of science, usurped the chariot of his divine father the Sun. Helios' steeds kicked over the chariot traces when they realized that a mortal's hands were at the reins. Only Zeus could then save the world from cinders. The Greeks sought explanations for the transmigration of souls in those two elements that carried on the most fervent dialogue throughout her history — the sea and the sun. "There is exchange of all things for fire and of fire for all things, as there is of wares for gold and of gold for wares," wrote Heraclitus (ca. 540–475 B.C.).[4] He saw the sea as turning into fire from an indigenous lightning flash, occasioned by fiery whirls over the water, what Seneca later named *igneus turbo*. From his *Cosmic Fragments* it becomes clear that Heraclitus — and, later, Aristotle who quotes the pre-Socratic in the Third Book of his *Physics* — related lightning and thunder, vapor and smoke, smoke and fire. The soul came to represent the dualism between these mutable forces — water bound to mist as vapor to smoke. "The soul originates out of fire."[5] Fire was craving, the soul aspiring upward, in the condensing vapors of the choppy Mediterranean.

These likenesses have been hailed an *empathetic* trend in philosophy. To most, this is downright confusion, looseness of simile. Abu Yazid al-Bestami, founder of Sufism, would adopt a more accessible image when speaking of his soul as "red hot in the flames of arduous endeavor." Richard Rolle was particularly fascinated by such allusions in his early fourteenth-century treatise *Incendium amoris*: "The heart that truly receives the fire of the Holy Ghost is burned all wholly and turns as it were into fire . . . that form that is likest to God."[6] In the sixteenth century, St. John of the Cross would speak of the "Living Flame of Love," the soul purified "like the gold in the crucible," in his *Dark Night of the Soul*.[7] Catherine of Genoa, St. Matthew, and Jeremiah were equally aflame with fire in their bones, heat in their loins. The analogies escalate from mysticism, sweat lodges and warrior wrath, to protest and political sacrifice. The Celtic Cuchulainn was only able to come down off his battle high by being plunged into three vats of cold water. Saints' blood is said to boil within glass reliquaries. Among the Andaman myths perhaps the most interesting is that of the flood: The danger to mankind was not in drowning, but in the near loss of fire.

Amid the glaciers of eastern Nepal, the Sherpa purify their dead on a pyre, around which Buddhist monks circulate, pouring butter over the flames out of respect to the Gautama's own body. The fire ceremony is presided over by one dressed as a Bodhisattva, the Mahayana saint. The fire portrays the illusion of life's permanency. It is a sacrifice and a reminder to those still bound by ego. Buddhism assimilated the practice of fire sacrifice from Vedic mythology, in which fire is said to have been brought from heaven by Matarisvan, the messenger of Vivasvant, the first sacrificer who specialized in fire, the gods' preferred medium. Traditionally, in the West, fire sacrifice was initiated first by King Mesha of Moab, King Ahaz of Judah, and his later successor, Manasseh. Innocent VIII issued a Papal Bull in 1484 sanctioning the burning of heretics; by 1750, 200,00 deaths had resulted. Throughout the history of Buddhism there are only twenty-five known ceremonial immolations. In 1963 Quang Duc, a 73-year old monk, took his own life with fire in protest of President Diem's repressive regime. A follower of Duc performed the same suicide 7 years later. In Czechoslovakia, several self-immolations occurred during the Soviet invasion, again by way of protest.[8] Throughout Vedic literature the terms *krodha* and *hedas* are enumerated with various meanings — heat, wrath, austerity, glow of devotion, the exultation of battle, fire. Those devoted to Rudra, forerunner of the great Siva, fear the ambiguity of such heat. But Rudra has it under control and extends a cooling, healing hand; unlike Agni, whose heat is always fiery. At the Harishchandra and Manikarnika ghats in Banaras (the ancient city of light, Kashi), the smoke of funeral pyres is continuous along the healing waters of the Ganges. In the late eighteenth century, with European contact, *kulinism* among orthodox Brahmans was revived and with it the practice of *sati*. Though outlawed in the 1930s, women are still known to occasionally burn themselves up on their husband's pyre. But Manikarnika is not merely a funeral place. It has always been associated with the ancient Chakrapushkarini Kund, meaning the Discus-lotus Pond in which the soul finds final peace and purification. Harking back to Heraclitus, fire and water mix to liberate the soul, as heat is liberated from the combustion.

In America, cremation has steadily grown in popularity. Today, approximately 13 percent of the population goes in for it, with higher statistics in metropolitan areas. During Passion Week throughout Italy, fires are gradually extinguished in each Church to make ready for the *new* fire. But this fire does not extend to the body. In Italy, cremation is illegal, despite the exhortations of its medieval saints.

THE ECOLOGY OF FIRE

Before teakettles, stewpots and water jugs, there were clambakes, dry and moist heat roasts, and cave broils. On a Paleolithic summer's afternoon, beach denizens on Cape Cod, Yucatan, or Greece might spread kelp over coals sunk into a pit on the sand. Atop the kelp might go layers of shellfish, leaves, and more kelp. Indians in Baja popped various seeds in hot sand and pit-steamed agave flower-stalk buds, along with tubers they roasted in ash.[9] American Plains Indians cooked venison over hot rocks and coal. To start their fires they used a fire stick in which a drill was spun about a hearth to produce glowing coal powder, which was placed into a tinder of cedar bark, buffalo chips, blue jay down, or birch fungi. Alternately, rawhide bow and drill were employed. Once the fire was stoked, it was transported in an airtight buffalo horn covered with punk and held by buckskin loops.[10]

Twenty thousand years ago, various animal fats and oils were used for lighting. Wicks came from moss, linseed oil from flax for use with hemp. In Britain, rush lights (*Juncus effusus*) worked fine; in Iceland, cotton-grass *(Eriophorum)*; in China, tallow-tree *(Sapium sebiferum)*; in ancient Egypt, castor oil. Torches were made of resinous wood faggots, rolled birch bark, fashioned vine tendrils, or papyrus. Oak or hazel would have made for excellent charcoal burning; tinder could easily have been gathered from the dried bracket fungus *(Fomes fomentarios)*. With the application of lichens, oils, fats, grass, and seeds to flame, some other transformational mechanism was engendered.[11]

Thirty-three California Indian tribes used fire to enhance the yield of seeds.[12] The meadows of Yosemite Valley attest to burning and suppression practices of the southern Miwok. In the Pennroyal, the first prairies arose as a result of such precise catalysts. With regrowth, the plants evolved higher concentrations of protein, calcium, phosphor, and potash. Throughout Mesopotamia this equation of fertility had momentous results on demographics and the domestication of plants and animals.

As early as 20,000 B.C. at the Shanidar Caves and at Zawi Chemi, the village 3 kilometers adjacent, is evidence of two virtually simultaneous occurrences: the domestication of sheep and of niche alteration. Repeated fires, over many millennia, accounted for the presence of cerealia, semidomesticated grasses, whose ash and pollen remains suggest a fire climax at Shanidar amid oak–pistachio woodland savanna. For tens of thousands of years, hunters–gatherers traded obsidian and asphalt across the Zagros mountains in western Iran. In the process, various hard-grained grasses (and the ungulates able to convert the cellulose into protein suitable to human consumption) were introduced to varying altitudes, where natural selection fostered radical shifts in genotypes. Marginal habitats absorbed spin-offs, creating reciprocal forces in the process of selection, and all shielded by the enhancing effects of transplant. During this botanical revolution, repeated mutations formulated hardy strains of wheat, barley, and oat; the frequency of such mutations occurred in concert with increased grazing of animals and ever-refined use of fire across the Assyrian steppe.[13] The rate of herbaceous plants coming into production could be calculated from season to season, thus spawning a veritable "folk science" of fire. A second fire-climax site, Proto-Neolithic Mureybit on the Euphrates in northern Syria, evidences the cultivation of Triticum boeoticum (wild wheat) as early as 8400 B.C. The sheep and goats feeding on these grains all year long at any altitude, would have affected the supply side of agriculture in two ways: typical predator–prey oscillations, given the differing speeds with which plants came back, as opposed to the ungulates; and a basic symbiosis between humans, other animals, and plants.[14] Fire would have exploited these associations so that successions of seedlings and harvest could be fueled without any actual farming. But as it turns out, such fire control may be the first swidden agriculture.

From 1967 through 1968, Richard Gould studied thirty-four Western Desert Aborigines 320 kilometers northwest of the Warburton Mission, and his research revealed the basis for this preagricultural fire harvesting.[15] Three types of fire making are prevalent in Australia among Aborigines: spear-thrower friction (rubbing a stick across another piece of wood), stone percussion, and drilling. Malcolm Calley observed stone percussion techniques among old people at Woodenbong, Tabulam, and Cabbage Tree Island in 1957. The technique consisted of two river pebbles, about 15 centimeters in diameter (flat, circular, white grey or yellow; known as *jalgun*, meaning sun, or weapon's raw material) – being hit hard in repeated bursts until they spark. The process is known as *dindabarng* and is purported to be the oldest technique in Australia for making fire.[16] Gould's subjects used the spear-thrower method. This involves splitting dry mulga wood *(Acacia aneura)* in the middle, laying it flat on the ground with a wedge anchoring the ⅓-centimeter splay into which bits of dried kangaroo dung have been chipped. One man stands on either side of the board, holding it down while a second person rapidly draws the thrower. In 15 or 20 seconds the dung is smoldering beneath dry grass or wet mulga. Every half hour at night, someone is expected to rekindle the fire. Women slowly burn down trees from the base; they burn out waterholes, the group constantly on the move to avoid ruining an area. The night fires on occasion serve as ceremony – dances and penis bloodlettings. In the daytime, the wild-turkey, dingo, kangaroo, and carpet-snake rituals accompany fire as well. The fire is used to drive wallabies and kangaroos over cliffs, to capture small rodents and cats. Smoke telegraphy between foragers is common. But the most important use of fire for the Aborigine appears to be razing spinifex ground cover prior to the summer rains (November–March). This practice insures the residual font of new grass shoots and accompanying macropods.

Gould interviewed many of the foraging women. They were completely cognizant of plant succession; of the fact that unless the spinifex was burned over upon complete foraging, such areas would never again see the profusion of succulents upon which their communities depend. Miles of spinifex are left to burn themselves out every fall. This is their science of farming, a most effective one.[17] In Brazil, burning over land has been shown, in some cases (depending on the precise soil type), to increase by 30 percent the produce, as opposed to unburned clearance. Fire can also rid the soil of deleterious bacteria, fungi, and nematodes, while fostering greater anaerobic nitrogen-fixing bacterial action and diminishing the acidity of the soil's pH.

We take for granted today our matches, natural gas jets, microwave ovens, electric stoves, prime movers. But throughout the world, other techniques of obtaining combustion are still in use. Fire-drilling is currently practiced among the Wedda tribe of Sri Lanka, the Tsarisen and Paiwan in Taiwan, among the remaining Ainu, Lamut, and Kamchadal in northeastern Siberia, and from the Nilgiri hills of India to the Caroline Islands. The !Kung at Lake Ngami use short fire sticks, some of them carrying the stick on a necklace. Fire saws have been turned up in Madagascar Antanala communities. In New Guinea, Bali, and Nagaland, thong sawing (rattan or raphia palm utilized like bow and drill) is still in vogue. In parts of the Cameroons and Philippines and ingenious fire syringe (cylindrical iron pistons, rocks, or hard wood nuggets forced against opposing surfaces) is put to use. Stone percussion techniques survive in the Aleutians and in Tierra del Fuego. In Borneo, pieces of Chinese porcelain are struck against bamboo. In parts of Central Asia fire steels are now and then discovered in use.[18]

HOMO SAPIENS WITHOUT FIRE: THE WILD MAN

In *Yvain*, Chrétien de Troyes cued up the Medieval personification of wilderness in the image of the wild man, of a beast yearning to be human, like a struggling embryo.[19] The term *wilderness* comes from the Anglo-Saxon and means *a place of wild beasts*. But no creature of the forest, save perhaps for the fire-breathing dragon of the Swiss Alps, exerted a greater fascination than wild men, women, and children: *"noir et velu com ours enchaines"* (black and hairy like a chained bear). Said to inhabit the highlands of Austria, Germany, and Switzerland, the creatures were portrayed as godless, giving free vent to a hypersexuality, innate irrationality, and occasional cooperativeness among themselves; they were even given at times to dancing, playing music, and farming. Typically, the wild man is aggressive, destroys the crops of farmers, is debased, melancholic, and naked. He crawls on all fours through brambly backbush, catching his food with his hands and devouring it raw, all in absence of fire. He is, in that respect, more primitive than the dragon. His woman is always occupied with suckling.[20]

Associated with that self-abuse that looks back to a pure, if pitiable, state of nature, the wild man began as madman in the literature of King Nebuchadnezzar, the sixth-century-B.C. Babylonian who went insane when cast into the wilderness. When Merlin went out to meditate in the woods, he too lost his wits. Categories of *ecological* madness personified ravenous desert satyrs (sacchani), demented gypsies *(hispani)*, wild devils *(pygmies)*, and demons *(diaboli)*. This lineage also accounted for the three wild saints – Mary Magdelene, Onuphrius, and John Chrysostom. Cast in accordance with religious awakening, unperfidious countenance and rustic ordination, the wild man became the Rousseauist prototype, and this upswing symbolizes a remarkable turnabout in feeling. In the manuscript illuminations to the *Book of Hours* of the Jean de Montluçon shop at Bourges (ca. 1500), 300 wild men are portrayed in 133 scenes. Not only has a shift in thinking occurred, but wilderness has seen its ultimate domestication. There is a unique, colorful glimpse of a wild man happily preparing a plump ox over a roaring fire. He is no longer wild. It was only 10 years earlier that Bosch had been the first to paint fire, making it all the more curious that a wild man should so quickly be endowed with it.

THE GAZELLE BOY

In the twentieth century, more than seventy cases of wild children have been reported, children usually said to have been reared by animals. Both the French Victor and the Indian Wolf Girls were of this breed. But there has never been so compelling a wild-child tale as that of the Gazelle Boy.

In the heart of the Rio de Oro, of the central deserts of Mauritania on the western fringes of the Spanish Sahara, the Swiss explorer Jean-Claude Armen discovered a gazelle boy living with a herd of white Dorcas gazelles in 1960.[21] In a swarm of surreality, heat, dust, and the skittish tumble of approaching hooves, Armen first caught sight of jet-black hair in the reddish glow of a rocky guelb at dawn: the boy – slender, darting through thorn bush, throwing himself at unearthed roots, cutting them to shreds with his expert incisors, quivering his nostrils, and fled with the herd on catching scent of Armen. After 9 days of successive gesturings, advances, and discretions, Armen made contact with the child. Exploratory sniffing enused, the proferring of gazelle codes (clicks and licks) until the boy sought Armen's body with his tongue. Over many weeks Armen watched the child climb cliffs in search of temen (jujube), steal catkins from sandwasps, roll in smectite clay to cleanse off grease, sniff hindquarters of his family members, drink the ooze of acacia gum, employ delousing shells, forage for wild berries on all fours, get into kissing rituals, take great bounds, sleep in irregular bursts, suck from his wet-nurse gazelle, and one day notice his own reflection in water without any apparent sense of alarm. The boy was unaware that he was human. At night, nearby, Armen built a fire. The child tried to catch it, entranced by the embers. It was apparent to Armen that the gazelle boy had never before encountered fire. Nor did it sustain his interest for more than a few moments. Having probed it sufficiently, finding it useless, the gazelle boy returned to the mounds of dried excreta where he slept with the herd.

In attempting to study gazelle language, Armen learned that the animals had nothing to signify a distance greater than 12 kilometers, the horizon, their infinity. The wind changed mysteriously one day, and the herd sensed something. They had been foraging over the seemingly barren sebkha plains for drinn and farinaceous zila, but were suddenly perked, jolted into dispersal, deploying at right angles and disappearing into the far-off empty quarters. Armen despaired, made his way out of the desert, only to return two years later. He succeeded again in relocating the herd. The boy was still in fine health, every bit a gazelle. And he seemed to remember Armen. Armen's military escourts got overzealous, invaded the explorer's reunion by jeep, terrifying the herd whose escape was clocked at 55 kilometers (35 mph), the boy in the lead! It was the last time that Armen ever saw his beloved wild child.

Later, military officers from NATO stationed at an American base at the Villa Cisneros learned of the boy and twice attempted to capture him using helicopters and suspended nets. They failed. The boy outran them. To this day the gazelle boy, now probably thirty years old, may still be out there.

THE YETI

The Mongolians term him *al-mass*, the Kazaks *ksy-gyik*, the Chinese *Yeh ren*, the Tibetans *mu rgod, lung-gum, gangs-mi*, or *mi bom po*: the Nepalese Sherpa, *Yeti*. He is a naked glacier man, a saintly madman, lord of the mountain, without fire. This abominable snowman has drifted from the yak-eating beast to an old rascal grandpa lecher. But he is wild indeed. Some legends claim he is strictly vegetarian. Chinese scientists from Wu Han University are searching for him in Hubei Province along the 1930-kilometer Shen Nong escarpment where *Yen ren* footprints have been found leading into caves more than 900 meters deep.

THE ARUNACHALESE

One of the most isolated of all regions in the Himalayas is Arunachal Pradesh, particularly its northern sectors. The northeastern State of India, surrounded by Burma, China, and Bhutan, the region consists of 150,000 square kilometers ranging in altitudes from 1500 to more than 7900 meters. At least half of the state is over 3000 meters high. There are twenty major tribes, seventy subtribes – a population of 500,000 people living in 2500 villages. Overall, the population density is two per square kilometer. With no written records, we lack precise data on the aboriginal and migratory history of these myriad groups. Most are Indo-Mongoloid, spinners, weavers, fanatical dancers and singers, husbandmen – at least in the lower elevations, where they have had steady contact with the outside over the last century. Though by outside I refer to a smattering of British and Indian Government officials. For in fact, authorities in New Delhi have successfully endeavored to protect their shakey control over Arunachal by stifling any hint of regionalism. Throughout the northeast of India, local tribes have frequently sought autonomy. To squelch these aspirations, the Indian government has a hard-and-fast policy of keeping out scientists, tourists, or anyone who might stir up renewed self-awareness among the indigenes. As a result of these travel restrictions, Arunachal Pradesh has remained the stronghold of extremely preindustrial aboriginal populations.*

In 1982, the *United News of India* received over teletype the scant details of a chance interview between one of their roving correspondents and an Indo-Tibetan military mountaineer, Captain H. C. Chauhan. Chauhan and his three compatriots had come down into the Northwestern Himalayan village of Josimath for supplies. They had been making the first trans-Himalayan traverse, exploring new territories of India (claimed by the Chinese) over a period of 13 months. They were now preparing for their last leg up into Ladakh, having come all the way from Arunachal, a year before.

Chauhan mentioned an event that had occurred in Arunachal: The team had been held up in a severe blizzard atop the remote, rarely visited Chetak Pass, not far from the Bhutanese border. The pass was more than 4500 meters high and had supposedly *never* been crossed during winter. After 3 days of effort, the mountaineers reached the top and started descending to the other side when they saw something: "nude men and women, long-haired, Mongoloid, who started running through the snow." The team gave chase, made contact, conveying no harm (in Chauhan's estimation). The four men followed them to their cave habitat and discovered bits of raw meat, flabs of uncooked flesh. The tribe lived in the icy cave, ate their meat raw, knew nothing of fire.

Various Yeti sightings are legion throughout the Himalayas and have entered into the mainstream of folklore and belief. But there was never such a report as Chauhan's; nor one offered by veteran Government people, the Tibetan mountaineers. Not unlike UFO sightings, the Chetak Pass incident may be difficult to ultimately verify. The U.S. Naval officer Charles Wilkes led the first exploratory survey, by authorization of Congress in 1836, to the South Seas. On Bowditch Island (Fakaafo) Wilkes reported in his nineteen-volume summary that the inhabitants did

*In India's Andaman Islands, three especially isolated groups do not know how to make fire, only how to keep it going. The Onge and Jarawas use slow-burning resinous wood to maintain their flame, a flame passed down from hundreds of generations. The Sentinelese wander their island's beaches in the nude, with a marvelous immunity: They have never been successfully contacted by the outside. Indian authorities have only managed to view them from a distant boat with the aid of binoculars. The first and last time that contact was assayed, the Sentinelese managed to discourage the intruders with poison-tipped arrows.

not know of fire. Accounting for nudity in the snow and the eating of raw meat pose their own special problems of adaptation. But being without fire is entirely baffling.

A month later I went to Calcutta and New Delhi, discussed the discovery with leading anthropological and military dons, conferred with ministers of state, with Arunachalese themselves, as well as Mrs. Gandhi, herself a graduate in anthropology. The military were hailing the top secret event as "stunning."

As of this writing (1984), we don't know where the tribe is, if they are nomadic, how many of them there are, their language, race, or true manner of survival. But the consensus of scientific opinion is encouraging among those familiar with the unique topography of upper Arunachal. The myth of such a tribe fits neatly with the myth of the very landscape.

The oldest known continuous habitation in Arunachal dates to about 2000 years. Remains from other parts of the Himalayas, most notably Burzahom in Kashmir, attest to older, Megalithic dwelling sites. In 1981, an upper molar of Ramapithecus punjabicus was discovered in a cave of south-central Nepal, 11 million years old. Part of the accelerated decay rate in the eastern Himalayas may be the result of the intense annual rain fall (nearly 15 meters!) and massive landslides. But the main explanation is the lack of accessibility and research. Throughout Tibet, Bhutan, Sikkim, northern Burma, and major portions of Yunnan, as well as the upper Salween, Irawaddy, and Mekong Rivers, there has been virtually no anthropological or paleontological inquiries. Arunachal has seen less than a few dozen Westerners in the past fifty years. Authorities in New Delhi speak with some pride of their government as being the most bureaucratic of any in the world. And it may be true. But it is equally true that the government takes little or no pride in its aborigines. Survival International, based in London, has had no maneuverability in India. From my many conversations with Indian officials, it became clear that they were not pleased to know of naked tribesmen running around their mountains.*

Northern Arunachal has long been held by Tibetans to be their sacred homeland, the true Shangri-la, adjoining Pemakod in China, across the disputed border, where Tibetans have long gone during times of trouble. Set beneath perpetual rain clouds, the immense, muddy Tibetan river Tsangpo (Brahmaputra or Dihang) twists feverishly around the highest unclimbed mountain in the world Namche Barwa. Picking up speed from then on, the mighty river plunges off the 4-kilometer-high Tibetan

*The enigmatic Sulung of Kameng, a western part of Arunachal, are known to go naked and live in trees. Hunter–gatherers of the Stone Age, they are nonetheless aware of fire.

plateau, slams through a narrow granite gorge 80 kilometers long and then falls even faster into the labyrinthine high country of Arunachal. Namche Barwa was explored for the first time by Chinese scientists in 1983. But the mountain has been kept out of reach to all outsiders.

A fabled waterfall somewhere in the river's course has been the romance of geographers for over a century. Early British explorers sought it out. These intrepid wanderers would think nothing of disappearing in dense uncharted thicket for a year at a time, re-emerging with a Tibetan blue poppy in their teeth and something of the Yeti's look in their eyes. But, happily for us, they never managed to fully reconnoitre the Tsangpo Gorges or find the elusive waterfall.

In 1911 the famous Tibetan prophet Rinpoche wrote a guide book to the area, for use by Tibetan pilgrims. He actually predicted the coming invasion of Tibet by China and stressed the need for Tibetans to retreat to their true homeland, Pemakod, where no Chinese could ever follow. This Tibetan *Pilgrim's Progress* was adhered to by a noteworthy family in 1954. The former Canadian Ambassador to Nepal, James George, interviewed a member of that family. The young man described how they had climbed into the deepest part of the gorge and there discovered a "magnificent waterfall" behind which lay a deep cave. The young man's father entered the cave and remained inside for three weeks, though upon reissuing perceived that he had been gone only part of a day. Inside the waterfall, the father had found true Shangri-la.

The mythological landscape is also a Pleistocene island in the sky, cut off from the rest of the world since the time of the eastern Himalayan syncline uplift and convolution. A nomadic, high-altitude breeding population might just live on in perfect ecological harmony without the necessity of fire. Were we able to find this group, a number of useful questions might be addressed. One of the more intriguing problems relates to their place in evolution. Benveniste and Todaro's studies, as we have seen (Ch. 1), indicate that the gorilla and chimpanzee evolved in Africa, whereas baboons, orangutans, and man originated in Asia, very possibly in northern Burma and surrounding areas. The basis for such claims lies with the careful profile of genetic data. The scientists wrote, "Our data suggests that man's direct ancestors probably spent most if not all of the Pliocene in Asia. The older Australopithecines found in Africa, though clearly hominids, were probably, therefore, not in the main lineage to man, but rather were unsuccessful offshoots whose progeny have not endured to the present."[22] As for other Himalayan paleontologists, David Pilbeam of Harvard conducted a decade of digging in the Siwalik Hills of Pakistan. Only in 1982 did it become clear that *Sivapithecus*, the fossilized local resident, was more akin to modern orangutans than to hominids, thus forging a "14 million year gap in knowledge of man's lineage."[23] Richard Leakey then told a private audience in London that a recent "reappraisal of fossil evidence" suggests to him that the origins of mankind are perhaps as much as 12 million years later than he always believed.[24] The C-Virogene theory, as it has been hailed, distinguishes between nucleic acid sequences in the cellular DNA of baboons, other monkeys, and man. Each sequence reveals itself to have been imprinted by identifiable environmental conditions. Living gibbons, siamangs, and orangs possess a gene sequence unchanged in millions of years that is most directly related to human beings. Such connections clear the ground. Is it possible that an Arunachalese group of mountain nomads, ignorant of fire and thus paleontologically more preindustrial than the 750,000-year-old Choukoutien (Asian) *Homo erectus*, might be a living ancestor of mankind?

FIRE, THE EYE, AND THE IMAGINATION

In his *Life of Apollonius of Tyana*, Philostratus (170–245) described the legendary statue of Memnon: "when touched by the first ray of the sun it uttered a sound as soon as the sunbeam reached its lips, and it seemed to raise its gleaming eyes joyfully towards the light, like a basking man. Therefore Apollonius offered sacrifices to the Ethiopian Sun, and to Memnon of the Dawn."[25] Powerful light affects the eye. This maxim, girded by physics and physiology, propelled the pre-Raphaelites and Impressionists into luminous zones of color meant to re-establish the primacy of light's reflection on the imagination. What the light may overpower, the imagination might revive with subtlety. Thus neither Homer nor Milton provides a single detail, luscious, lurid, or otherwise, regarding their Helen and Eve. But the human eye does all of the work, regardless of whether such nymphs are illuminated or not. Frenzied activity from cells, collectively called the *raphe nucleus*, signals the beginning of dreaming and image making when eyes are closed. The cells send out fibers to the visual cortex. In the secure fastness of their fetal darkness, prenatal rapid-eye movements occur, indicating a dream state. The kinetics of light response can be broken down into major categories: an equation of hyperpolarization, light intensity, flash duration, sensitivity constant, rate constant, time after the flash, and number of reaction stages to it.[26] All this determines, for example, the response in a red-sensitive area (the cone) of a turtle's eye to light, of our reaction to a meteorite shower.

Four kilograms of human brain contain about 100 billion nerve cells, or neurons, possibly with 10^{14} (100 trillion) synapses going on. The neurons are charged: 70 millivolts negative with respect to the outside. Sodium and potassium ions travel back and forth across the cell membranes, inhibiting, exciting, firing the synapses, and in concert with at least thirty chemical transmitter substances, terminating on gland or muscle cells with an exact piece of information. When a pupil responds to light shining on its retina and begins to shrink and constrict, that reflex is occurring by way of some five synapses. Over time, the synapses select from their reflexes. Memory is just such an example. Within the optic nerve are nearly 1 million neurons that connect the retina directly to the brain. Within each neuron are, again, 1 million sodium pumps busily conducting more than 200 million sodium ions per second across a membrane. Writes Charles Stevens, waves of voltage sweep the axons "much as a flame travels along the fuse of a firecracker."[27] The basic structure of the fuse and flame is no different in man than in squid, snail, or leech. Only the firecracker itself differs. The human reaction to stimulation is extraordinarily difficult to predict, because initially all such stimulation to the neocortex rapidly proliferates, going to the limbic system, the striatum, superior colliculus, and pons – regions of the brain where neurons are often so densely packed that it is impossible to formulate fate maps for them (the geography of axons); cellular distribution during early development. Mapped onto a mere 15 square centimeters of the primary visual cortex, the human visual world involves deceptively few cells (four retinal cells) and synapses (two to five). But numerous fibers pass to the lateral geniculate nuclei in the transmission of synapses. Messages arrive from rods and cones to ganglion cells, down axons to the primary visual cortex and other cortical areas. Many kinds of mathematical theorems have been put forth to explain how so much precise information can be condensed in the optic nerve prior to visual cortex expansion. But, writes Francis Crick, "we do not yet have any description of conscious perception that illuminates our very direct experience of it."[28]

The "eye of the imagination" *(oculis imaginationis)* according to the Renaissance physician and mystic Robert Fludd (1574–1637) is found in the frontal brain cavity. Five residual images perpetually emanate from it: an obelisk, the Tower of Babel, Tobias and the Angel, a ship, and the vision of the Last Judgement.[29] Fludd believed that all creation involved the separation of active light from passive darkness. Through such acts, all creatures – man, vegetable, mineral – may achieve immortality and truth. For Soren Kierkegaard (1813–1855), the imagination takes man to the edge of reality, "to get (him) far enough out, or in, or down in existence."[30] From Plotinus to Spinoza, the imagination – by dint of images, icons, the Greek *phantasia* – was associated with lust, became a religious issue of wet dreams and fractious physiologists. It kept pagan and Christian at hot-and-heavy loggerheads for 1400 years. It was the Renaissance Hermetic wisemen Giordano Bruno (1548–1600) who re-established the credentials of the imagination, believing such faculty best suited to discerning truth. But alas, he was burned to death by edict of the Church. Fire and imagination can be equally as dangerous.

The Egyptian book *Am Tuat* symbolizes the "secret wisdom" Bruno was seeking: Shen is the hieroglyph relating the *buto* (midpoint) of the Sun God Ra's journey into the underworld of Nut's body of night. Buto also refers to the Cobra Goddess Nedjoyet meaning "the third eye" and containing light in perfect balance. When such a vision was obtained, that person was said to have become immortal, to be possessed of a "fire-splitting eye." Osirus attained the highest immortality by becoming the God of the Underworld, spitting fire across Isis, bringing the Nile Valley into bloom from the depths of night.[31] A Gurage's tale from northern Ethiopia indicates the power of fire in darkness. The young servant Arha is offered his freedom by the master Hasei if only he can persevere without fire throughout a cold windy night atop a mountain. The servant pulls off the feat by having a sage confidant build a fire on an adjoining mountain. All through the night, standing on Mount Sululta, Arha keeps his eyes glued to the meager light miles across the blackened sky and this vision in turn keeps him warm.[32]

Curt Richter's work on the 24-hour clock and fire light (see Ch. 2) localized the biorhythms of a rat in the suprachiasmatic nuclei, in contact with the optic chiasma. Thickness of optic nerves, number of rods and cones, changes in neurology from light-active to dark-active firing, served over millions of years to condition and control the clockwork. The squirrel monkey's 12-hour fixed phase restricts the animal to a life in the tropics, where the light accommodates it. A move to any other locale, and it would risk oversleeping, thus becoming prey to animals with different sleep cycles.

In a study of Eskimos at Wainwright, Alaska one winter, Joseph Bohlen and his wife determined that electrically charged calcium production in the Eskimo's neurons was ten times that of summer production. Yet it was not enough, in some cases, to prevent the onslaught of what Eskimos call "arctic hysteria," an imbalance resulting from the midnight Sun playing havoc on the biological clock.[33] Migrating fish orient themselves partly by maintaining an angle to the Sun or to its polarized light. It is light, first and foremost, that enacts synchronicity, hormonal circadian rhythms, the 24-hour oscillator in cells. In drosophila, the circadian rhythms have been analyzed, their source of timing identified in a single chromosome. Gorillas cannot keep themselves awake at a certain point; the drosophila hatch their eggs on strict, lighted schedule. Plants open and close according to the light; hens the same with their eggs. But we seem to be free-wheeling, liberated into the fantastic hours, the after-dark moments of tenacious, manufactured light; and all because of fire. Electricity stirred Prometheus in Shelley's version, not fire. But if electricity is modern man's fire, then fire itself—the archetypal desire, the flame, the explosion—has a more prominent place in the psychology of *Homo sapiens*. Not only the body, the culture, the technology of our species, but the art as well. Note the warm hearth in pastoral literature, Samuel Pepys' hyped chronicle of the Great London Fire — coming oddly enough one year after the Great Plague, as if to purify the rotting city in 1666, awkward preface to the Romantic passion for cities "radiant with light and fire" *(Endymion)*. Joseph Mallard Turner had a profound, if scatophiliac, attraction to fire at sea. Blake was attuned to the same impulse when he spoke of a fearful symmetry, a tiger burning bright "in the forests of the night." Blake asks us what immortal hand or eye is behind such flashes of illumination and form. The Bible reproduces in Genesis the first, the most important issuance. Let there be light! Conversely, mankind plays with darkness, as instanced in the hideous blinding of poor Gloucester by the Duke of Cornwall atop a cliff: "Out, vile jelly!" *(King Lear)*. The eye, and with it, the firelight affording man his own reflection, becomes our ego, our surest obsession to the point of blindness. Notes and diaries by Pliny the Elder, Mark Twain, Goethe, Alexandre Dumas, and others all attest to the alluring incandescence of Mount Vesuvius. The giant spewed 1.5 cubic miles of debris in 97 A.D. and has continued to do so intermittently. Yet indefatigable locals persist in living near the volcano, fatalistic, empowered by the hubris that gives mankind to believe in its sure-fire domination over nature. Though in fact the only sure-fire is that of old Vesuvius. "Fire, FIRE. I insist on the absolute word . . . the sun-principle," wrote D. H. Lawrence.[34]

PATHOLOGIES OF FIRE

There are at least 50,000 fires per year in the United States. Every 45 seconds a fire breaks out in a city, either accidentally or by arson. Arson is especially pronounced in New York and Los Angeles, where nearly 16,000 suspicious fires eruped in 1981. Most arrested arsonists are back out on the street in a few days.

The Frenchman J. D. Esquirol in 1833 first coined the term *pyromania*.[35] It means one who regularly "watches" fire and may relish preparing for the fire, accumulating fire paraphernalia, giving in to "intense fascination," being "sexually aroused by fires." That at least is the definition given in the third edition of the *Diagnostic Statistical Manual of Mental Disorders*. More common in males, no predisposing factors are cited. Among schizophrenics, setting a fire may be their only possible response to a delusion or hallucination. Curiously, pyromania was not mentioned in the *second* edition of the *Manual*. Apparently, fire starting is on the rise.

The clinical, anthropological, and theoretical literature of pyromania confirms certain widespread associations between fire, polymorphic sexuality, and violence. Carl Jung's powerful study of fire boring, psychosis, sexual symbolism, and religion is perhaps the best-known analysis, encompassing Aborigine, Greek, German, and Hindu responses.[36]

In 1937, sixty pyromaniac children observed at New York's Bellevue Hospital displayed one or another sexual problem. In her groundbreaking essay, Helen Yarnell suggested that "adolescents set fires when they felt emotionally trapped by either sexual or aggressive impulses and feared being overwhelmed."[37] A broad range of other sexual polymorphism has been reported. In one case study, a 20-year old had the maladaptive habit of setting fires to buildings and squatting toward the beginning of the blaze to masturbate. His treatment consisted of the physician having him masturbate day after day before pictures of naked women, rather than before fire. Eventually, his heart rate increased with the application of nude imagery, and he was finally cured of his problem. In another case, a schizophrenic male who set fires in female restrooms was coaxed from his proclivity by the application of matches and toilet paper. Every time he went to light the toilet paper, little electrodes clamped to his hand shocked the hell out of him. This is known as "aversion therapy." Developmental problems concurrent with puberty, the experience of hallucinating ghosts, skeletons, of parents who hate the child, ego weakness, limited-impulse-control mechanisms, castration and devouring confusions, wanting revenge, wanting to kill — all of these facets of pyromania are ineluctably re-enforced by the very look and heat and smell of fire. Ultimately, Freud believed that "primitive man could not but regard fire as something analogous to the passion of love." In Zambia, the Bemba tribe's *chisungu* rite furthers this particular association. A menstruating woman may not touch fire, nor an adulterer, lion-killer, or generally aggressive person. At the time of the chisungu, a girl is consecrated in a secret hut for her forthcoming marriage. Her own internal fire is carefully stoked by a process called *namushimwa*.[38]

But the seduction of fire is not merely the invitation to reinvent sex. Rather, as we saw in chapter 2, it is the conferring of awesome power, of a transformed landscape, culture, and politic. Cooked food would predigest our foodstuffs. Over time, the facial anatomy, possibly the pharynx, was altered. But these were events without specific conditions or exclusive bias. The technology of power was the important evolutionary advance, bolstered by an equal storehouse of unexplainable emotion.

The invention of gunpowder truly embodied the rage for fire. With it the Mongols would conquer the Sung Dynasty. Edward III would use cannon fire in the Scottish wars of 1327. At Cambrai in 1339 and Crecy in 1346, a newly refined powder mix was employed with fulminant effect. The Moroccans, in possession of the new firearms, crossed the Sahara in 1591 and conquered western Sudan. At the same time,

Cossacks crossed the Urals and carved out their own niche. Ottoman rule in Algeria, Tunisia, and Tripolitania won out as a result of the weapons. The power was again visited upon the Osmanlis, Muscovites, and Timurid conquerors of northern India. The Japanese accepted the weapons and clouded their civilization by them. The Aztecs and Incas did not have the chance to.

By the eighteenth century the rate of firing speed and magnitude of explosiveness had become its own mountain to climb. Better gunpowders, ballistics, and machine metaphors brought the art of directed firepower into the vogue of science. At Silesia, the Austrians were wiped out because Frederick William I of Prussia invested in techniques, which increased his infantry's firing rate by 100 percent.[39] Romantic color, pomp, majesty, military marches symphonically arrayed, rapid-fire hysterics, factories mass-producing artillery — war was helped along by the rubrics of fire fascination and its endless applicability to conflict just as sure as it had revolutionized smelting, forges, bellows, blast furnaces yielding large amounts of iron, and, eventually, the heat theory of mechanical refrigeration and the internal-combustion engine.

The dualistic apotheosis of fire aggressiveness comes to us in a testimony of canvasses, works that include John Trumbull's "The Battle of Bunker Hill" (1786); Benjamin West's "The Death of General Wolfe" (1770), Albrecht Altdorfer's "The Battle of Issus" (ca. 1528), Pieter Breughel's "The Triumph of Death" (ca. 1564), the weirdly fire-lit "Third of May, 1808" by Francisco Goya, and, finally, Pablo Picasso's remarkable "Guernica" (1937). Altdorfer painted an aerial splendor of mountains, silver linings, irradiated, sea, blue fjord, glorious sun, and proud armies in a proud twilight pinioned from an altogether new and vertiginous perspective, the banners blowing below in celebration of Alexander's defeat of Darius. But Picasso seized on a different energy: the psychoanalysis of terror. Beneath a bald and scorching electric light, raving, reductionist mayhem clutches to the very radiance that burns the hand. Bodies cavort frantically, reaching from one fire into another. In a cancer cell's perspective, the hellish, cinematic light condemns the whole world. The light is a spot, given to criminal identification and oddly searching out anguish and a single flower. Here again we have the final contradiction of fire: As it brings life, so it brings death. To relish such paradox would incite the schizoid, or so it would seem. But in fact, recent research shows that even acute schizophrenics, when presented with fire, will think more clearly and run for their life.[40] Given today's workings of fire (HEAT rounds, charged-particle beams, nuclear stockpiles and dramaturgy), where is there to run, but (perhaps mercifully, a fast death) into more fire, as Picasso's "Guernica" intimated?

I never saw such a beautiful light in all my life.
Hiroshima survivor describing the bomb, in "Survivors,"
PBS-WGBH Boston, 1982

NOTES

[1] K. Oakley, "The Earliest Fire-Makers," *Antiquity*, June 1956.

[2] James George Frazer, *Myths of the Origin of Fire* (London, Macmillan, 1930).

[3] Ibid.

[4] G. S. Kirk, *Heraclitus, The Cosmic Fragments* (Cambridge: Cambridge University Press, 1954).

[5] Ibid, Fragment 28.

[6] As cited in David Knipe, *In the Image of Fire: Vedic Experiences of Heat* (New Delhi: Motilal Banarsidass, 1975).

[7] St. John of the Cross (John of Yepes), *The Works of St. John of the Cross*, 4 vols., ed. and trans. by J. Wiseman (London, 1906).

[8] D. O. Topp, "Fire as a Symbol and as a Weapon of Death," *Medicine, Science and the Law*, April 1973.

[9] Carol O. Sauer, *Seed, Spades, Hearths, and Herds: The Domestication of Animals and Foodstuffs*, 2nd ed. (Cambridge: MIT Press, 1969).

[10] Thomas Mails, *Mystic Warriors of the Plains* (Garden City, NJ: Doubleday, 1972).

[11] Geoffrey Dimbleby, *Plants and Archaeology* (London: Grenada, 1967).

[12] Joseph G. Jorgensen, *Western Indians: Comparative Environments, Languages, and Cultures of 172 Western American Indian Tribes* (San Francisco: W. H. Freeman, 1980).

[13] Henry T. Lewis, "The Role of Fire in the Domestication of Plants and Animals in Southwest Asia: A Hypothesis," *Man* 7 (1972): 195–222.

[14] Eugene Odum, *Fundamentals of Ecology*, 2nd ed. (Philadelphia: W. B. Saunders, 1959).

[15] Richard A. Gould, "Uses and Effects of Fire Among the Western Desert Aborigines of Australia," *Mankind* 8 (1971).

[16] Malcolm Calley, "Firemaking by Percussion on the East Coast of Australia," *Mankind* 5 (1957).

[17] Gould, "Uses and Effects of Fire."

[18] Sture Lagercrantz, *African Methods of Fire-Making*, Studies in Ethnography, Vol. 10 (Uppsala: Uppsala University, 1954).

[19] Timothy Husband, *The Wild Man: Medieval Myth and Symbolism* (New York: Metropolitan Museum of Art, 1980).

[20] Ibid.

[21] Jean-Claude Armen, *Gazelle-Boy* (New York: Universe Books, 1974).

[22] Raoul E. Benveniste et al., "Evolution of Type C-viral Genes," *Science* 190 (1975): 886–888.

[23] David Pilbeam, private meeting of the Leakey Society, Bel Air, Calif., 1982.

[24] Richard Leakey, *The Making of Mankind* (New York: E. P. Dutton, 1982).

[25] As Cited in John Cohen, *The Lineaments of Mind* (San Francisco: W. H. Freeman, 1980).

[26] Werner E. Reichert and Tomaso Poggio (eds.), *Theoretical Approaches in Neurobiology* (Cambridge: MIT Press, 1981).

[27] Charles Stevens, "The Neuron," pp. 54-65, *Scientific American*, September 1949, Vol. 241, #3.

[28] Francis Crick, "Thinking About the Brain," pp. 219-232, *Scientific American*, September 1979, Vol. 241, #3.

[29] Robert Fludd, *Utriusque cosmi majoris scilicet et minoris metaphysica, physica atque technica historia* [An account, metaphysical, physical, and technical, of both worlds, greater and lesser (Oppenheim, 1617)]. Fludd's theory is examined in Allen G. Debus, *The Chemical Philosphy: Paracelsian Science and Medicine in the Sixteenth and Seventeenth Centuries* (New York: Science History Publications, 1977).

[30] Cohen, *Lineaments of Mind.*

[31] L. Bellingal, "The Third Eye and the Cults of Osiris," Ph.D. diss., University of California – Santa Cruz (forthcoming).

[32] Harold Courlander and Wolf Leslau, *Fire on the Mountain and Other Ethiopian Stories* (New York: Henry Holt, 1950).

[33] Joseph Bohlen, as cited in Robert Evans Ornstein, *The Psychology of Consciousness* (New York: Viking, 1972).

[34] D. H. Lawrence, *Fantasia of the Unconscience* (New York: Penguin, 1971).

[35] J. D. Esquirol, *Mental Maladies: Treatise on Insanity* (London: 1833).

[36] Carl Jung, *Symbols of Transformation: An Analysis of the Prelude to a Case of Schizophrenia* (Princeton: Princeton University Press, 1952).

[37] Helen Yarnell, "Firesetting in Children," *American Journal of Orthopsychiatry* 10(1940): 272-1286.

[38] Knipe, *Image of Fire.*

[39] Sue Mansfield, *The Gestalts of War: An Inquiry into Its Origins and Meanings as a Social Institution* (New York: Dial Press, 1982).

[40] Gene Usdin (ed.), *Schizophrenia: Biological and Psychological Perspectives* (New York: Brunner/Mazel, 1975).

[32] Harold Courlander and Wolf Leslau, *Fire on the Mountain and Other Ethiopian Stories* (New York: Henry Holt, 1950).

[33] Joseph Bohlen, as cited in Robert Evans Ornstein, *The Psychology of Consciousness* (New York: Viking, 1972).

[34] D. H. Lawrence, *Fantasia of the Unconscience* (New York: Penguin, 1971).

[35] J. D. Esquirol, *Mental Maladies: Treatise on Insanity* (London, 1833).

[36] Carl Jung, *Symbols of Transformation: An Analysis of the Prelude to a Case of Schizophrenia* (Princeton: Princeton University Press, 1952).

[37] Helen Yarnell, "Firesetting in Children," *American Journal of Orthopsychiatry* 10 (1940): 272–286.

[38] Knipe, *Image of Fire.*

[39] Sue Mansfield, *The Gestalts of War: An Inquiry into Its Origins and Meanings as a Social Institution* (New York: Dial Press, 1982).

[40] Gene Usdin (ed.), *Schizophrenia: Biological and Psychological Perspectives* (New York: Brunner/Mazel, 1975).

Chapter 14

The Shipwrecks of Dawn

I see water, I see fire and air and earth and all their mixtures come to corruption, having little endurance; and yet these things were created, so that, if what was said is true, they should be secure from corruption.

Dante's *Paradiso*, Canto VII

He was a valiant lad;
With his dull gold buttons and his pistol
With a manly air in his stride
With his helmet, a glittering target
(They reached so easily into his brain
He who had never known evil)
With his soldiers to left and right
And revenge for injustice done before him

—Flame on lawless flame!—
With blood above his eyebrows.
The Albanian mountains thundered
Then they melted snow to wash
His body, silent shipwreck of dawn
And his mouth, small songless bird
And his hands, wide plains of desolation
The Albanian mountains thundered
They did not weep
Why should they weep
He was a valiant lad!

Odysseus Elytis, "The Lost Second Lieutenant," from *The Sovereign Sun—Selected Poems,* translated by Kimon Friar (Philadelphia: Temple University Press, 1974).

Whatever choice he makes will be wrong. To stay or to go; to fight or to be idle . . . Achilles is trapped in a "double bind": he will die young, or he will be forgotten. He will die, or he will die. The choice, ultimately, is no choice, and Achilles sits in his tent, increasingly bitter, increasingly isolated.

Paul Zweig, *The Adventurer: The Fate of Adventure in the Western World* (New York: Basic Books, 1974).

More than 100 million people have been murdered in the twentieth century. Sociobiologists are fond of claiming that this behavior is merely that aggressive residue forever lingering in the reptilian portions of the triune brain. But we have no special evidence that reptiles are particularly aggressive. And the fact that at least 40 percent of the world's human population is always engaged in peace further demonstrates the somber reality of schizophrenia as it is experienced by our species.

SCHIZOPHRENIA

Julian Jaynes has remarked on the schizophrenia he believes to have been normal 3000 years ago. At that time, man's brain evolved the means of orchestrating left and right hemispheres in the neocortex, and 200-million fibers of his corpus callisum came triumphantly together. Before then, says Jaynes, feeling and information fought fierce battles, came out eruptively in the visitation of voices and the edifice of hieroglyphic incantation.[1] What balance, unmasked, offers the rebirth that the twentieth century wrestles to invoke? The imagination, bridled and hesitant, falters to admit it. It can see disaster to the point of premonition. It can turn its back on all knowledge for the sake of color, whim, sex, sudden form; can pull any trigger, stack bricks higher and higher.

But can it accept its beginnings with any wonderment? We can kill ourselves in one hand and paint cobalt ceilings upside down in the other; enlist the myriad sounds around us and fashion musical cogency, that greatest of all beelines to the paleomammalian brain; re-create the fantastic, the ironic, the counter and contemplative, the comment; elicit unexpected action, lunge from bed, from church spires; cast spells, conjure phantoms, disappear, delight and demonstrate with fire in the triune brain, in the groin, aorta, pyneal gland, in the untamed iris, marathon legs and Horowitz hands, and penitent stand. Sensual, whipped into shape by the summons of that all-unknowing trajectory we call imagination, human beings can make themselves into angels. The kernel is our very own first day of life, and it repeats its true miracle with every baby born. We are all perfuse with stormy weather, somersaults, with luscious alpine streams, with wanderlust and heavy-breathing exasperation, each; with undecided premises, spitfire zest, aversions, with the sorrows of every young Werther, the primordial feasts of Dionysus, Appalachian Springs, Ninth Symphonies, Ming Dynasty tomb arrangements, Tudor mansions, Inner Mongolian yurts, Varanasi brocades, Yemenite jewelry, ascetic annals – we have it all – from Falstaff to the Hesperdes; from the elegant legacies of Cro-Magnon to such noble laureates of compassion as Gandhi and Chaplin, Okakura and Jack Benny. We have Leonardo's women and Giacometti's solitaires, all of Greek tragedy, Bengali love poetry, Carolingian tapestries, twelfth-century Cordova, the Fujiwaras, the Taj Mahal, Keats and Frisbies, and Telegraph Hill. And we have the bomb. A child knows such dichotomy.

This awesome conjunction of good and evil, of original and derived cultures in the same brain, is the true and terrible balance of power of which the nations speak; a struggle within the imagination, with all of its own throw-weights and megatons.

If we disabuse ourselves of selection theory (survival of the fittest, child weaning, predation), then mankind's abiding dream, his hierarchy of volitions, appears easily satisfied, in absence of dualism and violence. We can enjoy beauty, security, all manner of comfort, boost our self-esteem on a planet so bountiful as this one. With a global human population of hunter–gatherers numbering in the very few millions during the Pleistocene, this equation was indeed real, though not without its nascent discrepancies. Indeed, a majority of our progenitors' food derived from the female forager's hand, but the male hunter catapulted the group mind into conflict and duality. The woman would ease her offspring into such a mind, but not without inflicting her own form of power play; and the male ascended to petroglyphic portraiture, numbers, calculation. He preserved in the earliest art forms the first and foremost angst. H. Kuhn as described some pictures from the Valltorta ravine: "his enemy's arrows have pierced him, his life is over. Where is paradise receding to? Is it a dream of humanity? Is war, fighting, the purpose of human life? . . . Primeval pictures antedating all memory, older than all myth and legend – and already we see the killing of man by man."[2] Raymond Dart, furthering this view of innate human aggression, has quantified the high frequency of bone injuries sustained by our ancestors, from the Australopithecines to fourth-glacial *Homo sapiens*. His evidence has been contested by others who insist that these "Paleo-fractures" are just as likely the result of animal scavengers who normally disinter and drag bone remains; puncture, consume, scrape – do all those things easily confused with human violence. Writes Raymond Dart:

> *The blood-bespattered, slaughter-gutted archives of human history from the earliest Egyptian and Sumerian records to the most recent atrocities of the Second World War accord with early universal cannibalism, with animal and human sacrificial practices or their substitutes in formalized regions and with the world-wide scalping, head-hunting, body mutilating and necrophilic practices of mankind in proclaiming this common blood-lust differentiator, this predaceous habit, this mark of Cain that separates man dietetically from his anthropoidal relatives and allies him rather with the deadliest of Carnivora.*[3]

What is the price of progress? This grim evocation, selfish mind, cynegetic ardor, guilt-ridden devotion, political facade, polished marble, hallowed horrors of every kind lays claim to divinations, exceeds all homeostasis, gives powerful chase across tormented terrains, exasperates and self-abuses, is naked, unrestrained, invincible – but it is not progress, nor is it instinctual.

War and weaponry are culturally derived, symbolism aching for praxis in its life. "Homo homini lupus; who has the courage to dispute it in the face of all the evidence?" wrote Freud. He measured aggression as part of our "instinctual endowment." But the definition of *instinct* is still debated, lacking sure biologic parameters. Freud's view of the death instinct being choked by more sexual and self-preservative volitions further compounds the allegory, inspires a unique idea of fruitful self-destruction, of fertility prior to desolation, and much like a malignant, uncontrollable swarm we call evolution. This is the love–hate archetype, the growth–putrefaction dyad of Heraclitus.

Konrad Lorenz, Erich von Holst, Karl von Frisch, and Niko Tinbergen have all asserted the Freudian gloom, borrowing from scenarios of ethological doomsday to assert the role of innate release mechanisms (stimuli detectors). Lorenz looks toward "rational selection" to transcending our hideous dilemma. "Conscience . . . rationally guided evolution . . . [the] . . . recognition that war performs functions that will have to be performed in some other way, without bloodshed," are the determinist claims to a peacable future, though not without an obstacle course of hubris and the latency of violence.[4]

The record of violence is in fact awesome, though there are interesting twists to its history. At least 40 percent of the human population has always been engaged in peace. Of those who do partake in front-line offensives, only 15 to 25 percent ever actually fire their weapon. Medical Corps psychiatrists investigating combat in World War II found that the "fear of killing, rather than fear of being killed (was) the most common cause of battle failure."[5] On headier echelons, with a one-in-three chance of surviving the Bay of Pigs (in President Kennedy's own estimation), "the only thing to decide was whether one had the *courage* to use all that might."[6]

Courage or paralysis: Nearly 100 million people have been murdered by warfare in this century alone. Approximately 250,000 large, hostile encounters have occurred since 1500 A.D. America has fought over 10,000 battles in its history. In fact, there have been only 20 years in which the United States was not engaged in some bloody contest. Since World War II, more than 100 military coups have occurred in the Third World; between 1945 and 1976, 25 million people died in 133 wars. What conceivable logic enabled Napoleon to coerce 500,000 men into wintry Moscow; for Alexander to march his troops halfway across Asia; and Augustus to goad 333,000 young men up and down Europe? Promises; salary; threats; glory; boredom; lousy homelife; unsatisfactory marriages; sexual fantasies; desire to kill; fraternal altruism; existential, generalized hatred of life; the allurement to fire power; of a powerful weapon in one's very hand and with it all the conceits of converse insecurity; castration complexes, hatred of women? War neatly pits the brain against the mind.

Schizophrenia means "split mind." One of three major functional psychoses, the disease has organic chemical elements only recently investigated. All four areas of schizophrenia (simple, hebephrenic [regressionary], catatonic, and paranoid-type) concern the etiology of warfare. But how best to characterize this association? In her extraordinary book *The Gestalts of War*, Sue Mansfield described the elation witnessed among participants in the My Lai massacre: "In gestalt terms, their explosion had unblocked physical repressions, acted out their hallucinatory defenses, and allowed the discharge of all the previously retroflected energy."[7] The history of such slaughters points to a climax within any battle, which—like a statistical J curve—predicts the inexorable resolution of dichotomy through a law perhaps best understood in thermodynamics: *entropy*.

Entropy has orders of magnitude, from the basic evapotranspiration of plants, to more complex heat exchanges and redistribution of energy, to the movements of whole animal colonies. Nomadic migrations are entropic, a form of dissipative structure which, chemically, looks like a rotating spiral, whose waves rhythmically lead to patterns of interference (the Belousov–Zhabotinsky reaction). In the nineteenth century, Great Britain's *pax Britannica* maintained central cogency at the expense of international balance-of-payments entropy. The uncontrolled consumption of limited resources leads to confrontation with constraints, to dissolution and anarchy—all qualities of ecological entropy and existential breakdown, whether it be the national debt, the Hundred Years War, or the spate of mental illness in today's world. Historian Arnold Toynbee has systematically chronicled history's rampages.Warrior barbarians (Kassites, Hyksos, Semites, Dorians, Achaeans, Etruscans, Celts, Hittites, Teutons, Bantus, Taishans, Mongols, Turks, Aztecs, Vikings, Huns) all moved as if in some vigorous choreography of increasing chaos. The results had widespread effect. The Hyksos, for example, rode into Egypt on horses, fortified a previously tranquil civilization and engendered a complete warring transformation of that society. At the same time, all the first Minoan palaces were destroyed by mysterious invaders, and Babylonia was ruptured by Kassites. By the fourth century B.C., the three contending Chinese states were building walls along their frontiers to keep out the Turkish-speaking Hsiung-nu pastoral nomads (the Huns) who were anything but pastoral. But over the next millennium, these high-altitude bareback equestrians would erupt five times, in every direction—to the Oxus-Jaxartes basin, to India, to Iran, to Europe—and conquer all of China. The Hsiung-nu, Tibetan, Tungus, and Mongol invaders would partitian the North China Plains, invade Japan (unsuccessfully), and remain in control of the world's largest empire until their conversion to Buddhism. Between 221 and 48 A.D., the Mauryan, Seleucid, Ptolemaic, and Carthaginian Empires and the Kingdom of Macedon would all be extinguished by the churning outreach of nomads on the move. Parthian and Kushan warriors occupied Mesopotamia, Babylonia, and Iran. In the West, Huns overthrew the Ostrogothic Empire on the Dnieper River in 375 A.D., driving the Visigoths to seek asylum in Roman territory. This resulted in Rome's defeat at Adrianople in 378 A.D. Earlier, the Parthian *Völkerwanderungen*—huge forays of migration supported by Bactrian *caravanserai* and massive arrow armaments, metal shields, and horse riders—had defeated Crassus' infantry at Carrhae. The Roman nightmare was all of nomads.[8]

What separates human atrocity from that of other species is the option of conscience, which is forsaken. The denial accumulates in the genetic repository and carries with it psychic numbing, an ineradicable angst—the Holocaust survivor, the *hibakusha* (Hiroshima survivor)—and that mad, hateful contagion among victims of less-publicized entropic outbreak. We can guess at the pathological carnage that persisted throughout the Assyrian Empire. King Ashurnasirpal II (883–859) set the tempo by mutilating captive kings, flaying his victims, impaling 700 prisoners one day atop towers surrounding the city gates, making piles of their skulls. The Palette of Narmer, the bas-relief of Eannatum of Lagash, the stele of Naram-sin, Trajan's column, the platform of Chichén which is decorated with a frieze of human skulls, are monuments attesting to unrivalled acts of horror with a purpose diametrically opposed to that of its victims. Parisians of the 1730's could savor the carnage of mutilated alley cats and delight in the neighborhood spectacle of animal tortures, disembowelled dogs and the likes. But from the larger view, it is the mind whose stricken dialogue we are tracking – from Iran, Lebanon, Vietnam, and Germany – to the more ancient archetypal landscape, which is as charged and darkly forlorn as that of *Beowulf*.

When the Toltec city of Tula was destroyed by barbarians from northern Mexico in the late twelfth century, its founder Topiltzin illustrated the staying power of revenge. He migrated to northwest Yucatan, founded the Kukulkán–Quetzalcoatal state, and instituted the practice of human sacrifice as a political policy. The Athenians had practiced this propitiation of gods every seventh year, culling from among their ranks pretty little boys and girls. But Topiltzin and the later Aztecs went wild with the spectacle. At least 20,000 victims were slaughtered annually, fifty-five a day. It is possible that over 4 million in all were burned, tortured, sexually devoured, and thrown dismembered into wells and off cliffs. The Bonampák paintings west of the Usumacinta River indicate the extent to which the Mayan culture also partook in such outrages. But ultimately, none of it survived. San Lorenzo, La Venta, Teotihuacán, Cholúla, Tikál, and Uaxactún were all wiped clean, their inhabitants lost to jungle rot.

The same obscurity masks Temüchin's campaigns in Turkestan and northern India in 1220, Batu's Polish and Hungarian campaigns of 1231, and Hülegü's invasion of Iraq in 1258 – the worst of the Mongol plunders. An equally hideous venom enshrouds the campaigns of Antiochus III, of Cornelius Sulla, of search-and-destroy missions accomplished against Amorite, Canaanite, Perizzite, Hivite, and Jebusite under the approving eye of Deuteronomy (20: 16–17) and Joshua (6: 21). "And they utterly destroyed all that was in the city, both man and woman, young and old, and ox, and sheep, and ass, with the edge of the sword."[9] Compare this latter injunction with a description of a night's madness along the Merderet River in Normandy, June 1944: "The slaughter once started could not be stopped . . . Having slaughtered every German in sight, they ran on into the barns of the French farmhouses where they killed the hogs, cows, and sheep. The orgy ended when the last beast was dead."[10] William Barry Gault has studied massacres by categorizing the perceptions of a soldier caught up in this tyranny. The mechanisms of barbarism include "the universalization of the enemy; the 'cartoonization' of the victim; the dilution of responsibility; the pressure to act; the natural dominance of the psychopath; and the ready availability of firepower."[11] Gault's principles contributing to slaughter endorse our mental predisposition to kill so as to exorcise a world of evil around us, which is the way the world presents itself to the evolutionary mind. In Vietnam it was the fact of being away from home, surrounded by mines, booby traps, swamps, malaria, and a ubiquitous enemy who might be the farmer's young girl or the old woman as easily as a hidden terrorist. With comrades dropping to either side, screaming horribly with ruptured insides, kill or be killed the only purgative, there is indeed this rational excuse for rampage, one that ultimately even sanctions international intervention, further complicating any black-or-white dictum. Letters from the bishops condemning on moral grounds the use of nuclear weapons fail to win approval from a body politic whose own morality plays victim–avenger with the rules, perceives battle *above* moral grounds. Second strike may be human nature. But first strike combines the activities of reason with the medieval provocations of history that will ever stay with us. This is hard to acknowledge when examining the remains of tortured prisoners. In Columbia in 1947, terrorists "ripped fetuses from pregnant women" and "chopped prisoners into tiny fragments before whole villages."[12] Prisoners of the German *Sonderkommando* were compelled to seek out valuables in the orifices of corpses in the Nazi camps. Innumerable instruments of atrocity have been wielded by the terrorist groups of history, be they Tupamaros of Uruguay, the Baader–Meinhof group, the Palestine Liberation Organization, the Irish Republican Army, or Roman Nucerians. In Bangladesh, Burundi, the Congo, and Nigeria, whole sectors of the population have risen to the techniques of cannibalism and plunder.

But the personality behind such acts is not merely one style over another. Great men and women have fought battles. Sophocles was a general in the Battle of Samos (440 B.C.), and Socrates fought against Athen's insurgent ally, Poteidaia. The great sensualist sculptor Pheidias created his "Zeus" at Olympia to celebrate the repulsion of the Persians in 480 through 79. Goethe swooned over the battle at Valmy, and Hegel declared the "world spirit" upon seeing Napoleon on a horse following the victory at Jena.[13] Every U.S. president has been commander of the military, many taking roles in the execution of field strategy. Sue Mansfield's analysis of double-bind impetus among the players of World War I is revealing. The Austrian chief of staff, Conrad von Hötzendorf expressed his excitement for armed conflict; and a letter from Churchill to his wife in August 1914 reads, "Everything tends toward catastrophe &

collapse. I am interested, geared up & happy. Is it not horrible to be built like that? The preparations have a hideous fascination for me. I pray to God to forgive me for such fearful moods of levity." Fear of losing the "prestige of [a] Great Power" ripped through functionary cadres of five major and two minor powers in a period of 10 days prior to the outbreak of the war. The paranoia led from "uneasy peace" to armed confrontation. The Czar, a Chancellor, a Premier and numerous foreign ministers were pawns to their own rash of half-truths, misconceptions, adamantine stances of pride, arrogance, opportunism and warring adventurism. This double-bind acted as a blind momentum, the very personality of war.

Hitler, with his Berchtesgaden mountain retreat and beloved dogs, might seem to have hidden well the inferno within him, that is until he opened his mouth. The Nazi leaders extolled certain virtues – the music of Wagner and Beethoven. But this only heightens the polarity of mind at work, the physiological lack of clarity. Hitler could speak of being "wrongly judged," of offering a "love of peace," of "patience" to the world.[14] Macho-conciliatory deception and Doublespeak are easy means of convincing oneself. A U.S. Public Information officer during the Vietnam war insisted that American forces were not "bombing" but lending "air support." Reagan, whose own ancestors defended Sliege Bloon Mountain in Ireland for generations, quotes scriptures from his 172-year old adobe house 2400 feet up in the Santa Ynez Mountains: "Look to the hills from whence cometh thy strength." And what *kind* of strength? Well, for one, the kind which would enamor one of guns, of certain favorite practices, like killing rodents with beautiful weapons, and systematically crushing the heads of rattlers.[15]

In 1789 the French National Assembly denounced all war. In 1792, the French Legislative Assembly declared war all around itself. Bismarck's *Zollverein*, the U.S. War of Independence, and the Swiss confederation all resulted in bloodshed for the purpose of peace. In 260 B.C. the Kalinga enclave within the Indian Mauryan Empire was massacred by Ashoka, grandson of the Mauryan founder, Chandragupta. Ashoka's own horror at what he'd done converted him to Buddhism. This true missionary remorse is recounted in Ashoka's Thirteenth Major Rock Edict. His adopted tenets are not only Buddhist but Taoist, pantheistic, amounting to true tolerance of other faiths. He engineered welfare projects, managed Buddhist monasteries, restricted the killing of animals, and established hospitals. Is there predictable human penance in history? While the United States was approving the 1929 Pact of Paris for peaceful aspirations, it was busily building up its naval defenses. Quincey Wright has cited this contradiction of democracies with reference to the League of Nations, which urged universal repeal of war while resisting the obligations to carry it out.

A former Assistant Secretary of Defense can describe NATO doctrine as that position by which "we fight with tactical nuclear weapons until we are losing, and then we will blow up the world";[16] the Reagan Administration can push massively forward for additional nuclear armaments while feebly negotiating for arms reductions. And a dentist-turned-U.S. Energy Secretary can happily view a nuclear explosion seven times the force of Hiroshima, detonated in Nevada on the eve of the Hiroshima commemoration, and hail the test as "exciting," thus underscoring the administration's sound determination to postpone comprehensive test-ban treaty negotiations. All of these irrational patterns describe the net effect of arrogance and madness infecting those vested with power. Seventy-five (known) chiefs-of-state over the past 400 years have exhibited mental illness while in office. Others suffered from debilitating ailments that reduced the effectiveness of their thinking. Leonid Brezhnev had amyotrophic lateral sclerosis that produced incoherence, absent-mindedness and disorientation accompanied by muscular atrophy. Caligula, Mad Ludwig, Joseph McCarthy, Nero, Rasputin — illness. Lincoln, Churchill, and Teddy Roosevelt suffered apparent lithium imbalances that led to perpetual manic-depression. During Nixon's final days, many of his friends were afraid he would "push the button" in his incoherence. Personnel reliability examinations exist in the military programs for persons with access to nuclear weapons (over 100,000 people with such means). Statistics buried in a 1979 report before the Subcommittee on Appropriations of the House showed that in 1977, 1289 such people were removed from duty for a "significant physical, mental or character trait or aberrant behavior, substantiated by competent medical authority, (which might) prejudice reliable performance."[17] Among Congressmen, the 204–202-vote defeat of the Nuclear Freeze Initiative was close enough to sound yet another region of sorrowful, clinical undecidedness. A Republican leader from Illinois argued that the freeze was "tantamount to national suicide," and that only more and more superior weaponry would constitute security. No television station allowed pro-Freeze advertising spots. The language and opprobrium surrounding the "suicide" issue bespeak of the "banality of evil" syndrome: Unlike the victim who has suffered to such an extent that he can no longer feel pain (the brain's endorphins serving to buffer us), Hannah Arendt's phrase refers to those who are capable of committing evil without wincing an eye; who have trivialized barbarism to the point of losing sight of differences.

Given the right conditions (and they are present in the mind), a person can "become schizophrienic," with all the resulting schizoid perceptions and behavior: loosening of associations, the loss of unconscious representation of objects and the replacement with words, a constant state of alertness. What follows is the activation of an "opposite" switch, a complete tuning out: the denial of reality, the inability to attend to details selectively, a fondness for generating metaphor, extreme self-consciousness, futile attempts to reduce the incoming overload of stimuli, difficulty in concentrating, insomnia, intense delusions, fear, hallucinations, the hoarding of food, vagueness, heightened sexual yearnings, aggression. So common is this condition that psychiatrists have considered it one more anomaly of the development process – predictable and generalized. Never has such agonism had the easy ability to destroy the planet.

The impulse to separate rather than coalesce – the origin of taboo, of difference in cultures – stems from the necessary split between that site on which blood is spilled and the buffalo sacrificed, and that land to which the eyes seek out their unsullied heaven. The human paradox invokes this peculiar, divisive correspondence. We cannot safely look back, but yet we do. And having done so – having peered into the twilight of a carniverous predisposition – are accustomed to all manner of continued transgression, conditioning our conscience to death in one hand and life in the other. No spread is large enough to contain so shifty a human god.

Qoeleth's fateful words from Ecclesiastes are precariously balanced. In the grim darkness there is grim light. In the apocalypse, where is to be lodged the final comfort, all of our contraries will have exhausted their contest, will have eloped to the Big Sur of abstractions. We still hope that the human lodestone – the buzzing of dreams, the yawning of deserts and Orients and high mountains, of deep forests and of love – will penetrate the scream, outlive all combat, until vanity of vanities has been undone in trauma of traumas. The brain has either option.

Religious schizophrenia has further exacerbated human compulsion to enact violence. Revenge plays a major role in Christianity. Heads, scalps, and foreskins may not be first on the list of meting out justice, nor are ancestors needing to be repaid. But the cross is indeed a sword throughout much of Christian art, as Sue Mansfield has brilliantly pointed out, a symbol blood-drenched with all the vengeance of an original-sin psychosis. The slave-master debacle imbues the *New Testament* as it does the *Koran*. The Papacy could easily justify its waging wars of extermination against the Hohenstaufens under Urban IV (1261–1264), as against Frederick II by the bulls of Pope Clement (1265–1268). Muhammad was a businessman among the Banu Quraysh at Mecca. Following his mystical rapport with the Archangel Gabriel, he began his bloody career by preaching self-surrender (Islam) among the wealthy. This did not set well, and the brash Muhammad was shipped out of town. But he had another job offer waiting for him. The neighboring Yathribite clan invited him to unify their discordant factions, which he did, with the aid of Abu Bakr. Continuing his success, with newly acquired forces, he invaded

Mecca, massacred unyielding communities, enslaved women and children, and from 622 to 632 subdued all of Arabia.[18]

Under various dictators, Japanese Buddhism is fairly larded with such contradictions. And once the Mongols had converted to Buddhism, thus alleviating China of two millennia of menace, did they sigh and embrace pacifism? No. Under the leadership of Ch'ien-lung in 1757, the Chinese invaded Burma, Vietnam, and Nepal. The violence of the Israelites in the wake of Egyptian slavery (the significance to this day of the Wailing Wall), and modern Israel's tragic plight beside her multitudinous Arab adversaries bespeaks of a whole people who – by torture and diaspora and genius – have been forced to become schizophrenic. Israel's Lebanese invasion, after many years of shellings and guerrilla actions from across her border, is the nebulous offensive–defensive, the first strike–second strike ambiguity that transforms religious into secular war. Any hint of sacred mission or Prince of Peace is eradicated for a more practical realization. Elsewhere, this shift in political emphasis began in the late fifteenth century, with the union of Castille and Aragon, the rise of the Tudors, the first secular Pope (Alexander VI), and the rise of the Orleanist dynasty in France in 1498. Gustavus Vasa in Sweden, Frederick I in Denmark, and Peter in Russia each furthered this separation of church and state with an ideology of war making. As war became secular, so did death lose its sacredness. Economics replaced hypocrisy.[19]

Guilt, anger, vengeance, scapegoating, self-hatred, boredom, the death of nature, and insatiability, the loss of symmetry, added assaults on all that is feminine, and the sublimation of those creative energies that struggle to redeem our true, blushing, biological value (but are squelched) accompany the rise of socialization and symbolic transport. The tragedy of schizophrenic violence is that the metaphor – at one time mankind's golden advantage over the surrounding environment – should be degraded, like the historical image of Shakespeare's Richard of Gloucester, "between a warped child and a demon."

The tragedy is Greek, *hamartia*, the flaw, Oedipus' "black rain of blood," Hamlet's "rotten" Denmark, psychic abundance becoming juvenile hysteria. For all of his prowess and strength, the hero makes a mistake. Ultimately, we realize that such error is the only solution to so much overabundance. The one special, rare quality that could arrest tragedy before it is played out is the Greek concept of *sophrosyne* or self-control.[20] J. Glenn Gray has intoned its quintessence: "Rendering oneself unarmed when one has been the best armed, out of a height of feeling – that is the means to real peace of mind; whereas the so-called armed peace, as it now exists in all countries, is the absence of peace of mind."[21] Sophrosyne has not been a welcome concept in the West. The civilizations that have followed its mandates (the Old Kingdom of Egypt, early Sumeria, ancient China, Fujiwara Japan) were indeed blessed with sustained periods of temperate life. But from the burning of the Alexandrian library to the hallowed war cries of iconoclastic Machiavelli and Nietzsche, the western intellect has thrived on flaws; has sought out nihilistic conflict to prove biological and social intuitions, to justify power elites whose blessings may be worth courting, to incite the proletariat, to redress rejection. I don't doubt for a moment other, nobler aspirations. Buddha, Mo-ti, Isaiah, St. Francis, Gandhi, Lao-tzu, Erasmus, and Aristophanes were genuine in their opposition to the Hobbesian state of nature, *bellum omnium contra omnes**, a condition in which mankind is portrayed as free and frustrated, subject to the phlegmatic, irruptive, outward viciousness of a dictator who has promised security where nature promised none. Such a dictator, and his politics, are thusly construed as outside of, better than, nature. These apostles of peace understood that the absolute source of illicit, dangerous power was this separation of human culture from the natural world. For Rousseau, man was everywhere in chains for having degraded the original state. And Freud similarly stated that "it is easy, as we can see, for a barbarian to be healthy; for a civilized man the task is hard."[22]

The rise of humanism, international jurisprudence and organization, disarmament enthusiasms, the economic interests of physiocrats and the morality of utilitarians, mercantilism and free trade have all mitigated, if minutely, our habits of aggression. But the powers in favor of war are overwhelming. Philosophers from Plato and Aristotle to Augustine, Aquinas, and Shakespeare have borrowed from the classical historians — Thucydides, Josephus, Livy — and from the military handbooks of Caesar and Aurelius to promote the justifiable and honorable codes of violence. The Declaration of Independence states clearly that "experience hath shown that mankind are more disposed to suffer, while evils are sufferable, than to right themselves by abolishing the forms to which they are accustomed." This was Nietzche's very position in *The Will to Power*, that "the natural morality, which recognizes as good everything that is bold, vigorous, cruel, and self-reliant" must win out over the "slave morality of nonresistance which calls good all that was cautious, humble, pacific and adaptable."[23] Sartre would speak in this vein ("irrepressible violence . . . is man recreating himself") and Bakunin would call destruction "a constructive passion." The artistic canons of Italian Futurism were the incarnation of brutality, the love of machines of progress, and the hatred of women.

Some of this illness, at least since the Enlightenment, comes from a literary obtuseness first explored by Edmund Burke.[24] Burke was the master of oxymoron, of phrases like horrible beauty, terrible joy, and

---

*Freely translated, "eat and be eaten."

this particular penchant would have an astonishing effect on Romantic métier. Etymologies aside, the clue to power lay with the unlikely juxtaposition, the shock value. Violent art could be bountiful in the same way that a violent avalanche might be wondrous to behold, at least from a safe distance. By the turn of the eighteenth century, never a summer's day passed when three-dozen British nature savants on grand tour did not stop in at Chamonix to thrill before the Mer de Glace, the pinioned, sacrosanct glaciers poised for tumult and down which Frankenstein's monster would descend. "Whatever is fitted in any sort to excite the ideas of pain, and danger, that is to say, whatever is in any sort terrible, or is conversant about terrible objects, or operates in a manner analogous to terror, is a source of the sublime."[25] The nobility of suffering gives rise to beauty. When merged with power (a conscious, manipulative fusion) all of the elements of war—"strength, violence, pain and terror" (Burke)—assault the mind at once. Burke believed in the negative sublime—that self-abusive pain and melancholia, the spiritual riot of a Baudelaire and De Quincey—he termed "swelling." The condition arose in the romantic hero as in the political tyrant and operated according to the mind's acceptance of those terrible objects and events from which it expected to derive dignity.

ANIMAL AGGRESSION

Quincey Wright has said that "the characteristics of protoplasm" explain, ultimately, the psychological causes of war.[26] All living animals are endowed with the same basic cells that prevailed during the PreCambrian Period, undifferentiated symplasm, with its rudimentary nervous system response known as *irritability*. Lorenz and Tinbergen have described three components of an "instinct" at least easily imagined: "appetitive behavior" (goal-directed readiness), "innate releasing mechanism" (the override of inhibition so as to act singularly), and a "final consummatory act" (relief of tension following the discharge).[27] By examining the earliest vestiges of cell fragments and their living counterparts (at places like Australia's Shark Bay, where the blue-green stromatolite mats still thrive, and among the microfossils that have turned up in Swaziland in eastern Transvaal) some general conclusions about extracellular interaction can be formed. *Heterotrophy*, the ability to feed on other life forms, and *sexuality* were the two essential accelerators of evolution. It takes roughly 1000 generations to mold a new species, and given enough eating and cell dividing, we can account for the nearly 500 million species that have at various times inhabited the Earth.[28]

The first consumption of energy occurred by fermentation. Later, with autotrophic photosynthesis, enzymes manufactured their own food within the cell. But this was rather costly in energy. The origins of eukaryotic sexuality date to just over 1 billion years ago. Two-hundred million years later, heterotrophy appeared. Having learned to sexually penetrate, full consumption of a partner was just a few million years away. Sexuality transferred genes, selecting for other heterotrophic organisms of which the earliest were small worms. Equating early heterotrophic behavior with aggression gets us deeply into a bind. The association is valid if we presuppose an alternate masterplan by which the lion and the lamb proverbially murmur love mutterings to one another. And in fact the biosphere offered that very paradigm in the ancient form of prokaryotic cyanobacteria, and then plants, the autotrophs. If we ignore fermentation, photosynthesis, and other processes of organic energy transfer and simply set our sights on the thorough indictment of violence, then any philosophy of pacifism is shot to hell long before the Cambrian Period. The genetic components of aggression are heritable, subject to evolution. The vertebrate endocrine system is finely tuned for such behavior, and at no more precisely fashioned threshold than that of the androgens secreted by the Leydig cells of the testes.[29] The confirmation of this specificity came as early as 1849 in Arnold Berthold's experiments with roosters. As Edward Wilson points out, the same association obtains for many other species, including painted turtles, night herons, and chimpanzees.

In evolution, aggression selects for itself in greater, increasingly inviolate ways. Clausewitz, describing the euphoria of battle to his pupil the Prussian crown prince, told him to "engrave your image in the hearts of your last descendants."[30]

Jane Goodall and George Schaller have detailed the gory, if infrequent, acts of murderous behavior exhibited by chimps and Serenghetti lions. Such infantacides and cannibalisms are rare, not obviously provoked. But most animal aggression *is* provoked, whether it be that of a coyote haplessly cornered in a city, or a racoon backed into a cul de sac. When an animal is on its *death ground*, that point of doubtful escape, it will do anything to kill its opponent. No animal will turn the other cheek, though certain defensive maneuvers may appear to have invoked such a gesture. Aggression which has been provoked is justifiable in most people's minds — an eye for an eye. Sue Mansfield has suggested that "men are angry because they are engaged in battles as often as or more often than they fight because they are angry."[31] The ratio of first-to-second aggression — legal debate over sufficiency of provocation — has preoccupied the juridical mind and tended to legitimize the philosophical groundwork of an aggressive concept of carrying capacity. In biological contexts, such rubrics encompass the extent to which a limited habitat can contain any variety of adulteration. To justify the "wilderness" designation, the region in question must satisfy the human, cultivated demand for purity – whimsical, instructive residue of past civilizations (i.e. fossil femurs, colorful ceramic fragments – the rotting remains of an aqueduct, notwithstanding). But even these relics – worked pieces of obsidian, mounds, a contrail overhead – can upset the lovely *idea* of the place that some adepts have invented. The *pure* state of nature is a relative notion, of paramount importance throughout the history of art, chemistry, and biology. Both the Taoist landscape vision and the spate of litigations brought before the EPA attest to such thinking. It is equally true that the mental paradigm of paradise – inextricably woven of the purest, most natural of urgings – takes as its pacifist model the normal, nonaggressive behavior of animals. If there are those historians and warfare zealots whose sense of duty and justice gains cosmic consolation from a misreading of the East African savannah, the grizzlies of Glacier Park, of tormented rats in a cage, or carpenter ants laying siege, few will misread the prejudice.

What this should tell us, what millions of protestors have been begging us to understand, is that the wilderness of aggression, under certain rarified conditions, supports limited conventional rationality. Beyond it, all aggression is suicide, often masked by sympathetic impulses, sublimation, chivalry, and a distorted range of sociological priorities that make war out to be natural, or at least inevitable. But by the conviction of protest (eloquently enunciated by Thoreau's "On Civil Disobedience") even the conventional logic – in an age that has attained the gift of forward thinking benevolence, of a partition in our evolution where the mind has wit and savvy, and the dream of universal peace – will not suffice.

Military power has been forever construed as true diplomacy, an acceptable means of settling disagreements. When the Inupiat Eskimos got guns, they soon killed twenty times more whales than they had prior, not because they suddenly needed twenty times more food or income, but because you can't *kill* a whale with a gun, you can only wound it, before it flees hemmorhaging into the deep, there to die slowly and out of reach. Weapons are inherently inefficient in their overkill, and the power of inefficiency has always been attractive to mankind. Why? Consider Murphy's Law: If more than two people are privy to a disaster, then no one need be a culprit. Warfare is the laziest possible means of settling dispute, an economic curve that parallels cultural vacillation from forlorn to fury. Neolithic villages seem to have been completely undefended: Mere migration was enough to settle conflict. Among Irish elk populations, fighting was built into their genotypes, granting them huge and lethal horns useful in sexual competition but terribly cumbersome when it came to foraging in the dense backbush. They went extinct as a result. Natural selection, like photosynthesis, like all energy exchange, is highly wasteful. But perhaps no more so than the act of conscious murder.

AGGRESSION AND SEXUALITY

The love-hate ambivalence infiltrates much of human self-consciousness. Gregory Bateson suggested a common form of schizophrenia in regard to its double-binding mode, although he was criticized for blaming the mother. Bateson described the mother's profound sexual and authoritarian relationship to her offspring, indicating great and terrible latitude for anxiety and the feeling of being trapped on the part of the child.[32] Freud considered the psychoneural conflict of sexuality in the infant to be among those very few character builders that would stay with the child all of his or her life. We do not know whether certain psychopathic behavior derives from sexual chromosomal disorders. But in the etiology of social distress, and the heritability of such malfunctioning psyches, we have no doubt. The glands, the brain, the full ambit of the hypothalamus are implicated.[33] Kinsey listed fourteen physiological changes common both to sexual arousal and anger. Sexually related aggression has been identified in the anterior hypothalamus, the ventromedial hypothalamus, and the tubero-mammillar complex. The initiation of soldiers into war involves training rituals where the men are showered with genitalian insults ("you cunt"),[34] and contradictory data suggest that aggression can be modulated merely by showing soldiers pictures of naked women.[35]

Among other animals, the male aggressive-sexual selection is an elaborate physiology of weaponry: horns, tusks, mane, and bright colors. These effects are ordinarily meant to defeat rivals, but in human society, most killing is cross-genderal. Men kill women, women kill men, often in circumstances where a rape or castration has additionally been perpetrated. The Roman rape of the Sabine women in the eighth century is a symbol for more wide-ranging sexual atrocities condoned in war-time and satisfying male sexual repressions that may have accumulated throughout the matriarchic Paleolithic era. The Trojan War was fought over a woman, and it is ironic today that male contraceptives should be named after that war. If anything, war has been ejaculatory. Woman stealing was common among Egyptians, Hebrews, Greeks, Arabs, and Nordics. The Byzantine emperor Alexius held out the bait of beautiful Greek women in his efforts to gain aid prior to the First Crusade. Scholars have argued the importance of sexual gratification over economic interests; Grotius included sexuality as a legitimate cause of war. Among Greeks, Romans, Carthaginians, Teutons, Franks, the Crusaders, and the British free sexual exploitation for the victors of a beseiged city was assumed a basic right of conquest. Though Aristotle believed that the love of women is consistent with warlike races, Quincey Wright's in-depth analysis suggested to him that "sex has played an important role in civilized war, often lying behind economic, social and political motives."[36] After the conquest of the Midianites, Moses had all male dependents, and all women who had had intercourse, slaughtered. Only virgins were to be spared.[37] At least 100,000 women were burned as witches by church and state. Mansfield cites a dozen tribes, from Jivaro to Hottentot, where the "desire for sexual revitalization" strongly motivated warfare.[38] Pondo women were said to accompany their army and encourage the warriors from adjoining hills by lifting up their skirts. Clausewitz defines war by recourse to an oddly associate sexual imagery not unlike diplomatic rape: "an act of violence intended to compel our opponents to fulfill our will . . . . a continuation of political intercourse, with a mixture of other means." The Roman Empire's *deditio*, their affirmation of "total and unconditional submission to the imperial will," furthers this logic.[39]

Among the Yanomamo Indians of northern Brazil – beyond chest-pounding, side-punching, and head-wrapping duels and the proud display of scars – the motive for abducting women provides sufficient cause for genocide. Axes, arrows tipped in curare, and lanceolate bamboo are employed in the raid. Every opposing man is killed, every woman captured.

It is not too great a leap to envision enemies clutching sexually at one another. The Nuba wrestlers of southern Sudan act out their aggression in a peace-loving sport that nonetheless has all the trappings of sensual but deadly combat. Naked, with cicatrization beauty scars, the wrestlers who succeed will have a better chance with the women. Sexual continence occurs prior to harvest, and before a wrestling match. Both areas of contest prepare the ground for fertility. The village sacrifices its honey and milk to the young fighters. The wrestling initiation takes place outside the village in the ascetic *zariba*. Young men sleep on stones, cover themselves with white ash and milk cream and prepare for their first match, which is merely a mock engagement. Meditative, with horns and drums, covered with strips of cloth, the young warrior fights, as if to the death, and then sits expressionless, pondering the sacredness. The young man's head is shaved so that he appears to be a ghostly visage from the ancestral past. The Nuba heroes lead their villages to the dusty, elevated rings on the real day of intertribal battle. Gourds hang from their lissome, carousing buttocks; they are decked in elaborate feathers, cowries and goat-hair headdresses; they wear belts of woven leather thong, furs, swinging carcanets. They carry a long spear, their bodies are mud-plastered. Acacia wood goes to the victors from which the holy ash is taken, an ash which gives meaning and expression to the soul. The Pan Hellenic games also set naked men against each other, rigors not easily dissociated from the real thing, as figured between warring hoplites on any of numerous vase relics. In other animals – monitor lizards, marine iguanas, cichlids, horned or antlered ungulates – chivalrous feints, and neck-in-neck stand-offs determine the victor, a non-violent accomplishment (usually) in accordance with an immemorial and sensuous logic.

This Nuba wrestling is not restricted to sport and ritual and dominance hierarchy. It extends to the more familiar negotiating tables, though without obvious clues to its psychopathology. But the image escalation of sex on the battlefield carries over to the styles of Cold War, the language of indecisive politicians pressured by conflicting data, personal careers, unrelenting constituencies, and the very pessimism most laymen share in the outraged or apathetic perception of the inevitability of doomsday. This sense of ruinous outcome hanging over each moment is attributable to the mechanism of normal sex, with its beginning, middle, and end, in that order; a veritable doctrine of climax lending to war the same energizing principles that infiltrate the workings of biology. Peace, premature withdrawal, and onanism, undermine the otherwise healthy metaphor of liege, family man, and patriot — the gift of progenitors. In its absence there is only extinction, defeat, the dwarfed canons of odorous monks and masturbatory legacy. Peace initiatives are construed as impotency, the invitation to be violated. At the negotiating crossroads is pretense, sham conciliation, the flaunting of feathers, a routine show — political strutting. But in fact the peace table further incites this exhibitionism. During a British-Tibetan treaty-signing occasion in the early 1900s, a British member of the negotiating team drew his pistol and killed the Tibetan signatory.* Beyond the blatant is the slough of similes, from amity to rape, from parity and balance of terror to armed conflagration, suidical reciprocity, orgasm, which take their turn from U.N. cat-calls to more pernicious slanders.

In the end, there is never to be the balance to which these actors — in whom the human species has feebly vested its hope — have for so long and expensively pretended. Only death, death that is famous for its randomness. More likely, between the various scenarios worked out by computer, one nation will *come* before the others. Every tactile stimulus up until that judgement day provides all the necessary incentive, lewd and vicious innuendo, intervention, all proffered with a kind of "cunnilinguistics," a sadomasochism. And not a single verity to speak of. Both partners to the crime are zeroing in on infinite generations that will never see the light of day.

These super powers are lumbering giants, irrepressible twitches and erections. What keeps the sullen monsters temporarily in check is the merest suface of history, a shimmering, baroque registry of time-worn

---

*This little known fact was told to me by the late King of Sikkim, 2 weeks before India annexed his country in 1975.

dialectics, party turnovers, formula deceptions, ballot-box hoaxes, jubilant boasts, atrocious Doublespeak – the derisive duping of a sovereign and gullible people, all people – fueled in the tickling maneuvers of foreplay; in a bed where two estranged lovers collect their willfulness, flirt with true desire, before consulting their generals and unleashing their pleasure.

In the deadly game of détente (untouching bed partners) fundamental attraction is at work between opponents, an attraction that sustains the military. If there were no enemy, eventually there would be one. "Once injured, the body never heals," said Michiko Sako, one of the Hiroshima maidens brought to the United States in 1955 for surgical treatment. We could compare this to Edward Wilson's idea of genetic aggression selecting an irreversible path throughout the course of evolution. The sexual metaphor is the truest *idée fixe*: the need to touch, to be engaged, locked in combat. Both parties to a negotiation may sit disarmingly civil, side by side, soberly calling off the tally of demands, a science of safeguards, while each is busy auctioning arms contracts at home. Brezhnev was generous with his hugs, and Reagan preached a Puritanical, antiabortion insecurity. Together, their brio applied to consenting adults, consenting madness. Reagan's hatred of his adversary had no subtlety about it. Just weeks before the U. N. disarmament talks in the summer of 1981, the president threatened to reduce the Marxist-Leninists to the "ashes of history".

The sexual motive has the tantalyzing force of tradition, of everyday life to recommend it. Recast in wartime, the power and authority of state-of the art technology and manpower gains on the deception, the seeming attractiveness of war by national fervor. Again, the sexuality is implicit. A big penis is like a big gun, and the allusions to this obvious cognate have their own cinematic, literary, and practical jokester traditions, from Freud's universalization of the phallus, to ballistic spermatazoa in Thomas Pynchon's *Gravity's Rainbow* and the hieratic traditions of Portnoy and the Marquis de Sade. These unisons ramify so ubiquitously as to overstate any epistemology of sex and violence. And yet there is the technology, the rape: from Roman siege engines that could throw a 600-pound stone as far as 900 meters, to Medieval pikemen, halberdiers; of cutlass and longbow; from the Sumerian bas reliefs featuring heavy infantry with spears, to Australian Aborigine waddy, boomerang and stone hatchets; from javelin, blowpipe, thrusting spear, battle-ax and sword, to MX missiles there is the terrible infatuation with overheated, throbbing projectiles, points, collision media. The price of such antics, other than heavy losses of life (which many have condoned on the sheer ecological grounds of balancing an overly populated planet) and the obliteration of animal and plant species (nearly the entire primate population of Vietnam now extinct) speaks in subtle moments of dolorous doubt.

Adopting the sexual metaphor to explain instances of aggression is not to suggest unilaterally that all men — and presumably the few Joan d'Arcs — are sex fiends, repressed infants who have suddenly been afforded the opportunity to rape with impunity. We cannot prove it. Sexual valency, manhood, machismo, protection of loved ones, the desire to find a wife, fears of homosexuality, revenge and jealousy — all play into the sexual court. But whatever the personality twist and concatenation of synapses, the weapon makes all the difference, exerts persuasion between the dominant and submissive partners. King Kamehameha, the "Napoleon of the Pacific" transformed the tranquil Hawaiiaan Islands of the nineteenth century into a military empire with the acquisition of European firearms. The New Zealand Maori intergroup slaughters resulted from the importation of the European musket. The same happened on the New Hebrides. The legendary Swiss mercenary force — the only truly autonomous military power in Medieval Europe — was easily broken under the firepower of Francis I at Marignano in 1515. Imperial, Papal, Spanish, and northern Italian armies were likewise routed by the new order of explosiveness. In 1494, Charles VIII of France invaded Italy with forty new cannons and obliterated every previously impregnable castle from Leghorn to Naples. One particularly impressive fortress, the Castle of Monte San Giovanni, fell in only 8 hours.[48] The Yuman Indians of the American River went off to war hallucinating on jimson weed. But the short-barreled 1873 Winchester carbine, turned them and many other plains Indians into the same impulsive desperados as those who hunted them down. The fact that technology supports desperation in the face of death (offering the last hope for salvation) further likens its importance to insemination.

Finally, we must acknowledge the ecological coefficient between sexual engagement and reprisal in war. Ritual altercation among the New Guinea Maring tribe has always relied on a homeostatic ecosystem. The killing of pigs, the burning and clearing of forest for new crops, and the replanting of fallow soil, as well as the crucial constraints against occupying fallow land, has ensured — in the height of conflict between subtribes — that a third, higher court of law prevailed. Among the Maring there is no "tragedy of the commons," either in peace or in war.[41] Among the Pimas, Mohaves, and Pueblos (sedentary farmers) acute adaptation to life in the desert precluded any exhorbitant feuds that might endanger fertility of the fields. Again, this higher ratiocination demanded that people be able to till, plant, weed, and harvest on time, without fear of invasion.[42] The conscious externalization of the sexual metaphor is best visualized by considering mankind's dependency on fertile land, an absolute that defines Cain's paranoia, works ravenously to usurp, grips the warrior in the belly as much as in the groin. In terms of entropy, the drive toward agricultural plains characterizes the search for the sun, for warmth, for the lowlands. Such fertility of the soil must conjure an analogous if deep and driven inspiration between the legs. The majority of nomadic peoples who have migrated throughout history have done so with agriculture and population constraints in mind; have pushed south and west from mountainous terrain, where exiguous yields meant a mounting food deficit and an increasing need for contraception or infanticide. Sunny lands would offer the dreamed of reprieve from the two most deadening circumstances. Gutaean barbarian highlanders occupied Sumer and Akkad (2230–2120 B.P.), inundating the rich hegemony of Tigris and Euphrates basins. The Old Kingdom of Egypt resisted wave after wave of northern Mediterranean invader precisely because it had firmly consolidated control of the Nile below the first cataract, where jungle swamp had been transformed into fertile field. But by 1750 B.P., the Hyksos invaders on horseback – the horse being perhaps the most formidable first weapon, with its own constellation of erotic tie-ins to terror, speed, strength, majesty, and a momentum allied to the groin's own victorious throb – managed to disrupt Egyptian civilization for the rest of its duration. As for the fertile lands along the Nile, they have remained inviolate.

Of the nearly 5000 spoken languages in the world today, every one possesses the argot of sexual relations. Our capacity for verbal and written embellishment of sexual pleasure constitutes one of the most compelling and powerful of human manipulations.

LANGUAGE AND VIOLENCE

Etymologically the word *tragedy* comes from the Dorian words *goat* and *ode*. Together, the original meaning was "lyric song." The Attic Greeks made such pieces for the stage and called them drama. Tragic heroes, once lowly shepherds, were stricken with the caprice of deities, their eyes gouged out. Poetry has rushed to string its laundry along the rusting barbed wire of human anguish. The literature that endeavors to fashion or reveal such moments is like a photograph. There is nothing enrapturing or realistic about stories that have invented the inevitability that drives them, and from which the audience – the reader of a novel, of the morning newspaper – need merely walk away, turn off the television set.

There is no way to be certain of our myths, parables, and history. Dante claimed that the entire written past was mere gossip. Much of Western man's suffering has come under the umbrella of Job: "And after my skin has been thus destroyed, Yet from my flesh shall I see god."

The aesthetics of violence confront our sensibility with a grave dilemma: How much can we be moved before contradicting the very message of such art, and thus impairing its remedial intent?

The museum, movie, and bookshop goer has been besieged by horrors meant to titillate or confuse. It is manifest in the writings of the fascist Celline and Pound, in crucifixion scenes, in the cosmic belittling, nasty, farting ogres from Grimmelshausen to Artaud's theatre of cruelty. The success of such films as *The Texas Chain Saw Massacre* parallels the high ratings of fancily paced news, of famine, assassination and war. The arts have been caught up in an axiom of death-by-salesmanship, an apotheosis of all that is dehumanized. This corporate reflex coins catchwords in a connoisseurship of atrocity. The critique of this trivialization has seen numerous proponents — from the Polish survivor of Auschwitz, Tadeusz Borowski, who ironically gassed himself to death in 1951 at the age of 27, to Samuel Beckett and Solzhenitsyn. All three men have sought to describe the end of the world. What provides some reason for their pessimism is the fact that their respective truths must be immersed in the same language pool of comedy, prates, banters, and the day-to-day disinterested. The victim of tragedy is left with one last link to sanity — his language. With an inconceivable delicacy he must pursue those words conducive to describing a monster. Desiring no solace that would only cheapen his despair, believing neither in mankind's redemption nor in the advent of the New Age, how does the victim instill a searing, indelible image into the vocabulary of those left unscathed; how to forge an apocalyptic impulse in our very syntax.

But the more inspired our radical obsessions with literary expurgation, the more our lives have plunged in proportion. The artist must now confront an ultimate horror about his own words and their unlikely effect on his neighbor; language — by which we would create ourselves — is irresponsible. It has no memory, is as deadly as the troops that dragged the young Borowski off to the camps. Does the victim howl from a paddy waggon or intimate in rhyme; does he remain silent? All of the paintings capturing warfare are silent — imagine; in the midst of bright flashes, bloody ejecta, and swirling battalia, no painting will every convey a sound. The heart of any depicture is the same death-still silence of immobile Wounded Knee at its aftermath. From the eleventh-century Bayeux Tapestry commemorating the Norman conquest, to the Flemish illuminated manuscript of Saladin's "Siege of Jerusalem" (mid-fifteenth century), to William Simpson's "Charge of the Light Brigade" and Tosa Mitsuyoshi's "The Battle of the Uji River" there is only an errie quiet; an uncomfortable, mournful color composite in the eye of a hurricane, the frozen hell, Polish lancers, flamboyant Mamelukes, screens, canvases, a codex or two, not a word. The outrage is intellectural, framed, forever softened in the zoom lense of artful purgatory.

Language promises a more direct confrontation. "The final Solution to the Jewish problem" was a phrase coined by Hitler and his cronies (Alfred Rosenberg, Herman Esser, Heinrich Himmler, Hermann Goring, Reinhard Heydrich, Adolf Gemlich, Adolph Eichmann, Josef Göebbels). This catch-all euphemism was one of many, and represented Hitler's talent for generating virulent ambiguity. In his early speeches he used the words *Entfernung, Aufräumung,* or *Beseitigung* – all meaning elimination *and* expulsion. In this way he never actually appeared to be saying outright that he intended to murder the Jews, only to rid Europe of them.[44] The diabolical verve of Hitler's vocables (first captured in Leni Riefenstahl's films) has often struck the passers-by of history since World War II as incredible proof of mankind's naīvete. For sociologists of mass behavior, the very language is the aggressive codon, amplified by so many Victorola-like repetitions.

The literary critic George Steiner, famous for his proposition that silence was the only dignified response to the Holocaust, re-examined the nightmare in his disturbing novella *The Portage to San Cristobal of A.H.*, published in 1980 and performed on the London stage in 1982. As a deep analysis of violence and how it can maintain its hold upon the imagination, this work of fiction is bold and picric, straining between good and greater evil.

Hitler spent the end of the Second World War in a shelter beneath the Berlin Chancellery, playing out his feverish histrionics to their rightful conclusion, marrying Eva Braun and then setting loose a conflagration in their bunker. But Steiner's theory postulates some other wantonness: a look-alike (the ashen carapace and shattered dentures nearly convincing), while the real Hitler is off and winging free in an airplane through a Wagnerian bluster of gutted skies – the cocoa, hammocks, and curare of South America on his mind. But five members of a relentless cabal stalk the Fürer who is now ninety, eyes undimmed, with singular purpose still (like Moses on Mount Nebo), surrounded by rotting trip wires, a thousand scummy miles into Brazilian outback. The five avengers come from Israel and Europe, sending out their progress reports on short-wave, in ciphers from Revelation and Malachi. They have inexorable Kaddish and nausea in their hearts. They are cautioned over radio by their liaison: "Gag him if you must. Words are warmer than fresh bread; share them with him and your hate will grow to burden!"

Through sulphrous airs, their eyes swelling, toxins in the lungs, noses to the breeding swamp, each Uzzi-toting envoy is consumed by holy conjuration: arcane testimony from Deuteronomy, the memory of torn nails, shaven skulls, yellow stars, identification numbers branded onto arm pits, children thrown from trains, Ruth Levin forced to clean the police station's latrine with her hair; of the Rabbi's tongue ripped out, and Rahel Nadelmann's legs tied down while a half-dozen SS rammed rifles up her vagina. But A. H. only gains in endurance. Though his captors keep him leashed, he grows stronger, comes alive; is ready to spring with millenialist ferocity.

Hitler challenges his escorts to Rabbinic psychologies and glowering polemic. Instinct with malevolence, antipodal to the Jew, he is connate with biting advocacy. Hitler proclaims *himself* a Jew. Later on, in discussing this boast, one of Steiner's characters writes, "How else could he have understood us so perfectly? Found us in the hidden places? Known that we would walk into the firepit, with a dozen butchers, lame men, dogs gone in the teeth, herding a thousand Jews, a hundred thousand? How else would he have known?" Hitler will defend himself with hammering persuasiveness, knowing all too well the "secret power" of the chosen ones, they who "invented conscience."

Steiner is all *vivace*, his resounding indictments nearly palpable, in mordant peels and strident arpeggios, and the macaberesque, same rush of wit that galls while giving Yiddishkite its fatal blend. Hitler marshalls his vindication with insane reasoning, projecting a thoughtful humanism. He predicates a single syllogism in defense, that "the Holocaust was the necessary mystery before Israel could come into its strength." That Jews thrive on adversity is fitful platitude, and Steiner is remorseless in his trenchancy. With a final hellish grammar, with *the* evil utterances that can, among a welter of 3 million and tweleve characters in Torah, destroy the world, Hitler exploits the limits of human semantic organization.

But the Jews are determined to bring Hilter back alive and try the bastard properly, send him branded throughout the countryside where, irony toppling irony, he may well die a contented tourist. Such is justice. The Jerusalem image of that glass prisoner's pew inheres as two West German lawyers charged with their government's confounding briefs, attempt to size up the very premises and principles of law, in Hitler's case.

The younger lawyer Rolf, from Oxford, opines a legal forum, international in appeal. Dr. Röthling is querulous, treading with doubt on the very question of responsibility. Can the guilt of Nazism ever be identified? Could *any* law procure a measure of justice? Would we have all acted like the Germans under similar circumstances?

But once stated, these injunctions lost their sting. The older Doctor halts his idle flow and marks his bigger meaning:

> "Like a terrier scratching for old bones. People who do too much of that go queer, they think they're making deep and terrible statements on the behalf of the dead. They aren't. They're puffing up their own little lives. Oh it was hell, we were in it up to our eyes — while it lasted. And for a few years more. You can't remember, you were a baby. But now, looking back, don't misunderstand me *Junge*, I can't help wondering whether it was very important."

Steiner strikes powerfully at the German psyche with its historic hells, and neo-Nazi revivalism. An elusive, spine-tingling cancer works an old psychosis — gathers grim, grey speed in the cortex — is enlivened, luminous in the personage of A. H. He comes back to torment the living with their own deadly flirtations, only better stated. In Japan, history books delete all mention of Japanese atrocities in World War II; in the Soviet Union, Stalin is again lionized, the favorite darling of Tiblisi. The late Andropov spoke a fine English. One assumes, one hopes, that a shared language would promote new gains toward peace and decency. But then the language of violence has also, for all time, been shared.

NURTURANCE

The psychological, sexual, and linguistic motives of warfare begin in fairly innocuous ways, at home, during the early years, when the accumulation of certain toys, friends, and relationships with animals leave their bias and predillections in the young adult, the conscript. The nurturance of the child within its community constitutes every hope. The anthropological literature is enticing, depending upon which observations, case studies, and prejudices we wish to believe.

Among the New Guinea Fore tribe, infants are kept in perpetual bodily contact with their mother, but at the same time are given the unrestricted freedom to actively explore their environment. In their uninhibited pursuit of interest, the children are encouraged to fool around with adult toys, to make use of the parent's own tools. "There was no need to play tricks or deceive in order to pursue life," writes Richard Sorenson. "Nor did they have to act out impulsively to break through subliminal fears induced by punishment or parental anxiety. Such children could safely move out on their own unsupervised and unrestricted."[45] Granting complete security to children among Fore clans obviated any need for the young person to sublimate and connive. Their external world was allowed to meet their internal. This harmony has of late been challenged by the conversion of their traditional protoagricultural lifestyle into a fully sedentary farming settlement. The influx of outsiders buying up land has mounted tremendous pressure. With the arrival of Australian rule, the Fore have actually welcomed outside arbiters in the newly arisen disputes over territory. As a result of this mediation, and of the particularly sensitive and open relationship engendered between parents and children, Fore society lacks nearly any violence.

Robert Levy's study of Tahitian society yielded a similar views. But among Tahitians the nurturance appears to be embodied in the very "gentleness" of their islands, and in their "adaptation to a sufficient, dependable, steady-state, delicate ecology."[46]

Colin Turnbull's intense 'upbringing' among the Mbuti Pygmies of the

Ituri rain forest in Zaire has been frequently cited to prove mankind's peaceful capabilities, given the right circumstances. From a child's first few days, when he or she is literally dancing in synchrony with the mother's movements, she is wrapped in bark, sleeps on phrynium leaves of which the forest leaf hut is composed, and is thus never out of olfactory range of the life-giving environment, *ndura*, God. The children quickly learn to climb trees, to meditate on the natural history of sylvan sensation, to imbibe the spiritual essence, in other words, of the forest, the *molimo*. More quickly than their Western peers, the Mbuti babies express themselves. Within a few years they may wander naked under the full moon, across the damp forest floor, transfixed with the skein of dew collecting over them, the breeze lolling their bodies into perfect equality with nature — and this is the peace they will give to others. Sexuality is neither stifled in the young nor overtly encouraged, merely taken for granted. Curiously, the Mbuti will use matches only for "profane" acts, such as lighting a cigarette. And at the end of every year, they will burn up most of their possessions as a cathartic means of maintaining their purity. Remarkably adept at dwelling in their unique locale, Pygmy civilization is older than that of the Egyptian.

Among the !Kung bushmen, there is minimal competition. Continuous lack of privacy — all the more remarkable given the 1-per-26-square-kilometers density of hunter/gatherers throughout Angola, Botswana, and Namibia regions — enhances the socialization of the young. Lacking surplus storage facilities, they find that they need each other. Patricia Draper, who claims that the !Kung are incredibly modest and funny, writes that "real anger frightens and sicken the !Kung."[47]

In 1966 a Philippine hunter discovered the Tasaday clan on Mindanao. He taught them how to hunt small animals. Previously, they knew nothing of hunting, a condition also prevalent among the Phi Thong Luang, first discovered in 1941. The Tasaday comprised thirteen children and thirteen adults. Journalist John Nance managed to spend 73 days with them and wrote, "their love was everywhere — for each other, for their forest, for us — for life . . . . They have no weapons, and no apparent aggressive impulses."[48]

Citing work in the early 1920s by A. Kroeber, Quincey Wright claimed a "remarkable peacefulness" for the Yurock Indians of California, a group who susbsisted on salmon and acorns, choreographed the deer and woodpecker ballets, kept strong legal codes, but by the 1950's had dwindled to below a breeding population size after innumerable assaults in the ninetheeth century. The Samoans, the Tanzanian Hadza, the West Malaysian Semai (with their numerous expressions of external and internal restraint), and the Utkuhikhalingmiut and Qipisamiut Eskimos, have all been looked to for supposed nonaggressiveness. Among still other clans, what little belligerency does prevail is reduced by the ritualization of restrained technology and impulse. Australian Murngin avoid blood shed almost entirely, says Eibesfeldt. Opponents dance up to each other with their spear tips deliberately blunted. The infrequent murderer is likely to receive no more punitive action than a wound to the thigh. The Australian Walbiri will aim their spears only at their opponents' legs. Among the American Indians, the ambush and murder of one person was considered the only true style of manly raiding. The Dani of West Irian forego feathers on their arrows to decrease the likelihood of serious injury during battle. Eskimos take out personal revenge in song duels. And even the Pentagon rhetoricians worry about certain in-house devices likely to confuse the Soviets, thus activating mutual hair-trigger computer trajectories when none were intended.

But Eibesfeldt, the brilliant student of Lorenz, has debunked much of the foregoing. He doubts that any culture is free of some form of warfare. The original pacifistic portrayals of the Ukusiksilik Eskimos by Fridtjof Nansen are more than reversed by sobering renditions of hunting boundaries and intergroup aggression between Cape York and northwest Greenlanders, and the Copper and Netsilik Eskimos, as provided by R. Peterson and H. W. Klutschak. Similar balance sheets can be drawn up between L. Marshall's empathetic study of the !Kung versus less complimentary ones (V. Lebzelter, and H. Vedder). Eibesfeldt compares Turnbull's work with the Pygmies to those accounts of M. Bicchieri and P. Schebesta, the latter actually describing a Pygmy ambush employing poisoned arrows, clubs, lances, sickle-shaped knives, and woven shields. Eibesfeldt's own field work showed Margaret Mead's perceptions of the supposedly loving Samoans to have ignored among other things, the fact that the word *malietoa* ("brave warrior") is stamped upon every Samoan coin. Mead's insistence that the Arapesh of New Guinea are peaceful Eibesfeldt rebukes by turning to the work of R. Fortune who actually detailed the Arapesh war-making facility.[49] And Finally, unrelenting, Eibesfeldt disparages of all the romance surrounding the Tasaday. "If one were to conclude from observations of the friendly relations in a Tyrolese neighborhood that Tyrolese are a particularly peaceful people . . . one would be wrong." His logic that twenty-six people (the number of Tasaday upon their discovery) would discourage aggression obtains. As does his personal interview with the Minister for Minorities in the Philippines who related the fact that the Tasaday were known to have attacked a neighboring tribe, the Tboli. The attack, admittedly, was not unprovoked. The Tboli had been hunting monkeys in the Tasaday forest and may have set a trap that caught a Tasaday. No one could say for sure.

What can be said with some resolve, I believe, is the flexibility of human nature, particularly in the early stages of a child's life. A loving environment will always prove crucial to a happy human being and may be the only unbiased argument of optimism to serve against mankind's penchant to destroy itself. When has there ever been a truly happy environment? We all need to give and take love, lots of it, endless profusions of it. And when so doing – when elevated to the position of intimacy – is not aggression irrelevant? No one would disagree, assuming that the whole world were swept up at once in this embrace, from the personal sighs of fond companionship and liberating passion, to the full acknowledgement of global nudity in which all of us, heart and soul, might renavigate the clashing rocks of our endangered humanity. The language, the sentiments, the supplications are as universal as music, dance, the drawing of stick figures, and making love. Communication and love are the only antidotes to war, and more often than not left untried in the shipwrecks of dawn.

NOTES

[1] Julian Jaynes, *The Origin of Consciousness in the Breakdown of the Bicameral Mind* (Boston: Houghton Mifflin, 1976).

[2] Irenaus Eibl Eibesfeldt, *The Biology of Peace and War; Men, Animals, and Aggression*, trans. by E. Mosbacher (New York: Viking, 1979).

[3] Raymond Dart, "The Predatory Transition from Ape to Man," *International Anthropological and Linguistic Review 1* (1953).

[4] Eibesfeldt, *Biology of Peace and War.*

[5] S.L.A. Marshall, *Men Against Fire* (New York; William Morrow, 1947).

[6] George Will, "A Word for the Wilderness," *Newsweek*, 16 August 1982.

[7] Sue Mansfield, *The Gestalts of War: An Inquiry into Its Origins and Meanings as a Social Institution* (New York: Dial Press, 1982).

[8] Arnold Toynbee, *Mankind and Mother Earth: A Narrative History of the World* (New York: Oxford University Press, 1976).

[9] Ibid; and Eibesfeldt, *Biology of Peace and War.*

[10] Marshall, *Men Against Fire.*

[11] William B. Gault, "Some Remarks on Slaughter," in Stefan A. Pasternak (ed.) *Violence and Victims* (New York: Spectrum, 1975).

[12] Louis Rene Beres, *Apocalypse: Nuclear Catastrophe in World Politics* (Chicago: University of Chicago Press, 1980).

[13] Mansfield, *Gestalts of War.*

[14] Eibesfeldt, *Biology of Peace and War.*

[15] *Los Angeles Times*, 30 May 1982.

[16] Beres, *Apocalypse.*

[17] J. Muller, "On Accidental Nuclear War," *Newsweek*, 1 March 1982.

[18] Toynbee, *Mankind and Mother Earth.*

[19] Quincy Wright, *A Study of War*, 2nd ed. (Chicago: University of Chicago Press).

[20] Angeleos Terzakis, *Homage to the Tragic Muse*, trans. A. Anagnostopoulos (New York: Houghton Mifflin, 1978).

[21] J. Glenn Gray, *The Warriors: Reflections on Men in Battle* (New York: Harper & Row, 1967).

[22] Sigmond Freud, *Outlines of Psychoanalysis*, trans. J. Strachey, reprint, 1940 (New York: W. W. Norton, 1949).

[23] Freidrich Nietzche, *The Will to Power: An Attempted Transvaluation of All Values* (London, 1913).

[24] Edmund Burke, *A Philosophical Inquiry into the Origin of Our Ideas of the Sublime and Beautiful* (London, 1757).
[25] Burke, *Philosophical Inquiry.*
[26] Wright, *A Study of War.*
[27] Cleveland Pendleton Hickman et al. (eds), *Integrated Principles of Zoology,* 5th ed. (St. Louis: C. V. Mosby, 1974).
[28] Edward O. Wilson, *Sociobiology: The New Synthesis* (Cambridge: Harvard University Press, 1975).
[29] Ibid.
[30] Cited in Wilson, *Sociobiology.*
[31] Mansfield, *Gestalts of War.*
[32] Gregory Bateson, *Mind and Nature — A Necessary Unity* (New York: E. P. Dutton, 1979).
[33] W. Green,
[34] Arthur Engendorf, "One Vietnam Veteran: A Study of Continuity and Change," *Dissertation Abstracts International,* October 1977.
[35] Eibesfeldt, *Biology of Peace and War;* and Luigi Valzelli, *Psychobiology of Aggression and Violence* (New York: Raven Press, 1981).
[36] *Politics,* ii, 9; and Wright, *A Study of War.*
[37] Numbers, 31.
[38] Mansfield, *Gestalts of War.*
[39] John Keegan and Joseph Darracott, *The Nature of War* (New York: Holt, Rinehard & Winston, 1981).
[40] Ibid.
[41] Peter Marsh, *Aggro: The Illusion of Violence* (London: J. M. Dent & Sons, 1978).
[42] Joseph G. Jorgensen, *Western Indians* (San Francisco: W. H. Freeman, 1980).
[43] George Steiner, "The Portage to San Cristobal of A. H.," *Kenyon Review* (1980). Steiner's novella was also published by Simon & Schuster (New York, 1982).
[44] L. Dawidowicz, The War Against The Jews, 1933–1945, (New York; Bantam, 1976).
[45] E. Richard Sorenson, "Cooperation and Freedom among the Fore of New Guinea," pp. 12–30 in Ashley Montague (ed.) *Learning Non-Aggression* (New York: Oxford University Press, 1978).
[46] Robert Levy, "Tahitian Gentleness and Redundant Controls," pp. 222–235 in Montague, *Learning-Aggression.*
[47] Patricia Draper, "The Learning Environment for Aggression and Anti-Social Behavior among the !Kung," pp. 31–53 in Montague, *Learning Non-Aggression.*
[48] Cited in Montagu, *Learning Non-Aggression.*
[49] Citations for Nansen, Peterson and Klutschak, Marshall, Lebzelter and Vedder, Bicchieri and Schebesta, Mead, and Fortune can be found in Eibesfeldt, *Biology of Peace and War.*

Chapter 15

Ecology and Existentialism

Since wealth and civilization breed as many causes of wars as poverty and barbarism, since the folly and wickedness of man are incurable, there remains but one good action to be done. The sage will collect enough dynamite to blow up this planet. When its fragments fly through space an imperceptible relief will be given to the universal conscience — which, by the way, does not exist.

Anatole France, *Penguin Island,* (1908)

All in all, there's no shortage of great Führer figures: a bigoted preacher in Washington and an ailing philistine in Moscow . . . . Of course we still have (as trademarks of salvation) good old capitalism and good old communism; but thanks to their tried and true enmity, they are becoming . . . more and more alike: two evil old men whom we have to love, because the love they offer us refuses to be snubbed. Big Brother has a twin.

Gunter Grass, *Headbirths or the Germans Are Dying Out,* 1980

The creature must, while there's still life,
Give all support to the spine he bears,
And on that invisible backbone
The wave will play for all it can.
As if a child's tender cartilage
Were the earth's own age of infancy,
They've brought the very crown of life,
As lamb, in sacrifice again.

Osip Mandelstam *"This Age"*

For Norman Mayer, who threatened to blow up the Washington Monument in protest of the nuclear arms race. Killed on December 8, 1982.

We are the species of conscience, giving life and taking life. Aware of this syndrome, the ecological and artistic sensibility has sought analogues throughout the biosphere. But these excuses of simile can not atone for the barbarism that, apparently, presists in our own nature. Our ability to feel suggests its own special place in biology, an island of hope in the tumultuous seas of evolution.

EVERY DESPERATE AGE

In Tertullian's second-century *Apology*, the record of Christians filling the ranks of military service is offered to a doubtful, war-peppered posterity. Pacifist doctrine in the West was uncommon before the Waldensian movement in southern France during the twelfth century, (though for a previous millenium in China, rudimentary convictions of peace were synonymous with the active codification of nature). Anabaptists, though many were to end up in Oliver Cromwell's Model Army, promoted nonviolence during the Reformation, as did the Mennonites. Quakers too have taken up arms at various times, though their creed is one of passive resistance and best epitomized by William Penn's unarmed settlement in the colony of Pennsylvania. The Plymouth Brethren and the Christadelphians refused to bear arms, and numerous elements in Judaism, Methodism, and Roman Catholicism have opted for conscientious objection during wartime. Nevertheless, the Old and New Testaments sanction violence, racism, as well as genocide. Even Gandhism can be perceived in a disappointing light. Twenty-six years of nonviolent struggle resulted in a free but divided India. Hindus fought Muslims, and an estimated 500,000 people perished.

Dozens of inspired peace plans have been advanced throughout history, visions of universal, benevolent monarchy during the Renaissance, the Hindu *Tat tvam asi* and Buddhist *Shikishin Funi* – calls for spiritual brotherhood, the Jainist nonviolent *ahimsa*, and *Jiva*, or life principle. Dante, Erasmus, Rousseau, Kant, Thomas Paine, and Jeremy Bentham proposed enlightened solutions to human self-aggression, which is what the conflict between nations always amounted to. International free trade, the effective distribution of greater shares of wealth, defense of one's borders with no further affronts, mandatory disarmament, neutral treaty zones, alliances of nonfortification, the League Covenant and U.N. Charter: despite such pledges at various times, an inexorable determinant in human relations has not stopped for a moment the deadly escalation of arrogance, hatred, and an aggressive technology, often under the galvanizing principles of Crusade, French Mission *Civilatrice*, Manifest Destiny, the German Place in the Sun, Pax Romana, Pax Brittania, America Love It or Leave It.

We have no solution to any problem of character; no salvo that can easily redress, no proclamation to make things better. But there may be an approach to problems, a true dynamic, which speaks the same language between parties, instructs on the vagaries of personality, and humbly, simply assays to find coordination. Without that language, no world consensus will ever manifest the power of restraint to cede its rancor to a higher authority. God has not had sufficient clout; the beauties and fragility of life on Earth – everywhere nude and blushing – have not apparently impressed the mind, though the heart quivers to the slightest touch. The soul begins to flourish when someone, or something, speaks to it. But amid the warrior's wrath, where is the soul's access point? At what crossroad does humanity speak out? And when it speaks, who might be listening?

These queries are those of every desperate age. In 1648 Europeans fell back sickened and exhausted after 30 years of redundant bloodshed. It took not one but two A-bombs to persuade the Japanese Empire to quit its own offensives. We are stubborn and have always been so. By chemical and biological drives we are the species of hedonists. The blood rushes in our brain. The adrenalin courses our veins. The awakening of conscience must compete with a hormonal makeup most closely related to the orangutans, but whose outlet of feeling holds on to arsenals far more reckless and cataclysmic than mere sticks and stones. A bias works in favor of our oblivion, we are lazy, our reaction times sluggish. Extinction patterns are closing in on our species. When the air begins to cool, you can be sure of a cloud overhead. Whether today's cloud is a passing one depends on no caprice of meteorology.

We can reduce a sequence of events in our mind easier than we can add to it. Our imagination determines our reference point, and it is tragically limited. Since 1945, we have had the statistics, seen the ashen after-images, kept relative tabs on the increasing number of warheads, and been hapless, apathetic or raging bystanders to a U.S.–Soviet collision whose purpose and inanity must cause the stoutest archaeologist of a phantom future to wince, to curse out in perplexity.

CONSCIENCE IN HISTORY: THE PARADISE COLLISION

The spectacular ascendancy of human aggressive curiosity in conflict with the subtler spheres of feeling, of delineations between good and evil, must take its place, in evolution. This latter word is apparently useful in describing chaos and regularity, a barrel of monkeys, a belly of microbes, the vast wastebin of has-beens, the launching pad for up-and-comings. Evolution, as experienced by *Homo sapiens,* is a polarity of forces perceived between being and becoming, – that is to say, conscience. From as early as the fifth-century, BC, Greek Stoicism described an instinct for self-preservation whose expression was termed *synderesis.* In Ezekiel (i.4-15) Jerome translated it as *scintilla conscientiae,* or "spark of conscience." Physically, our species has remained the same for tens-of-thousands of years; culturally, the changes have been mechanically expedient, rapid, though we have no way of properly assessing the dimensional shifts in self-perception or basic need since the early Paleolithic Era. We have ample evidence of postglacial Aurignacian religion from a cave in Spain; of a Chinese (Abbevíllean) world view replete with fancy dinnerware and a pronounced preference for pleasing views 600,000 years ago; of European Acheulian campfire girl indulgences on the Rivieras of prehistory. Human *being* has remained the same, undifferentiated, wisely stable, devoutly happy, and integral. Human *becoming* introduces what may very well be an opposing pace, out of chorus with the older, more fully guaranteed niche. This quandary has obsessed twentieth-century anthropology and ecology.

As beings we had managed quite charismatically to enjoy the tubers, berries, occasional rabbit and venison; the full moon, the seashore, our mates, offspring, and whole community; the swoosh of adrenalin, rapture of sudden silence. Eight million strong, in global clusters, following the Würm fourth glaciation we were the birthright of a 40-million-year-old strategy, having arranged our simple hierarchies, animism, stone cultures, and migrations in accordance with circumstances deemed optimal by Earth. The planet had contrived through diversity and patient manipulation, the fondling of pearls, and the sensualization of mangroves to ensure the modest livelihood of our curious and ungainly forebears. The species built on a record of substantial investments in time and attrition. The sub-Anthropoidea branched out in trees and dispersed across savannah. Like scat, our myriad tools, charred seeds, and pollen remains tell a long, unarguable tale of being at home in the world. That's *my* interpretation.

But subtle change was underway, a grain, then a pebble, then a fist. Mind, like the Chesire cat enamored of itself, seized the opportunity of its own malignancy. We are only now wondering whether or not the Earth has been keeping tabs on this swelling of consciousness; if it *feels* what's going on and can or cares to cope. But the battle lines are assuredly drawn up, and a checks-and-balance system is at work, if only residually. As our ideas populate the planet, a new order of attrition is engendered. Nearly 3 billion microorganisms per 6 gram of fecund soil are annihilated; the reorganization of vegetal and inorganic mass and the patent provision for murder on a scale equivalent to some unthinking asteroid come into collision with the whole Earth. Our powers of abstraction and engineering finesse have not catered to the paradise originally granted us, but to some idea, an alien thing, dumb matter from cold space, brute rock burned and hollow and going nowhere and reverting to the indivisible glints of mute element that compose it. Not one to be stunned by a warning, the mind carries on in its conversions. Human cognition is prone to betting on bad odds, even in the face of stronger alternatives. The condition is both left and right handed, (hope and forlorn), and the two opposite peripheries are distinguished by the advantage of infinity and idealism in the left, and gross, final ultimatum in the right. The mind in a state of becoming gives promise to the human being while tempting its surfeit. War, robotics, artificial intelligence.

What questions does this paradox raise? The questions themselves are weighted by the languages we wield and the circumstances each of us individually can refer to. Without a joint god or recourse to common intimacy, even to shared information, we have little willingness to trust a stranger's sources of judgment — his facts, his perceptions. By remarkable degrees, we each know a different Earth. We raise our nostrils to the night firmament and sniff a different world; touch ourselves to a different rhythm; inflect, taste, imagine a differing drama. We each possess private *logical types*, cybernetics stamped with our own personality; our feelings and biases can be said to be *stochastic* — given to certain outcomes allowed to endure despite prevailing, outer selection mechanisms in nature and society. Our beliefs are *tautological*, which means that they comprise propositions we tell ourselves repeatedly (and live by) and whose truth cannot be doubted — whether it is correct or blatantly fallacious — because of the logical validity of the links, the person enjoining those beliefs; and this inviolability of self, of being, is easily translated, missed, likened to *topology* — that branch of impersonal mathematics treating formal relations, all of those characteristics that remain unchanged despite continuous quantitative distortion. In planetary biology, topological insights tell us that nature prefers no one species over any other; that the British navy existed because of British bumble bees (navy-meat-cows – clover – the pollination by bees).

As early as the Twelfth Dynasty under Amenemhet I (twenty-second B.C.), a genre of dialogues had been inscribed in stone. "The Dispute of a Man with his *Ba,*" probably written by one Khety (Papyrus Berlin 3024), reveals the life and anxiety of a "seeker" who asks to speak with someone in the universe capable of redressing the wrongs which are at large in the world.[1] That someone is the man's soul. By arraignment, reply, argument, rebuttal and parable the ba convinces this ancient melancholic to "be adamant about life . . . to make a harbor for the occasion." The seeker, who has seen his wife and children die, has sat all day under sails, face into the breeze, "longing for home, the clearing of the sky, a rain-washed path . . . the smell of lotuses," is rejuvenated through his reflections. He has held the problem in his hands, figured it out, and restored equilibrium to his heart. We can *feel* for this man in a manner that, conversely, elicits only scorn for the seeker's contemporary, Naram-sin, pinioned atop the victorious, marauding shoulders of his blood-thirsty comrades; the ugly vision of a 3500-year-old, 6-foot Bronze Age warrior buried near Thebes with a gleaming sword, spearhead, a helmet made of boar's tusk sewn to a leather skull cap. Interred near the home of King Oedipus, this corpse is a clue to that "sanguinary dementia" of the *Iliad,* which so contrasts with Old Kingdom Egyptian peace and moderation, heart-to-heart commiseration and the love of family. Even the Old Kingdom Pharoahs conceded to an equality of genders. Note the statue of Fourth Dynasty Mycerinus, hand-in-hand with his wife, both of them very much in love, steadfast, the exact same height. "Happiness is the same height as man," said Confucious. He, too, was referring to woman, to a universal absence of arrogance.

The Greek word for peace was *irene,* and it figured in the earliest philosophical essays, histories, and drama, as chronicled by Gerarado Zampaglione.[2] It was spellbinding reality in Hesiod's conception of the Golden Age, the source of spirit in the earliest Greek elegiac poetry of Callinus, in Theognius' maxims. Theognius was forced from his home of Megara in wartime, and spent much of the sixth century B.C. wandering throughout the Hellenic mainland extolling the virtues of peace. His passion signalled the advent of peace cults in Greece, dominated by statues of the Goddess Irene, which were erected in markets and forums, the most famous having been scuplted by Praxiteles' father Cephisodotus. Aeschylus and Euripides, and particularly Aristophanes, were the first *radicals* in history for their literary expressions of these cults. The notion of *victory,* they said, was a lie. "Why does war ever have to break out," asks Euripides in *The Suppliants.* "Leave others in peace. Life is short." The career of Aristophanes (448–385 B.C.) paralleled that of the Peloponnesian War, and suggests the painful interface of artistic sensibility and mass violence. Aristophanes' life was marred by the failure of his own earnest efforts to draw attention to the immorality and senselessness of armed contestation. Following the invasion of Attica by the Spartans, he produced his play *Acharnians* (425 B.C.) in which an Athenian farmer, Dikaiopolis, concludes his own treaty with Sparta, eludes military services, and lives peacefully and privately according to a plan better suited to his age, – but rejected. A classic in pacifist literature, it was soon followed by *Knights*, in which Cleon is accused of illicitly pursuing war against the notorious aggressor, Brasidas. A year after the play was first produced, both leaders were killed at the battle of Amphipolis. The playwright celebrated the ensuing Treaty of Nicias in his character Trygaios, who liberates Irene from the war god Polemos in his play *Peace*. But the euphoria was short-lived: Seven years later, the Sicilian expedition resulted in the loss of 50,000 soldiers. In *Lysistrata* (411 B.C.), Aristophanes' women sexually refuse their husbands until a lasting peace is assured. It never was. Zampaglione's analysis is trenchant.

The crisis of conscience among commentators of the Old Testament produced the prophetic literature, the fear that Jews had been abandoned by their God and the millenialist rallying of Isaiah, Ezekial, Jeremiah, and Micah. The capture of Samaria by the Assyrians in 722 B.C., the later military campaigns of Nebuchadnezzar II, and the two-time conquest of Jerusalem with subsequent deportations into Babylon – would all confuse self-defense with the concept of everlasting peace during the wars of the Maccabeans against the Seleucids and the insurrections against Rome's second-century dominion. "The Lord is a man of war," Moses had said. (*Exodus*, 15:30) He is "mighty in battle" (*Psalms*, 25:8). Justice, divine wisdom, eternal dominion, and the washing away of sins was apt theology when conducted with the weapons of self-preservation. Paul exploited the military metaphors (helmets, breastplates of faith) toward the hope of salvation. In his *Epistle to the Ephesians* (6: 13–17) he advised a Christian to don the armor of God and be a "good soldier of Jesus Christ." (Quoted in Zampaglione.) Canon XII of the Council of Nicaea (325 A.D.) discouraged military desertion. And writing during the period of Emperor Valens' defeat by aliens, John Chrysostom conveyed desperate belief in Christ, the Prince of Peace, though not before recommending upon the virtues of waging massive war against the corruptibility of the body, whose immoral consequences he classified as more odious than civil war. If the body was sinful, then the massacre of other human beings could easily be condoned, as it later was in the Crusades. Again, Zampaglione is compelling.

But of all the early religious speculations on the nature of aggression and the human potential for morality, St. Augustine's theory of *just wars* is perhaps the most inciteful. Zampaglione has brought selective evidence to indicate the extreme dichotomy at work in early Christianity, a metaphor implicating a pattern of reality hard to dissociate from the appearance of human nature. In mathematics, Hindus, Mayans, and Babylonians invented the *zero* which, like the discovery of oxygen, plunged human morality into competition with desperate proportions, the exponential and the combustible versus nothingness, embers, humility. In his *Contra Faustum*, Augustine analyzed the inevitability of conflict. His model was the thermodynamic rush of catastrophe, Alaric's Goths. He asks, "Who can really say whether, in time of peace, it is a good or evil thing to reign or to serve, or to rest or die, or in time of war, to command or fight, to vanquish or get killed?"[3] Augustine held that Christians were the best fighters, and that those whom they killed would die anyway by God's will. But he did condemn some aspects of war, namely cruelty, revenge, "uncontrollable rage," and the "desire for dominion." Recommending military duty for all disciples of Christ, Augustine went firmly against the grain of the leading Donatist and Pelagian pacifists of his age, Christians looking to a remote state of nature ruled by benevolent angels.

Despite its unprecedented record of violence, the Roman Empire produced an equal testament to restraint, of an original condition in nature when men lived in caves or trees, surviving on fruit, milk and honey, in absence of all greed and manufacture. Peoples of the Mysian coast were said to live on in that state, opposing all war; vegetarians, the inheritors of Arcadian ethos. Virgil was of a like mind, proposed simple gardening, the rural life of *Georgics*. "Then shall wars cease and the rough ages soften . . . . The gates of war, grim with iron and close-fitting bars, shall be closed; within, impious Rage, sitting on savage arms, his hands fast bound behind with a hundred brazen knots, shall roar in the ghastliness of blood-stained lips."[4] Cicero went on to deny morality in any action of expediency. But these were regrettably thin, little doted-on sentiments.

A man could expect to live but 32 years at the time the Stoics were flourishing in the Roman Republic: Reason in itself to be philosophically minded. By the fourteenth century in Europe, society had gained some reprieve, matching in longevity the achievement of the Neolithic, or 38 years. Our ancestors had struggled for at least 6500 years to establish commercial empires, imposing edifaces, reliquaries of one divinity or another, all suffused — so many words, deeds — all lost in the shuffle for dominance and added life. It was Dante who first called for unity, a world order under the Holy Roman Emperor.[5] Pope Boniface VIII proclaimed himself global peacemaker. The idea was revolutionary, enlisting cartographic balance-of-power ploys, guidelines, an evolving *shape* to human communications that transcended the every-day; made commoners more feverishly aware of stepped-up potential, be it in the cities or in some Orient. Europe in the fourteenth century was beset with birth pangs, the oozing wobble of protoplasm, what Barbara Tuchman marvelously hailed a "calamitous" time.[6] It was an era weathered by the welter of fatal judgments, bad luck, tragic heros, villains, monsters, insane kings, grappling dukes, knights errant, philosophers, recklessness. There were louts behind cassocks, prostitutes forced to wear their clothing inside out, pet monkeys, ubiquitous flower venders, legless beggars with wooden stumps strapped to their hands; a continent without knowledge of potatoes, tea, coffee, and tobacco; of heights measured by somber monks lying prone with trigonometrical devices. The *chanson de geste* were aswoon with the lurid details of romantic love, of which children received little, more normally beaten for sport, as were cats (quartered in village games). Rape was common, hardly a crime. Money deals were conducted in public steam baths. All the while Giotto was painting *feeling* for the first time, and Marco Polo was lending more land to the human imagination than any monarch. It was a time seemingly ripe for a message, and that injunction came murmuring on Genoese trading ships putting into harbor at Sicily in 1347, carrying dead and dying seamen aboard. The Black Plague struck literally overnight, originating on caravan routes in Central Asia. It was to diminish the population from India to Iceland by 50 percent. Scholars at the University of Paris insisted that there was no religious cause but rather the rare triple conjunction of Mars, Saturn and Jupiter in the fortieth degree of Aquarius. Holy flagellants danced themselves to death. Others wandered from village to village, setting fire to Jewish quarters, where soap was invented. Thousands of knights and peasants joined up in Free Companies to pillage towns already in decay. Whole cities were raided. In France, the only protection to be had was in the dubious charge of these *routiers.*

We might expect a disaster so all-pervasive as the Plague to bring people together, drawing from the experience of community zealots banding together in dungarees with sandbags in hand to save a town. This species of world-threatening disaster – armageddon, the awakening trumpet blasts of Ishmael on Judgment Day – has been approvingly counted on in much science fiction to congeal nations in a unified front. But unfortunately, the machines of an H. G. Wells, fast-approaching meteors, or the Great Flood have had no more systematic purchase on the sense of internationalism than the equally fatal and certain threat of nuclear suicide. There is indeed a psychology to mass panic, and it looks very different than the ordered altruistic effects of small-town catastrophes, where people can relate to the problem in the company of acquaintances. We might then assume that local victims of the Plague inspired the collaboration and goodwill of their neighbors. Nothing of the sort. Highwaymen taunted the swelling ranks of refugees, and the

Albigensian Crusade furthered the decimation. Everywhere the Plague struck was levied the license to loot, murder, to act in keeping with a God who had lost faith in man.

Curiously, certain pockets of mankind were momentarily spared ruination: a region of Bohemia where not one citizen got the Plague, the residents claiming that their gaiety kept the Devil away. Others were less happy about the century into which they were born: Margery Kempe, for example – a Saint famous for her weeping. Those who survived went on about their business. More wars, more killing, no sense of human vulnerability, or not enough of one to bring people together. If anything, the Black Plaque reaffirmed the idea of evil, justifying an intensification of weapons to conquer it. Moreover, nature was devalued, so that all romance, all images of Eden were either forced back in time or inaccessibly away – to Abyssinia, the Himalayas, remote cliffs in the North Atlantic. The belief structures in the applicability of a golden age were eroded to an accompanying resolution that violence begot and befitted mankind, now construed to be a victim whose only hope was martial strength. God was dead. It would take an army to resurrect him.

Erasmus cried out against this trend in European culture.[7] He urged the clergy to refuse burial to all who were slain in battle, remarked on nature's law of love and cohesion as man's truest reason for alliance, and asked that the "unbelieving Turk" be coaxed by the example of Christian innocence, "friendly treatment," and beauty and gentleness over massacre. But Erasmus' appeals were muted by a blast of new artillery introduced in European battlefields: infantry muskets able to kill a horse at 400 paces, the bullets powerful enough to render body armor ineffective. Three years after Erasmus' treatise, the old defenses of the Landstün fortress in the Rhineland were demolished in a single day. Unthinkable! Bullion from Mexico and Peru, meanwhile, were proving to the Spanish crown just how lucrative genocide could be. A Dominican friar of Salamanca, Francis de Victoria, following Cortes' Mexican rampage, delivered lectures stressing the need of international law. And 50 years later the judge advocate general of the Spanish armies that invaded the Netherlands (1566), Balthazar Ayala, called for restraint through law and the exchange of diplomats. These were feeble hopes.

Shakespeare's plots give ample evidence of dramatic ambivalence and pathology during the period of escalating firepower that followed Francis Bacon's research on gunpowder. Europe witnessed cannibalism and the murder of more than 7 million German-speaking peoples from religious wars. And this tragedy is reflected throughout the body of Shakespeare's works, redundancy of grief. The examples are too numerous for easy digest, but a few will make clear the symptom. Note the murders in Macbeth; Richard II's treacherous murder of an uncle and his banishing of his cousin Bolingbroke, only to be murdered himself; the too-hideous career of Richard III (still contradicted by fan clubs), that bloodiest of princely protagonists, whose imbroglios and Tower travesties involved no less than nine slaughters, preceding Richard's own death; the rivalling Montague and Capulet families ending in the suicides of Romeo and Juliet; a dinner that turns into a massacre complete with the cutting off of heads, tongues, and hands before General Titus' son Lucius finally assumes the throne in *Titus Andronicus.* Horatio at the end of *Hamlet* is persuasive and unflattering about this sickness of his age:

And let me speak to th'yet unknowing
world
How these things came about. So shall
you hear
Of carnal, bloody and unnatural acts,
Of accidental judgments, casual slaughters,
Of deaths put on by cunning and forced cause,
And in this upshot, purposes mistook
Fall'n on th' inventors' heads.

During Shakespeare's generation, the Dutch jurist Hugo Grotius (1583–1645) responded to the mayhem of the Thirty Years War with a call to global codification of ethics and laws diminishing the possibilities of uncontained conflict. His treatise *On the Law of War and Peace* laid the groundwork for a mightly presumption that "there is a common law among nations which is valid alike for war," a notion based on his belief in the universality of rational thinking.[8] With equal verve, Grotius' contemporary the French monk Emeric Crucé was writing with the idea of a world community guided by "common sympathy" and the hidden alchemy of economics. The young Pascal similarly sought such a body, a "neutral third party". In absence of arbitration, argued Locke, men were pitted against themselves in a state of nature. For Montesquieu society falsely bolstered human confidence, inspiring complacency, inequality, and inevitable warfare. Pacifist jurisprudence had to set us straight, put us more in touch with our beginnings in nature.[9] This was no easy task, as Count Buffon concluded. He considered our species the most unhappy of all animal groups, frail and failing, born too soon.[10] Development of the human foetus is oddly cut short, perhaps by as much as 12 months, whereas we attain our puberty at twice the body weight of most other mammals. This situation if reversed, would endow us with smaller, less lethal size and a far wider, more purposeful humanity. Shakespeare's dramas would have been free to explore and encourage more gleeful play beyond the horror. Between all of these sinister clashing rocks, the existentialist precognition grew up in Europe. Following the War of the Spanish Succession, the Abbé Saint Pierre called for a mutual security pact between nations, relying on the idea that perpetual peace required constant amendment of treaties in keeping with the fluctuating imperfection of human nature.

But the Enlightenment produced a wave of optimism that moved from anthropology to economics. Adam Smith and the French Physiocrats reinstated the laws of nature with an aim towards promoting laissez-faire biology, invisible hands, the *four stages* theory of human development: hunting, pasturage, agriculture, and commerce, this idea embodied in Locke's celebrated adage "In the beginning all the World was America". Rousseau's profound interest in the evolution and corruption of mankind in any society and his alterability in the social contract is only the most romantic and revolutionary of treatises written at the time. But whatever the lead-in, equal distribution of wealth was resoundingly perceived among liberal philosophers as the obvious antidote to war. This was not a cry for socialism, but the heart of industrialist drive, the "False Consciousness" against which Marx was vituperative. Thomas Paine's pamphlet *The Rights of Man* actually provided a preamble to the Marxian critique by asserting that "man is not the enemy of man but through the medium of a false system of government".[11] Jeremy Bentham called for the abandonment of colonies and alliances, the reduction in size of Britain's naval forces, and the opening of all free trade.[12] In America, Benjamin Franklin published another universal peace plan, this being the purported work of a "former Galley-Slave," Pierre-Andre Gargaz, who suggested that militiamen be converted into an army corps of engineers to help society.

Of all the late eighteenth-century peace proposals and theories of human nature, Immanuel Kant's *Perpetual Peace* is the most thorough and engaging. "Within the limits of pure reason," says Kant, peace is the only purpose and end of legality.[13] The Americans were good at this. In 1781 Ambassador John Adams told the Foreign Minister Vergennes under Louis XVI, that "the dignity of North America does not consist in diplomatic ceromonials or any of the subtleties of etiquette; it consists solely in reason, justice, truth, the rights of mankind, and the interests of the nations of Europe." Kant's work detailed the formation of an international federation of seperate nationalisms founded on the imperatives of cool logic and survival. He recognized individual preferences of behavior in any civil array, but insisted on a body politic that would fruitfully inspire checks and balances, calm public relations, and the abolition of evil. Kant did not look so far as altering human morality. Rather, he sought out a variety of social sciences — from transcendental aesthetics to human geography — for insight into the "mechanism of nature," that perfect regulation of conflicting interests. By adopting the "irresistible will of nature" human beings could realize peace and this "guarantee" would bypass languages and differing religions. Like his Scottish predecessors, Kant proclaimed the "power of money" to be mankind's easiest solution. It wasn't. Napoleon's meteoric dash from Madrid to Moscow to Leipzig created profound national deficits, with 500,000 men needing amenities to keep them convinced of the battle rationale.

Historically, an economic curve could be drawn in which no monarch has ever been able to seduce – by compulsion or otherwise – more than 2 to 3 percent of the total population into war readiness, with the exception of the samurai standing armies. Nevertheless, the Prussian Chief-of-staff Count von Moltke relied on the doomed Napoleonic ethic to overwhelm Austria in 1866 and France in 1870 with massive armies. Though no shot was fired between European nations from the time of the Prussian alliances to 1914, the cause of strained alliance was not wealth but the exhorbitant cost of battle. Maintenance of vast armies and munitions kept factory workers enslaved throughout Europe. America's Civil War, the Lopez War in South America (1865–1870), the Chinese Taiping Rebellion (1850–1864), the wars for Italian and German unification, and the Crimean War suggested nothing approaching universal peace. As some indication of the armaments consumption, the United States contracted the supply of 1.5 million rifles during the Civil War. Cost-benefit contradictions were to escalate.

There was no better portrait of revulsion, disappointment, and futility for these incessant conflicts than in the eighty-five etchings, "Los desastres de la guerra" (The Disasters of War) by Francisco José de Goya y Lucientes (1746–1828), published posthumously in 1863. Like Beethoven, Goya's work took on additional outrage, irony, and punch following a mysterious illness in 1792 that left him completely deaf. In 1808 Napoleon planted his brother Joseph on the Spanish throne, Wellington's forces invaded the Iberian Peninsula, and the horrors of occupation ensued, augmented by the French paranoia. Goya had written below one drawing, "I saw this," referring to the depiction of rape, mutilation, murder, and madness. In another, dated 1814, he wrote painfully, "One can't bear to see such things" beneath his rendition of men and women begging for their lives before being blown away and bayonetted. Only the bayonet points come into the picture. His protests were beyond Romanticism, concerned with real horrors and the speechlessness of the victim. Eugéne Delacroix's "Scenes from the Massacres at Chios," based on accounts of the Turkish atrocities during the Greek War of Independence for which Lord Byron gave his life, sustained the growing artistic passion to depict such events while championing the isolation of modern man. It is extraordinary that the two artists who singularly gave expression to the power of crime and its modernist rebuke should have been physically isolated from the world. It is as if their being deaf was their initiation, enabling them to step into purgatory and recount the descent without distraction, only the eyes burning, the conscience rended; a chord, a few words, strokes of a faint brushwork. The pain of impotency, by our own inventiveness. Again on a Spanish battleground, Ernest Hemingway reiterated Goya's crisis of conscience: "You see the murdered children with their twisted legs, their arms that bend in wrong directions, and their plaster powdered faces. You see the women, sometimes unmarked when they die from concussion, their faces grey, green matter running out of their mouths from bursted gall bladders . . . . You see them sometimes blown capriciously into fragments as an insane butcher might sever a carcass. And you hate the Italian and German murderers who do this as you hate no other people."[15]

To this day and onward, despite 10,000 years of yearning for paradise, for private, confined love; of families and friends and laughter and mortality; of discussions with the soul, the heart's rambunctious meditations; of frescoes and operas, of delicate sonatinas and solemn occasions; of laborious legal formulations that describe, but are unable to dictate; of awe in discerning our smallness, amid ever fancier magnitudes – despite all of this, each one of us is more deeply naked before the eternal, frustrating fact of our species' depressing ailment. For all of the nineteenth century's delight in nature – from Francis Danby and Dante Gabriel Rossetti, to Pierre Bonnard, Melville, Van Gogh and Sir Leslie Stephen – there was an immense, unspeaking gap which knew no salutation, had no lineaments for commentary, was simply there; malevolence, awful necessity, survival of the meanest, the richest. No parkland legislation could mask the increasing range and effectiveness of firearms – 550 meters by late century; of ironclad ships and railroads capable of quickly moving whole regiments; of anchored mines and torpedoes and the first submarine. Industrialization meant technological war. It was a disgrace more fantastic, more alluring (to some) than anything in our history. If recession fueled it, and the balance of powers demanded it, where was the human conscience to impound it? The retreat to ecology and existentialism would enhance the tragedy of that gap as scientific revelations, social realist fiction, the discovery of origins, and the first hints of a doubtful future amassed on the frontiers of every thoughtful denizen. These hints raised the spectres of the unconscious, brought the entire gallery of medieval demons, superstition, and then panic upon the unnerved sensibility of every land.

"Without communion with the dead a fully human life is impossible," said the poet W. H. Auden. *Homo erectus* in China arrayed the crania of its dead as ponderously as the monks of Mount Sinai have maintained a charnel house of bones watched over by St. Stephanos since the seventh century. According to Shinto doctrine, a child is not to be named until the end of its first month, when all the ancestors have been acknowledged and the child's survival taken as a sign of their propitiation. That life is inseparable from death – that we are unable, other than as distant, unmoved spectators, to observe, even to imagine our own dissolution – is at once the core of all human dramaturgy, of the future tense, an angst that fights foreboding, even in the thick of existence. Given our rudimentary cognition of death, the greater enigma may be our ability to tune out from death, to live our lives with little or no effort in preparation for the inevitable, about which we have made no sense. If birth is a gift, death, too, is a gift. As there are ill-timed pregnancies, so too, are many terminal illnesses premature, by which I would cite those of youth or middle-age. We do not dictate the duration of a pleasant sound's Dopler effect, nor count the seconds of a shooting star; we feel no intolerable grief for the collapse of an old tree or the fading of autumn, but rather know such cycles for what they are.

And for our fellows we mourn but go on living. And sometimes there is added courage and tenacity in painful memory, a nearly happy outcome of the dead issuing in our own declarations to ourselves, the seizing of days, the powerful tainting of vital signs, the honor that we pay to fragility. In the wake of tragedy and tears, we are, if embittered, yet more emboldened to our quirks, brief joys, and requisite life in the coming days. Our language is not inconsiderable on the occasion of death. We have the means of full-blown tribute, of surrounding pain with passion. On Greek islands, in high Ladakhi yak-herding camps, among mountain tribes of Ecuador, at Black funerals in the Transvaal threnodies are cacophanous and prolonged. We can absorb the life of a deceased, live it alongside our own, set placemats, or intern articles of good use in the afterlife – a penny, a candle, gold lockets. However we accommodate the *idea* of death, we have no choice but to live, idea or no idea. And this rendering of thought is what has inspired in us the hated, sacred "other," that ineffable beyond words and thought, the flight of the mystic; no special category of thought, but rather of the delimiting, the extinguishable in thought; the realm where all words and dreams pale beside action and the moment. The incantations of today will always subsume disasters of yesterday, regardless of pyramidal funearies, alabastor memorials, resplendent adagios, or the cold, lopsided slate, deep-etched with the wrath and misgivings of survivors in Puritan cemeteries throughout New England.

For all our efforts to improve death – from Edwin Lutyens' architecture of national pride in the cemeteries following World War I under the Commonwealth War Graves Commission (gravesides adrift in dwarf lupines, alyssum, and candytuft) to flamboyant funeral parlors (the American way of death), we do what is necessary to ensure business as usual. This genuine capacity of the viable-in-nullity works against us, degrades our language, infects humanity. Its zenith is war. I want to quote an extended passage from Jonathan Schell's catalytic work, *The Fate of the Earth*, which is particularly apropos here:

Meanwhile, we are encouraged not to tackle our predicament but to inure ourselves to it; to develop a special, enfeebled vision, which is capable of overlooking the hugely obvious; a special, sluggish nervous system, which is conditioned not to react even to the most extreme and urgent peril; and a special, constricted mode of political thinking, which is permitted to creep around the edges of the mortal crisis in the life of our species but never to meet it head on. In this timid, crippled thinking, "realism" is the title given to beliefs whose most notable characteristic is their failure to recognize the chief reality of the age, the pit into which our species threatens to jump; "utopian" is the term of scorn for any plan that shows serious promise of enabling the species to keep from killing itself (if it is "utopian" to want to survive, then it must be "realistic" to be dead); and the political arrangements that keep us on the edge of annihilation are deemed "moderate" and are found to be "respectable," whereas new arrangements, which might enable us to draw a few steps back from the brink, are called "extreme" or "radical." With such fear-filled, thought-stopping epithets as these, the upholders of the status quo defend the anachronistic structure of their thinking, and seek to block the revolution in thought and in action which is necessary if mankind is to go on living.[16]

But this paradigm of suicidal consensus is not "anachronistic"; there was *never* a time when such structures of thought were not lethal; when a "sluggish nervous system" did not incur extinction. The lime burner Ethan Brand can not decipher his own incineration, and all the meditation, self-knowledge, restraint, and reverence of charitable folklore has not calmed our method. One good person, it is said, can save the world. But *will* he? The canons of biology, physics, and art condemn us to drinking our cups dry after letting them spill over. We're a sloppy species. Samuel Beckett has pithily dissected this paralyzing intensity of volition and genetics. In *Texts for Nothing* he wrote, "Suddenly, no at last, long last I couldn't anymore, I couldn't go on . . . . .How can I go on, I shouldn't have begun no, I had to begin."[17]

In *Some Thoughts Concerning Education*, John Locke commented on the penchant of children to "torment" animals "with a seeming kind of Pleasure." This custom, he went on, would work to repeat itself in later adult interactions.[18] Today the United States has more than 20 million hunters and 150 million public firearms. Each of us has carnivorous fangs and vegetarian molars. "When one is hunting," wrote Ortega y Gasset, "the air has another, more exquisite feel as it glides over the skin or enters the lungs; the rocks acquire a more expressive physiognomy, and the vegetation becomes loaded with meaning. All this is due to the fact that the hunter, while he advances or waits crouching, feels tied through the earth to the animal he pursues." This is disheartening ebbulience from a man of some insight who might have more humanely thought about what he was saying. Who *cares* what the hunter feels about the rocks and the air? Certainly not Shelley.

> *Let the advocate of animal food, force himself to a decisive experiment on its fitness . . . tear a living lamb with his teeth, and plunging his head into its vitals, slake his thirst with the steaming blood; when fresh from the deed of horror let him revert to the irresistible instincts of nature that would rise in judgment against it, and say, Nature formed me for such work as this. Then, and then only, would he be consistent.*[19]

Shelley actually traced the derangements of the nineteenth-century mind to the unnatural habit of meat eating, citing the passage in Milton between Raphael and Adam – "moon-struck madness, pining atrophy, marasmus, and wide-wasting pestilence, dropsies, and asthmas, and joint-racking rheums." For Shelley, the human carnivore is at the "root of all evil."

It is true that the wolf prefigures the myth of meat consumption. The animal moves vitally at brisk clip through deep snow, charges the moose, rips into its coelomic cavity thereby transmitting the pathogen of peritonitis. The moose flees, but over the coming weeks weakens until the same wolf can easily bring it down.[20] I fail to see the particular romance in this scenario, and wonder what person's conscience does not shy away from such a masterplan, sterling or not. As we have unnecessarily forged complex technologies of the kill, so must our consoling faculties balance lethal ledgers with a moral code as extreme, even counter, if need be, to that system of predation deemed necessary in the natural world. If we have truly sailed beyond most need, than this one as well can be dispensed with. Our cousin the gorilla, that he-man vegetarian, that hero whose musculature may be bristling but is fully ungiven to demonstration, he has rejected "tooth and claw" for a world of foraging. There is no sure estimate of how many species actually chase their food down, as opposed to picking it off the ground, and off trees.

Hunters claim that they are *concerned* more than most meat eaters with the quality of the kill, and of its wilderness habitat. Indeed, concerned to preserve it for continuous killing. In an age of overshoot weaponry, the only educational precept urgently applicable to all morés and gradations is the manifest cessation of all killing; to discover the megatonnage in conscience, and wield it.

The Indian pre-Buddhist Nirgrantha sect, today known as Jain, spearheaded this commandment. But unlike the dictum of Moses, then made it mandatory for all living beings, from gnat to elephant. Today, 10 million Digambaras and Swetambaras protect all life. The Jain metaphysic urges forth a soul that can never be extinguished because it has neve annihilated – a verb for a verb. The Jain soul has six colors – three bad, three good – and may enter various stages of fresh karma (body). But ultimately it is destined to live on, immortally, for it has even preserved vermin, abstained from all flesh, suicide, from abortion. The trangression of Eden, in the West, was the confusion of murder with curiosity. Indeed, Genesis (1:29–30) is explicit in the conjunction: "the fruit of a tree-yielding seed; to you it shall be for meat." We make our own myths. The wilderness reflected that complex of forces against which a conscience was first enlisted, the frown and temporary guilt, sorrow proceeded by impunity. This pitfall of Western hubris is the source of redundant protein, the hypocrisy of penance: "Make sure one's worst forebodings coincided with one's dreams of glory," wrote Yukio Mishima. .[21]

The semiotics of our death entanglement has its own charismatic syntax, a deep structure which has eluded philologists for lack of fossils.

The existentialist victim must determine his own priorities beyond the contradiction of his setting, society, loved ones, and inner laziness. Risking oblivion, taking the hard path, his analysis is full of suffering and bound by heredity and wrathful sociology. His choices and fantasies are conditioned, as is his very set of beliefs. If once he has killed, is there any sense in discontinuing the process into the future? The virgin precedes all transgression. Once defiled, the image-maker revels in the liberation to defile himself over and over. In his passion to escape massacre, the condemned lingers undecided, craves an equal freedom to rape and kill, loses heart. This futility is what Splengler termed "the decline of the West." In "The Second Coming," Yeats prepared its death ground:

Things fall apart; the centre cannot hold:
Mere anarchy is loosed upon the world,
The blood-dimmed tide is loosed, and everywhere
The ceremony of innocence is drowned;
The best lack all conviction, while the worst
Are full of passionate intensity.[22]

Sartre's conception of transcendence encompassed this terrible cycle of moral enervation, exhorted *responsibility* over the vertical occultism of escape (Christianity), and strove to condition our heart through a re-education on the street. In *Literature and Existentialism* he asked what medium might work the needed miracle in the Third World,[23] and again it was this principle of the *street* that surfaced, as it did in Paris in 1968, Argentina in 1983, in the character of the noble doctor of Albert Camus'

The Plague. It also accounts for the street gang, street smart, the street of anarchy.

Sartre was born in a compelling year, 1905: Not only did it witness the birth of Cubism in Picasso's "Les Demoiselles d'Avignon," but Einstein's special theory of relativity. Both events typified the coming concerns of existentialism, however futile it may be to pin down the term: humankind's odd role in an insensate universe, and its tragic effort to outwit dumb nothingness with machines of the future. The effrontery was characterized during both world wars in such works as Wilfred Owen's poetry, Maxwell Anderson's and Laurence Stallings' *What Price Glory*?, E. E. Cumming's *The Enormous Room*, the works of Hemingway and F. Scott Fitzgerald, E. M. Remarque's *All Quiet on the Western Front*, Jaroslav Hasek's *The Good Soldier Schweik*, Charlie Chaplin's film *Shoulder Arms*, Brecht's *Mother Courage*, and painter Franz Marc's "Fate of the Animals," which tragically anticipated his own death at the front. This painting, as Roger Rosenblum has described, merges the pastoral and apocalyptic chaos of conscience, "kinetic sensations of a modern, mechanized world, transformed by Marc into bolts of spiritual lightning that wreak havoc on a primal universe of landscape and animals."[24] The animals haunt us. Their torture we can not tolerate. Vietnam, American prisoners in Iran, and world wars will pass many viewers by. Yet the clubbing of baby harp seals ignites a nexus of feeling so fulminant as to thrust us into gear, to reawaken what Jung called that "undiscovered vein within us."

But Cubism – with its brutally confining definition of time and space – and the spate of social manifestos coming in its wake, added another dimension to the fragmentation of that Jungian vein, and this was the analysis of its very ore, the breaking down of impregnable substance after a long, glorious, untouchable history from Democritus to Dalton. J. J. Thomson's discovery of the electron in 1897, Lord Rutherford's netting of the proton a decade later, the first acquaintance with half-lives – the very jargon is dismal – chrono photography, the strobe, the motion picture – yes, the *excitement* of dissolution and reconstruction, – gave over a new world view, all in a cache of fragments. What individual could begin to unravel it? So as to retain the old world integrity, artists sought the personable. Finding little in World War I, with its 50 million enlisted men and 10 million dead, many went into exile, or upstaged an absurd era with absurdist art: the auto-destructive sculptures of Yves Tingueley, Marcel Duchamp's signed urinal presented to the 1917 New York Exhibition of Independent Painters, the feckless Initial Manifesto of Futurism (1909: "We wish to glorify war . . . . contempt for woman"); the Bauhaus mavin Moholy-Nagy's "Machine of emotional discharge," Tristan Tzara's Manifesto of 1918: "We are furious wind, tearing the dirty linen of clouds, and prayers, preparing the great spectacle of disaster, fire, decomposition . . . . knowledge, knowledge . . . . boom, boom, boom!" Whether in the push valves of Picabia or Max Ernst's gas refinery, it was an age indebted to a well-earned cynicism.

By the spring of 1927, 15 million Model Ts had been sold, coming off the assembly line one every 15 seconds. Yosemite already had an emissions problem, the same haze of distortion characterizing the fracturation of solid form in Braque and Picasso. Reality was no longer the exchange, merely, of concrete elements in a composition; it could ask of two plus two some other conclusion, change convex to concave, turn cylinders into bottles, cogs and propellers into altars of sacred space. Léger, for example, had been in the trenches of World War I and had had a self-proclaimed mystical experience with a gun barrel, then expressed in a series of works that transformed warm-blooded figures into eccentric geometries. Juan Gris would concentrate on billboards. The mechanistic bias and denunciation of nature, armored trains and weeping women, provided the confused prelude to modern abstract art.

Sartre's book *Nausea* was published in 1938. Many consider it the long-lived philosopher's greatest work. On a Monday at 6:00 p.m., Sartre's antihero Antoine Roquentin goes berserk: "the root, the park gates, the bench, the sparse grass, all that had vanished: the diversity of things, their individuality, were only an appearance, a veneer. This veneer had melted, leaving soft, monstrous masses, all in disorder – naked, in a frightful, obscene nakedness." This might have been a description of Hiroshima, offered up one year before the war got underway. Sickened by his connections to earth, unable to flee the onrush of ecological awkwardness, he is literally crushed from within by the inability to cope, and this inverted frenzy, this carefully protected twentieth-century vision of ecstacy, Sartre aptly named and exorcised during Roquentin's famed museum tour. "Farewell, lovely lilies, elegantly enshrined in your painted sanctuaries, good-by, lovely lilies, our pride and our reason for living! Good-by, you bastards."[25]

NIKOS KAZANTZAKIS

In the autumn of 1940, Mussolini's forces attacked Greece. The following May, German soldiers descended on the country. Partisans ripped the German flag from its Acropolis mast, food supplies were cut, and the outlying islands were plunged into famine conditions. Three hundred and fifty thousand Greeks would starve to death. Nikos Kazantzakis was living on the island of Aegina, one hour by boat south of Athens. His food had run out. A spoon of olive oil each day was his sole sustenance. Lying in bed to preserve energy, the 57-year-old bard wrote *Zorba the Greek* to pass the time. Two years before, his masterpiece *The Odyssey* had been privately published. Greek intelligentsia had been outraged. Only boatmen, squid fishermen, and Cretan montagnards could understand all of its 33,333 lines of iambic hexameter.

Sixteen thousand German parachutists landed on Crete. With a single antiaircraft battery, the Byzantine muleteers down from their mountains managed to kill a handfull of the Nazis. These bronze highlanders were accustomed to heartache, other foreign overlords, and doomed insurrections. They had celebrated their heros in memorized epics from the Renaissance like the *Erotokritos* of Vincenzo Kornaro. Fighting the Turks and the Venetians, this Cretan spirit – armed with all the vigor of Africa, Europe, and the Near East (Crete being the original crossroad) – moulded Kazantzakis. He was a self-proclaimed Jew, a heretic, a Christ figure nearly excommunicated. Above all, he was the modern Prometheus: "Set fire! This is our great duty today amid such immoral and hopeless chaos . . . . Sow fire to purify the earth! Let a more dreadful abyss open up between good and evil, let injustice increase, let Hunger descend to thresh our bowels, for we may not otherwise be saved."

Eighty-thousand Jews were exterminated in Thessaloniki. The Nazis put young Greek girls on supply trucks to dissuade the Partisans from attacking their convoys. Kazantzakis was nearly assassinated. At various times he was suspected of being a Communist, a German spy, a traitor. When police came to arrest him, he invited them in for tea. When his island was liberated at the end of the war, he received for his troubles 100 rolls of toilet paper from the Red Cross.

As the audience for his *Odyssey* grew, Kazantzakis assumed the presidency of the Greek Socialist Workers' League. In this capacity he visited Crete to assess the damage done, marking all of the black crosses on the doors signifying executed males. In the hills he encountered an old shepherd jumping from rock to rock who shouted down to the poet, "How is Norway getting along, my child?" To which Kazantzakis replied, "It's doing better grandfather . . . . Do you want a cigarette?" The old man smiled, shook his head, made the sign of the cross and started up the mountain. "I don't need anything since Norway's doing better."[27]

Kazantzakis went to England to write a book about postwar sentiments. Speaking on the BBC radio he addressed the "pure and honorable men throughout the world," citing the critical moment at hand and invoking the planet as a "unified organism, one nation that cannot be saved unless all are saved." Elsewhere he wrote, "Yesterday the Americans proudly announced the discovery of a poison so horrendous that with nine grams of it they can in one fell swoop kill all the inhabitants of the United States and Canada [i.e. Russia]. The gorilla, you see, without taking the time to become a human being, has discovered fire and is going to burn the world."[28] Athens refused to renew his passport. He was now officially in exile. It was dangerous even to utter his name in Greece. He was sixty-four when Julian Huxley appointed him director of translation of the classics for UNESCO. After living with Eleni Samios for 20 years, he decided to marry her. After one year, he quit the UNESCO job, moved to Antibes, travelled incessantly as he had his whole life, and wrote one novel after another: *The Greek Passion, Fratricides, Freedom or Death, The Last Temptation of Christ, St. Francis.* His many plays, *Zorba the Greek, The Odyssey*, were appearing in dozens of languages. Jules Dassin was preparing to film *The Greek Passion.* He met with Albert Schweitzer in Gunsback, was given the International Peace Prize in Stockholm, and then continued to the Dalmation Alps where he met with the American poet–translator Kimon Friar who had been commissioned by Max Schuster to translate *The Odyssey*. Kazantzakis finished his autobiography, *Report to Greco*, and then left for China. In Canton he was mistakenly vaccinated against smallpox. Kazantzakis had leukemia. His arm swelled, he raced to Tokyo, to Copenhagen, finally to Freiburg, having flown over the North Pole. But it was too late. Schweitzer came to visit him in bed a few days before he died, October 27, 1957, aged 73. His last letter from China had foretold all: "I force my body to obey my soul, and thus I never tire. We shall return to Europe via the North Pole. I am saying farewell to all things, all things are saying their farewell to me. Nevermore. The fairy tale is coming to an end."[29]

But let's go back towards the beginning and examine Kazantzakis' particular insights into freedom, and its expression for him in Nature, his tonic. Though anticipating the existentialists, Kazantzakis renounced guilt. In his fifty-two books, his politics, his asceticism, and world-roving observations he was to create a world as luscious and profound as that of Picasso's and to provide a dazzling portrait of peace and tranquility amid all the sorrows of the Holocaust. Kazantzakis did not deny the horrors of his age: He recast them in the context of an altogether new belief in evolution, the role of art and aesthetics in human culture, and the profound clarity by which the wilderness critiqued human immorality.

He was born in 1883 in Megalo Kastro, beneath the limestone crannies of the Cretan alpines, overlooking the startling blue sea. His city had always been besieged, under garrison. His and his father's enemy was the Turk. As a child he had been prepared to kill his mother and sisters rather than let the Turks at them. After a particular night of bloodshed, his father took him to kiss the feet of Cretan heros hanged by the Turks in the public market. His grandfather had been a pirate, his mother a saintly woman. Kazantzakis grew up a self-proclaimed "scribbler," to his father's dismay. He was early obsessed with martyrdom, pictures of the saints. Later, at Mount Sinai and in Toledo, this passion would be consummated with his discovery of El Greco – his new Grandfather, also a Cretan. El Greco's "Healing of the Blind," "Assumption of the Virgin," "St. Peter," "St. Francis," "Crucifixion," and "Virgin with the Saints" were imprinted on his soul. From the earliest age he was determined to steal away on a boat and visit Mount Athos, there to become an eremite.

He was schooled by French Roman Catholics on the island of Naxos. One night he was lowered down from the school's turret in a basket into the arms of his waiting father, an irate, wounded man, just back from liberation fighting at home where he had gotten wind of his learned son's immanent departure for Rome at the urging of the Fathers. So he threatened to burn the school down, and Kazantzakis was thusly relieved of a religious career. He went to the University of Athens, wrote a thesis on Nietzsche's concept of moral law, obtained his degree, and went on to Paris where he studied under Henri Bergson. His first novella, *Serpent and Lily*, was published in 1906. In it we can see the first outpourings, the advance notices of firebrand, the way a wanderer in high country hears the wind before such wind is upon him.

> *O to love nothing, to hate nothing, to go far away from men and near the beasts – away into the wilderness. And there alone with my untameable soul confront the heavens. To strengthen my thought with the spectacle of the endless wilderness and to become an element of the tempest and a gust of the simoom and to unite with the silent spirit of the wilderness and baptize my soul in fire, in the colors that revel and carouse every night out there in the west.*[30]

This kind of provocation and announcement is prevalent in all of his early work, the plays *Day Break, Fasga, The Overseer, Comedy* (remarkably like Sartre's *No Exit* but 50 years anticipatory), and his essay "Science Has Gone Bankrupt." The material grew in him insatiably, was unprescribed, unchannelled; it was simply there, a gigantic desire. There is an archetype in all of his early writing synonymous with forceful gales, labyrinths, flying fish, and high mountains; of the bull rings in the Cretan frescos at Knossos. This was Kazantzakis' deep feeling for nothingness, the abyss that was to haunt all of his works, but that gave his characters a compelling dignity and hope. He fled, embraced, craved all, craved nothing, in a vortex of sweet and sensual desire identifiable in the masks of the women he adored, and, especially, in the one who came to him when he was already the mature philosopher, Eleni (Helen), who stayed by his side until the end, a freedom fighter herself.

By 1910 Kazantzakis was furiously translating everybody – Nietzsche, Eckermann, Maeterlinck, Darwin, Bergson, William James, and Plato. He met the Greek poet Angelos Sikelianos, and together they spent 40 days and nights exploring Mount Athos, communing with cave-mournful monks, combing the aeries to "discover the consciousness of their history." Christ, Odysseus, and Don Quixote were wrestling inside Kazantzakis to emerge. "Here in my impregnable solitude I sensed that even the most insignificant of God's creatures – a grain of wheat, a worm, an ant – suddenly recalls its divine origin, is possessed by a God-inspired mania, and wishes to mount step by step."[31] Back down from the mountain on Christmas Eve 1914, he approached an almond tree, asked it to speak of God, and the tree blossomed. It was the beginning of the Kazantzakis legacy.

In 1916 he hired Alexis Zorba to work with him at his inherited lignite mine in the southern Peloponnesus. The mine collapsed, Kazantzakis gave up business. But Zorba taught him a great deal, most of all to love life unstintingly, not to fear death. He accepted these beliefs with savage optimism. Zorba went off to Serbia, and Kazantzakis headed back into Europe, equipped with prolific paper and ink. He had nothing to offer a Zorba, a man who could charm and plumb every oblivion with his santir music and moonlit beach dancing. One day later on, Kazantzakis would receive a telegram from him: "Found most beautiful green stone. Come immediately. Zorba." He taught Kazantzakis the difference between thought and action.

Nikos followed the itinerary of Nietzsche through the Swiss Alps, searching for a single rock somewhere between Sils-Maria and Silvaplana where his acknowledged master had had a revelation of bitter joy. But even his new Golgotha dissolved: The Superman was another bribe of the hereafter, a mirage. How did Kazantzakis interpret him then? He got only the business about struggling, for that's what he needed to find; to stare with juvenile bravura, or romantic melancholy, or foolish abandon, straight into the abyss; to burn with joy and renew all life on the very edge of disaster, thus elevating nihilism to an ingenious euphoria. Nietzsche, the Alps, and poverty were working like nettles on Kazantzakis, urging him outward. The heavy casualties and atrocities of World War I fully disclosed, Kazantzakis drew up his own League of Nations blueprint. At the same time he was busy firing proposals off to editors. Nothing worked to bring him an income. Until came a miracle – his first opportunity to put theory into practice.

Venizelos, a fellow Cretan and first President of Greece, appointed Kazantzakis as Director of the Ministry of Public Welfare, with the specific mission of restoring 150,000 uprooted Hellenes of the Caucasus to their native land. These Greeks were being harassed by the Kurds and Bolsheviks, horseshoes nailed to their feet. For 18 months Kazantzakis traversed the Caucasus, making four trips to Russia. But Venizelos was forced to resign, and the Greeks whom Kazantzakis managed to resettle in Thessaly and Macedonia suffered yet more torment. Kazantzakis considered his one period of politics to be a failure, resigned himself to the written word, and headed for Vienna, where he wrote *A Year of Solitude*, burned it, and then prepared the first draft of his monumental play *Buddha*. At the same time he wrote his philosophical credo *The Saviors of God*, the very germ of his *Odyssey* to come. At Alserstrasse 26, he had his first bout of hideous eczema on the very eve of a new romance. Freud's disciple Dr. Wilhelm Stekel diagnosed it as Saint's Disease. The 39-year-old had only recently finished his play *The Christ*. Around him were the inflamed signs of the rising National Socialists, flagrant anti-Semitism, Communist worker strikes. A newspaper cost 200 billion marks. "An appalling mask of flesh was glued to my face . . . . I was not a man, I was a demon," he wrote. "Get out of Vienna," Stekel warned.[32]

In Berlin in 1922, surrounded by five devoted Jewesses, but having renounced the flesh, troubled by the Nietzschean Overman ethic as Hitler seemed to be embracing it, Kazantzakis immersed himself in Buddha, the prophet that was to have the most long-lived hold on his spirit. The play works on many levels. The Yangtze River is on the verge of flooding, and the diverse residents of one particular village are in a frenzy of propitiation. Revolutionaries, old wise men, whores, warlords, children, their mothers, magicians, poets, the chorus, political mandarins – they all must confront the inevitable. Suicides, sacrifices, engineering chimeras, ashes of ancestors, the clangor of gongs, and the dowsing of dreams commingle in the resignèd pageant of obliteration. Two characters, Mei-Ling, sister of a revolutionary, and Old Chang will attain their salvation by recognizing that the river *is* Buddha, come to deliver them. But the Buddhistic acceptance of death beyond the vanity of all effort is not the end. Old Chang persuades the wailing chorus that necessity can be transformed into freedom. Chang opens his arms to welcome the river. Full of color, music, bold language, mythic in scope (but local in detail), set in twentieth-century China, the underlying assertion is that of a radical ecology. For Kazantzakis the entire planet comes alive, makes value judgements, ruthlessly determines its future. And Man is witness.

He clarified his views more succinctly and spectacularly in *The Saviors of God*, envisioning that ultimate freedom like this: a man must first be a body, then a heart, and then must free himself of the illusory hopes these dualities inspire. He must recognize bravely the necessity of nothingness in this world. Having come to the dizzy vantage, the man must cry out, "Yes!" before plummeting into his own soul, there to touch the delicate, fearful void – no empty space, but an organism itself, palpable, yearning to be touched. God. This God is wounded, says Kazantzakis, wounded and in desperate need of man. Aware of his responsibility, the savior must then lunge beyond ego, straining toward racial origins; must then follow his sixth sense beyond race, into all of mankind, and then, finally, beyond mankind to all of nature. In this way human imagination takes its valliant place within the biosphere.

His philosophy of evolution (he had already translated *The Origin of Species*) was, with the American George Santayana's reading of the nineteenth-century Transcendentalists, a spiritual ecology. He had mounted a tiger — all of life — but there was more, the concomitant summons of Lenin.

In the fall of 1926 a Greek newspaper sent him to Moscow on the first of three journalistic forays. His spate of articles appeared in the Athenian periodical *Elephtheros Logos* and would be consolidated in a two-volume history of Russian literature and in the book *What I Saw in Russia*. But again, Kazantzakis saw and felt a different world than other newly converted intellectuals:

> *What moves me in Russia is not the reality that they have achieved, but the reality which they long for . . . all human value roosts in this chimera; for this intoxication, it is worthwhile to act and die. For the human masses — workers, peasants, women — to enjoy themselves, to eat better, to become enlightened is certainly a goal worthy of man. But I regard all these things as petty, practical illusions, crumbs for my heart, which when it has eaten is hungrier than before - - - - . Russia was being crucified.*[33]

The newspaper *Kathimerini* then sent Kazantzakis to Palestine. He interviewed the first Kibbutzniks, including Ben Gurion, and idealized the Jewish destiny, committing the gravest fallacy by assuming the Hebrew greatness to lie in Diaspora. This was in keeping with his own sense of flattery concerning death and transcendence. Accordingly, he hoped that the Arabs would drive the Jews into the sea, to vitalize the Jewish race, keep it on the move. The Jews wanted no such philosophical attribution. In 1927 Kazantzakis went by camel to Mount Sinai. There, Father Joachim was to reveal a dream that the Greek poet would many years later refashion into Christ's own story in the novel *The Last Temptation of Christ*. Instead of saving the world — a fantasy — Christ would be cured of his longings and become the best carpenter in Nazareth (even going so far as to build crucifixes). From the summit of Mount Sinai Kazantzakis seethed: "I felt they were not mountains but the fossil remains of an antediluvian cerebrum and that high above me, on my right, an immense cross was embedded in a boulder, with a monstrous bronze serpent crucified upon it."[34] Years before, in the chapel atop Mount Psiloriti on Crete, Kazantzakis had lost his virginity to an Irish girl who lay beneath an icon of the Virgin. Now the ravine of Er-Raha opened beneath him and he saw the race of Israel ("hungering, thristing, blaspheming") emerging from the God-trodden desert, sulphur in the air.

Kazantzakis returned to the cold nocturnal haze of Russia and in winter travelled through Siberia, working on his extended poem of *The Odyssey*. He noted the Eskimo's single prayer, "God, do not kill me!" Lenin and Stalin had become slogans. Lenin was the light, Trotsky the flame, but Stalin was the soil. Kazantzakis gravitated toward such myths, but not to convince others nor to rise up in revolution himself. He tired too easily of solutions, was forever concerned to break every equilibrium. This was not an easy man to live with. "It is not human beings that interest me," he wrote from the Arctic, "but the flame that consumes human beings . . . . . Russia does not interest me, but the flame which consumes Russia."

Commuting between France and the Greek islands, Kazantzakis, aged 50, was writing terza rima cantos dedicated to all of his heros: Lenin, Don Quixote, Buddha, to himself, to Nietzsche, Mahomet, Moses, "To Rhyme," and to Helen. He translated Cocteau's *Infernal Machine* for the Greek National Theatre, wrote more children's books, finished a fourth version of his *Odyssey*, then set off for the Orient for an Athenian daily. Down the coast of Egypt, to Colombo, Sumatra, Singapore, Nara, and Kyota, he fed his *Odyssey* with everything – the Goddess of Mercy at the Horyuji Monastery, erotic Indian women, fine eyebrows, slithering snakes, Chinese silks, the Forbidden City, the Deer Park. "I am not made for Europe; it's too poor and the plunder too meager. It's the Orient I long for, a stroll along the Tigris and the Euphrates, an ascent into Tibet, an expedition to Central Africa."[35] Pursuing the penultimate saga, a nakedness nearly lavish, and the topographic twilight of idylls down which he marauds, Kazantzakis, like his protagonist Odysseus, is a modern sunburst. His motivations are complex. Courting the whole world in his harsh cry and reckless solitude, this hero plunges in and out of stylized iconography, bludgeons icebergs, witnesses his African Erewhon disintegrate in earthquake, and sees himself reduced to a mere god in the minds of men. The epic breathed new life into myth: to come at the contemporary inferno with ancient impulses and details; to insist on the absolute at each junction, and never to compromise; to love Nature entirely. Kazantzakis' excesses were not Proust's nor Joyce's, nor Mann's, though the comparisons have been made with each of these writers. Kazantzakis' frenzies were more dated, precedented in the provinces of lore and the baroque. His true elevation derived from his sheer will power, but his stubbornness was all masochism, a love of humanity that he dared not overindulge for fear of total dissolution.

Odysseus was searching for harmony, rearing and out of breath, litigious, all powerful, a Moses of an ascetic renowned throughout Africa. He fought bulls, abducted Helen of Troy, seduced Emperors, monsters, his own people, was a volcano. Unlike other twentieth-century heros born of empty space and ambiguities, pithy and slight, epigrammatic, dour, given to the neuroses of their time by way of commentary, hacking out more mediocre finales, in tune with defeat, or a defiance destined to disappear, Odysseus stirs us to *remember,* inculcates the impossible, raising his eyes to the sky and soaring like a giddy bird.

Odysseus returns home to Ithica at the beginning of the epic, as in Homer's original song; he subdues his wife's suitors "like a lion" but rather than giving into the contentment of old age, sets off again with a motley crew. He sails to Sparta, convinces Helen to run off with him to Crete where Odysseus will manage to set fire to the incestuous court of King Idomeneus. Helen will migrate with a blond barbarian and Odysseus continue to Thebes, another decadent empire, where the bronze-muscled man lands smack in the Pharaoh's dungeon with other revolutionaries. He will spawn the collapse of Egyptiann civilization, flee southward into the desert where salvation and destruction merge into an all-encompassing vision of God. Odysseus describes his planned-for Utopia to his faithful followers Granite and Kentaur. They descend to the source of the Nile. Granite captures a female leopard cub and makes a gift of it to Odysseus who takes the cub with him up a mountain where the sage communes for seven days with God. Contraries are resolved for Odysseus on hot slabs of rock, in caves swollen with the odors and phantoms of primordial pasts; the light plays upon his mind, nature rushes in, and good, evil, soul, mind, heart, and hands come alive, can be allocated, traded, reapportioned. Odysseus re-creates the world. "Who are you, virile voice? Pirate, what is your name? I am that dark beast, God, who mounts eternally."[36]

Odysseus comes down off his mountain; lays the foundations for a socialist state that merges the craftsmen, warriors and intellectuals; and outlaws marriage with the end being to rear all children in common. He sees his principle expounded in the fate of male termites who, upon mating in air, fall to earth and there are ravenously consumed by others, scorpions, beetles, birds, and snakes. This is pure bliss to Odysseus who says convinced, "Whatever blind worm-mother Earth does with no brains, we should accept as just with our whole mind, wide-eyed. If you would rule the earth, model yourself on God!"[37] The city is built upon maxims culled from Plato, Augustine, and Thomas More. But on the very day of inauguration, an ugly chasm opens, the African jungles tremble, and the city vanishes. Odysseus' troops manage to escape north, but Odysseus heads south, lost in inner thought. He becomes deathless, evolution incarnate, encounters Prince Motherth who has come seeking wisdom from Odysseus. "Death is the salt that gives to life its tasty sting," the Grecian tells him. Odysseus meets the prostitute Margaro, whom he celebrates, hails as his "ascetic fellow-toiler" when she confides to him her own secret knowledge: that only in sexual love, amid a wretched world, can two become one. To which Odysseus replies, "Even this One, O Margaro, even this One is empty air."

He journeys onward, meeting a blind hermit; the schizophrenic Prince Elias; crazy Captain Sole who broods wildly on Everyman's salvation, running perpetually into the stew; a gentle Negro fisher lad who comes in the guise of Christ to haunt Odysseus; and then a host of demons. Odysseus builds his last skiff with the aid of spirits and sets sail for the South Pole–"God is wide waterways that branch throughout man's heart." He crashes into an iceberg, climbs aboard at sunrise, wanders its wide margins, and discovers a desperate human settlement of igloos. He lives with the Witch Doctor until spring, while many perish. Invoking the sun, he sets off again, bidding life farewell, embracing Death with a harmonious cadence. A mountain of ice crashes into his boat, and he flips into the gelid Antarctic waters, naked, the ax having slipped from his waist, his memory volplaning into pure fire, as all of his old comrades and lovers come to join him.

The twenty-forth Book of *The Odyssey* is a feast of ancestors cavorting in the august glow of ice floe, shuddering, swirling nature and the all-expressive Sun, with which Odysseus has been enjoined throughout the long poem. Burning, soaring, fusing, this enraged melee of life – even Odysseus' old friend, a cricket accompanying him – bolts from ultimate insides to black-hole nothingness. With a final laugh, thrusting his hands once more into a plentiful horn of pomegranates, figs, and grapes, Odysseus departs his last cage, freedom itself. In the Epilogue, the Poet rejects his mother's offering of a meal, lamenting "that his beloved one has vanished like a dwindling thought." Matter, to fire, to spirit – such is the extravagantly prolonged theme of *The Odyssey*.

Kazantzakis published the book by hand with a $1500 gift from the American Josephine MacLeod. Buyers specified how many "grams" of the book they wished to purchase. It weighs several pounds, had taken him 14 years to write. Its speech and consciousness was that of *Demotiki*, the peasant's tongue, scarcely understandable to those high-Greek (Katharevousa) wielding bureaucrats of Athenian academia. Its seventeen-syllable verse startled everyone, for the normal Classical Greek form consisted of fifteen syllables. Greek is fiercely polysyllabic, enabling the poet to string millipedally six, seven adjectives in a row before the noun. With Kazantzakis' addition of two syllables, the poem took on unprecedented, out-of-breath richness. Ironically, Greece had been under the dictatorship of General Kondylis for three years when Kazantzakis' masterpiece emerged. The island of Aegina, where he and Helen had built their house of stone adjoining the sea, was under strict military censorship. Yet he could still swim naked every day, out past the sharp coral reef. He believed it a mortal sin to ever enter the sea with any clothes on.

"Faithless generation, defiled, ungrateful, doomed," he stated in his play *Christopher Columbus*. "This generation mentions the earthly Paradise and bursts out laughing!" In all of his letters and prolific creations, there is very little connection between Kazantzakis the man and those actual events occurring by the time of World War II; virtually no mention in any of his exposition of the Holocaust or Hiroshima. He had, in essence, experienced a massive hemmorhage of feeling in *The Odyssey*, gone to heaven and back, explored all of Asia, the North Pole, the South Pole, resurfaced cloaked in seaweed at his ancestral hearth, and systematically erected a monument to the ageless, whose fundamental concern was the transmutation of energy, the spiritual evolution of consciousness. The Russian Revolution had merely confirmed his urgent believe in *all* change, and World War II, as well as his own near starvation (which accounted, ultimately, for his protracted ill-health and death), merely reaffirmed his basis for claiming that Evil and Good were voracious bed-partners; that only superhuman willfullness could rightly choose the one over the other. The choice was intrinsically vested and self-propogating, in all of us – *Home sapiens sapiens* – potent representatives of the entire Earth. We are not alone in our perplexity, though, Kazantzakis assured readers. Never forget the worm that becomes the butterfly, the flying fish that transcends its nature, and the silkworm whose very entrails are golden, subject to God's apotheosis. "Even if I were certain [of] heaven," the poet remarked, "I would pray to God to let me go by the longest possible route."

At his funeral (he was refused permission to lie in state by the Greek Orthodox Archbishop of Athens), legend has it that a colossal mountain peasant came from the hills beyond and single-handedly lowered his coffin into the earth. I have studied the photographs from the funeral. There appear to be dozens of such figures mingling stoically in the huge procession, all likely candidates, clinging to the simple wooden casket. Kazantzakis' tombstone, atop the windy Martinengo Rampart above Heraklion (Crete), reads,

> *I do not hope for anything*
> *I do not fear anything*
> *I am free* *[38]

THE MEETING OF MINDS ON AN ICEBERG

> *Solovya basnyamee ne kormyat* *

There are myths, religious biological rules, that intercede from time to time in matters of destiny. The discovery of America saved Europe from

---

*Russian proverb meaning "Actions speak louder than words," but more literally, "Let's stop talking and get something to eat."

overpopulation and entropy. Americans are the very product of entropy, and their spirit is still roaming, seeking new horizons. The Russians have been, for most of their recorded history, consumed by self-doubt and wintry obsession, anxious for the spring. When Peter Romanoff smashed Charles XII of Sweden at Poltava, when Napolean was lured by Kutuzov and De Tolly into Moscow, and when the Russian soldiers retook Rostov and salvaged Stalingrad in 1942, it was the very winter that rose to fame, defining the ardent spirit of Leninism and the frigid, genocidal archipelagos under Stalin. The Russian landscape reveals something of its impenetrable nightmare and power; the long enduring reaches east of the Urals with their unassailable volcanos, energy resources, literary metaphor; the wild, untrammelled Pacific coast and polar north; Pamir country with its countless unnamed massifs beyond Mount Lenin and its lovely valleys of longevity hemmed in by apricot trees and unknowable nomads and lusciously described by Chingiz Aitmatov; Crimean Summer dochas for the elite party members and the ever-growing middle class; desert elements of Iran and Turkistan, northwestern Himalayan backcountry, poplar groves stretching to infinity, huge fresh-water lakes, pieces of Europe, the Near East, the Balkans, Central Asia, the Orient, the Arctic – half the climate of the globe is Russia's flavor.

I have been embraced by Soviet generals on a windy street in Kiev; they unabashedly declared their unanimous love of Eisenhower, whom they had met at the Elbe River in 1945. Inebriated, they repeated loud pledges of universal brotherhood and wept upon hearing of Ike's "untimely" death. I was followed in Leningrad. The kindly fellow, – wearing true Richard Burton *Spy-Who-Came-in-from-the-Cold* drag, later shared an ice cream with me on Nevsky Prospect when I blew his cover. Most Russians I got to know well were far more familiar with American goings-on (literature, statistics, an opening at the Met) than most Americans could claim about the Soviet Union, a ratio overwhelmingly weighted in favor of Soviet discipline and homework.

Russians are given to emotional fatigue and embrace, whereas Americans more vigorously blazon with embarrassing gab. American's have a penchant for instant photographs, the unsubtle; they'll fire querries, exhibit themselves where any Russian would suffer pause. Yet American manners are usually the more lively, honest, open in any international exchange, like the wheat-thick thought and down-to-earth imperatives of five midwestern Ranchers for Peace who took off for Russia intent upon solving the arms race, ready to talk to anybody.

Whatever the differences in etiquette, outlook, reasonableness, or history, these two styles of thought and thirsting must annex themselves, recognize how slight their differences are – neither has any claim against the other. They must own up to the spectacular possibilities of their alliance.

The Chukchi Sea in winter south of Point Hope across the Bering straight – gauzy, glorious, imponderable – reflects the mating ground of symbols, a pristine border shared beneath aurora borealis. The polar bears and salmon move freely between the two lumbering giants. I have walked along this border in winter and felt the pull and allure of our Soviet neighbor. If only we could relinquish our greatness long enough to feel how *small* we are in the Arctic; hands pressing in its sleeves, face beaten down by wind, the long night coalescing over Yupik hamlets of reindeer antler and aluminum on both sides of the border. The Arctic would make a fitting site to celebrate the ecological truth of this U.S.-Soviet dependency; no summit, but a bottom-line coming to terms, there surrounded by the ineluctable flow of ice; a landscape so elemental as to tempt our sluggish leaders into fresh air, clear thought, one-on-one simplicity. Let the two or three leaders meet in solitude, without the charade of limousine trains and armored helicopters. Let them come in down parkas and woollies, throw a few snowballs, horse around, and be marooned there, no back up. Let them sit and sit for days until the revelation of their mortality assails them. These ephemeral potentates, little old men with blistery convictions, both will be dead within a decade or two, have no reason whatsover to interest us, command no spot in history, except that they have managed to appear in the limelight of the disaster zone. For their brief regnancies they can be counted on to savor fully the illusion of power. Their subordinates claw at one another for such glamor and would destroy the world in their ague.

In accordance with the planet of life on which we live; a planet that has delivered at least 500 million other species and ideas before Man, I think it time for us to stop taking ourselves so seriously. It is not Communism versus Democracy, I versus Thou.

I know of only one gesture that would, in effect, preserve future generations on Earth: let us meet on the ice, having renounced, defused, all of our bombs. If the opposing side then conquers us, considers us pacifist fools, makes us learn to read Tolstoy in the original, alright, so, at least we can begin to live again. Americans will be remembered for having turned the other cheek. And that is a humility which may not exist in nature, unless we have the courage to insist upon it. If we fail to re-kindle that endowment which is our unique gift, namely conscience, then we will surely die out there amid the drifting floes.

The history of Homo sapiens is surrounded on all sides by 20,000 years of previous glaciers. For millennia the climate of Earth has been more mellow than usual, has allowed us to prosper and experiment.

But in the crevassed peripheries of current civilization, there are aquamarine gapes, glowing cold and immutable; holding secrets long-forgotten, worth remembering.

NOTES

[1] Jans Goedicke, *The Report about the Dispute of a Man with His Ba: Papyrus Berlin 3024* (Baltimore: Johns Hopkins University Press, 1970).

[2] Gerardo Zampaglione, *The Idea of Peace in Antiquity*, trans. Richard Dunn (South Bend, IN: University of Notre Dame Press, 1973).

[3] St. Augustine, "Contra Faustum Manichaeum libri triginta tres," in J. P. Migne (ed.) *Patrologia Latina*, vols. 32-46 (Paris, 1841-1842), quoted in Zampaglione.

[4] *Aenied*, I: 291-296, quoted in Zampaglione.

[5] In *De monarchia*, 1311.

[6] Barbara Tuchman, *A Distant Mirror: The Calamitous Fourteenth Century* (New York: Alfred Knopf, 1978).

[7] In his *Querela pacis*, published in 1520. English edition published as *The Complaint of Peace* (Chicago: Open Court, 1917).

[8] Hugo Grotius, *De jure belli ac pacis libri tres* (Paris, 1625). English translation under the same title by F. W. Kelsey (Oxford: The Clarendon Press, 1925).

[9] Baron de Montesquieu, *De l'esprit des lois* [The Spirit of the Laws], trans. by G. Truc, 1750 reprint ed., 2 Vols. (Paris, 1945).

[10] Buffon, *Histoire*.

[11] First published in 1791. Reprinted in *The Writings of Thomas Paine*, edited by Moncure D. Conway (New York: 1894-1896).

[12] Bentham outlined this proposal in his *Plan for a Universal and Perpetual Peace* (London, 1789; reprinted in *The Works of Jeremy Bentham*, 11 vols., edited by J. Bowring (Edinburgh, 1838-1843).

[13] Immanuel Kant, *Zum iwigen Frieden: ein philosphischer Entwurf* Königsberg, 1795). English translation by C. J. Friedrich, *Perpetual Peace* (New Haven, CT: Yale University Press, 1948).

[14] M. Howard, *War in European History*, (New York: Oxford University Press, 1976).

[15] Ernest Hemingway, "Humanity Will Not Forgive This," *Pravda*, 1 August 1938.

[16] Johathan Schell, *The Fate of the Earth* (New York: Avon, 1982).

[17] Samuel Beckett, *Texts for Nothing* (New York: Grove Press, 1962).

[18] Cited in Tom Regan, *All that Dwell Therein: Essays on Animal Rights and Environmental Ethics* (Berkeley: University of California Press, 1982).

[19] Percy Shelley, "A Vindication of Natural Diet," in *Shelley, Selected Poetry, Prose and Letters*, ed. by A. S. B. Glover, (London: Nonesuch Press, 1951), pp. 900-913.

[20] John G. Mitchell, *The Hunt* (New York: Alfred Knopf, 1980)

[21] Yukio Mishima, *Sun and Steel*, trans. by John Bestor (Tokyo: Kodansha International, 1970).

[22] William Butler Yeats, "The Second Coming," in *Michael Robartes and the Dancer* (New York: Macmillan, 1924).

[23] Jean Paul Sartre, *Literature and Existentialism*, trans. by Lloyd Alexander (New York: New Directions, 1949).

[24] Robert Rosenblum, *Modern Painting and the Northern Romantic Tradition: Freidrich to Rothko* (New York: Harper & Row, 1975).

[25] Jean Paul Sartre, *La nausée* (Paris: Gallimard, 1938). English translation (*Nausea*) by Lloyd Alexander (New York: New Directions, 1949).

[26] Nikos Kazantakis, *The Odyssey: A Modern Sequel*, trans. by Kimon Friar (New York: Simon & Schuster, 1958).

[27] Nikos Kazantzakis, *Report to Greco*, trans. by Peter A. Bien, (New York: Simon & Schuster, 1965), p. 453.

[28]Helen Kazantzakis, *Nikos Kanzantzakis, A Biography Based On His Letters* (New York: Simon & Schuster, 1968), p. 456.

[29]Kimon Friar, *The Spiritual Odyssey of Nikos Kazantzakis: A Talk* (St. Paul: North Central, 1979).

[30]Nikos Katantzakis, *Serpent and Lilly,* trans. Theodora Vasils (Berkeley: University of California Press, 1980).

[31]H. Kazantzakis, *Nikos Kazantzakis.*

[32]Reported by Kimon Friar in his Introduction to Kanzantzakis' *Saviors of God* (New York: Simon & Schuster, 1960), p. 3-11; H. Kazantzakis, *Nikos Kazantzakis*; and N. Kazantzakis, *Report to Greco.*

[33]H. Kazantzakis, *Nikos Kazantzakis.*

[34]Kazantzakis, *Report to Greco.*

[35]Ibid, p. 382.

[36]Ibid, p. 441.

[37]Ibid, p. 465.

[38]"Kazatzakis," a film by Michael Tobias, KRMA-PBS, March 25, 1984.

Epilogue

Ice Bird

Sunt qui dicunt locum illum esse umbilicum terrae nostrae habitabilis. (Some say that place is the navel of our habitable earth.)

FOREBODINGS

In some ways it was easier like this, destiny less wide open, less room for heartbreak. No generosity of distance taunting the eye, no confusion of horizons. There was, to be sure, a certain comfort in living so close, so cozily in touch with the end.

The signs of imbalance had emerged over a period of a few bent decades. In the year 1994, for example. An old widow on a remote end of the mountainous Greek island of Ios, having displayed no evidence of bigness, uttering no discomfort, suspecting nothing, one day, in the candescent heat of an afternoon – not a murmur of meltemia through the olive grove that was her providence, on whose warm loam, and in whose shadows she reclined, mending, daydreaming, remembering, a mere kilometer from Homer's legendary burial ground – suddenly was thrown backward. In a vomitous rending her guts were strewn. An enormous black swarm of bees burst from her vagina and rolled away into the higher reaches. A goatherd lad found her gauzy corpse some weeks later. The disintegrating sloughs of skin were enflamed, deep in globules of golden honey.

Six months later. Lavoris, one of Gaia Company's experimental voltaxis communes in the Punjab. An eleven-year old information retrieval technician awoke at 3:26 a.m. screaming convulsively. He continued in such manner. Tests revealed a terminal brain condition: His left and right hemispheres were devouring each other. The boy was, apparently, unable to counter with his emotions the kind of data he'd been gathering. Other technicians throughout the Alliance were to come down with screaming.

Northern Ethiopia, the year 2020. An army of white ants swept down the slopes of Ras Dashan, devouring every living plant and animal all the way into Gondar, before cannibalizing their own corps, until one bloated ant, size of a Cambodian water buffalo, wobbled clear at the zenith of devastation, was captured, subjected to intensive scrutinies and nervously hailed the product of some new evolutionary process at work in Africa.

Toward the end of the twenty-first century, following a series of tremors, Iceland's nuclear diameter exploded, igniting a volcano. England and part of Belgium and France were flooded. Body count around six million, sufficient incentive to the USSR, which occupied all of Europe without contest. At the same time, the Soviet premier was heavily drugged with cortisone to counteract a painful tennis elbow, the effects of which, scholars later subscribed, may have augmented both his backhand paranoia and pugilistic pyrotechnics.

Mount of Temptations Hermitage, above Jericho, year 2103. Father Makarias had a disquieting vision of the end of the world in which was revealed to him the face of the Lord. White. "Like a snowstorm," he relayed to Father Tomaine, an epileptic evangelist who stood perched year after year waiting atop a thirty-cubit column of rosy schist.

Los Angeles County. Forty-eight hundred Oriental children, all under nine years old, died within thirty-five hours, as did every gerbil in the city. No reliable explanation.

Like a serac ready to collapse, the two-hundred-foot Tower of Symbols rose medievally, beyond the pale of memory, near the outer zone of fusion domes that encinctured the last colony of animals on earth, Point Mind. In the cyan mist of dawn, the tower's rotting bricks and eroding crenelation shone carnelian, even prettily, one might venture. A triangular superstructure of cells, all blackened, housed the 2937 inhabitants. Black and insulated. No snow got in. The tower stood here, the cells there.

The circle of whirring reactors had no beginning and no end, forever and reputedly keeping the horizons in check; horizons of grey, slatty glacier ever threatening. The inhabitants were largely prevented from seeing the outside. The machines were more than machines, known collectively by an unnameable name, euphamistically referred to as NO. This injunction effectively summarized the populace's all-embracing sense of adventure, perception, and jurisprudence.

One day per annum, male and female of each caste, according to an order long upheld, gave obeisance to NO. The explanation for this one indecorum was lost. Not even the tireless archivist could trace legislation giving cause to such anachronism. The unenthusiastic, the inept ones were induced. The remaining few maneuvered clumsily on their own, circling the concourse by foot, squatting expeditiously — with a year's slow build-up of fever — to have done with their ejaculation, a form of expression out of tune with the twenty-sixth century.

The stones spiralling to Hardanger's aerie were smoothed down, palimpsests. With primitive leftover bones in his face, he presided over, dwelt within, the museum atop the Tower. No one in his lifetime had ever come into the Tower, save for Hardanger's deceased father, and father's father. And Margo.

Its architecture, purpose, and very contents — Hardanger himself — were throwbacks that held no interest to contemporaries. The truncated arc of sky admitted a merest thirty degrees of light to the senses; only the high, exaggerated diffusity of cirrus, seasonless. Cloud, sun, winter, summer had been *fixed* to eliminate all need of concern or speculation. The moon glowed cold and partially; the Sun had no other function than to feed NO, which in turn, by machinations no longer inquired after, kept the ever-advancing ice from sweeping over this final vestige of breathing man and woman.

Other than those manufactured, no more children were delivered unto this paradise. The very longing had been gallantly overcome. The earth, admittedly, was exhausted. No plants grew in there, other than an algae — *Spirulina* — easily cultivated for its sixty-percent or so of protein. And a gourd, like Assamese plantain, a breadfruit cooked, baked, prepared in myriad ways to satisfy the oligophagous hunger and expectations of a people that had miraculously survived. All other animal species had been extinct for centuries.

Hardanger descended from a northern race. This was his secret. That he had descended at all. He was thick, bedevilled with forethought. His body conjured legacies of kayak malaise amid ice floes, of prison sentinels in winter hunched against electric fences, cigarettes pinched between tumid lips. He sweated much, breathed heavily, had the additional weight of blondish curls crowding his forehead and double chin. Like Michaelangelo's long lost "Brutus," Hardanger looked unfinished, dangerous, stout as an oak tree. For all his lugubrious show, his cobalt eyes bore upon others with a gladiator's lunge, a Kazak's horsemanship. His wrists were hairy and expansive, neck bristling, the total picture prognathic. At forty his thickness moved, dominated the svelte, unsullied Margo. But that night, Hardanger refrained, sat troubled before the fire he had kindled from his own wastes in direct violation of the Complex. And in his hands, a book, Spinoza's *Ethics*.

He had an awesome responsibility no one else avowed, save dimly for Margo, the woman drawn to him. He was in possession of the only testaments: a rubber tube of one-half-inch diameter, soil samples from several planets, a used thread of dental floss, a tuning fork, the dried leg of a tarantula, a decal from Mt. Ararat, a nuclear detonator long ago rusted out, a jar of sugar. In addition, a map, unguessable.

His sullen urgings this night were not uncharacteristic. Hardanger was, it is true, *heavy* and not easily enjoyed. For years he'd wondered, done something, in other words, of extreme risk and injustice. But why not? The monitors were not able to beam into the Tower or feared to, fearing that Hardanger himself was one more artifact not safely toyed with.

Hardanger first learned of this rare deliverance, this freedom to think, when think he did, nestled in the woolen blankets of the museum, high atop the refuge. His father Nils assumed all the early duties of childcare, in absence of Hardanger's mother, Inge, who'd perished during childbirth. He thought about the ticking of a cuckoo clock, the wooden mantle on which it rested, the only wood; a coiled rope, a doll. He thought, read all the precious volumes — Hon chi-yueh's *Burnt* trilogy; *The Landing*, by Bowra; Cinshee's celebrated eight-volume *History of Nervous Endings*; Roger Bacon, De Vaca, Peter Carter, *The Springs of Marah* by Eric Davis. There were old astronomical instruments of mysterious purpose. Paleontobilia. All Hardanger's companions in youth for lack of other playmates. He was reared on phonetics, memorized San Jose's elaborate *Silicon in Antiquity*, counted fibers in the woolen remnants at night to put himself to sleep.

"You're different," Nils had told him. But Hardanger *knew* for himself. He looked different: less waxy, hair unshaved, bones less sleek. But he was a man, among 1500 men. What truly distinguished him was his Tower. Without it, he wouldn't be Hardanger. And indeed, the Tower would long ago have fallen to ruin were it not for Hardanger's loving attentions. He was a supreme builder. Where chinks needed caulking, bricks relaid, seepage cleared before it could undermine the weakening stereobation altogether, Hardanger was there, all industry. Though the past few years (Hardanger consciously measured them; no one else did) were increasingly futile. Something was happening beyond the Outer Zone. From his Tower, he smelled, felt stronger winds, and heard throughout each day the rumblings of ice on the move. As never before he was alerted. No one else was. No one else had ever *seen*, or been told of, the ice. The six reactors blocked the view. But the Tower reared high, high enough to lend a vantage. Having grown up with ice out the adobe fissures, Hardanger thus possessed a second secret: that of its beauty and allure. Oh, the others knew something, of course. That beyond was aftermath, the end, post-Finality. From what the Council had told them.

The Council. Eccentrics. A hackneyed governing. Reduced to utter efficacy. No complaints, no questions, socialized in all things, safe, as long as NO continued to whirr. The protein was cultivated in large cisterns, extracted simply, hardened into loaves. Those impervious gourds, weighing thirty-five kilograms, provided additional meat. A spring issuing in central Point Mind acquitted all thirst. It emerged from a well rumored to have connection with the Earth, access to inner labyrinths. But only rumors. Hardanger had often stared down into it. With Margo by his side.

Margo. Nearly six feet, common height among the others of her gender. There were five genders in the twenty-sixth century. But she had uncommon hair. Never clipped. Because of its color, flaxen. Below her waist, lascivious, full, rampant. Margo *felt* things, also an injustice, subject to monitoring. She wanted—this is awkward—she wanted to open, to quiver in her buttocks, to break out in some fashion still dark and obscure to her. To see something other. Believing there was more. But no. Such talk, hope, illegal, all.

She was twenty-one, eligible for an education, or tapping, as it was donned. A simple operation, undergone in the calms of the parents' home, and by which, with the aid of frequencies, the neurons in her neocortex, within moments, were lent new dimensions, loftier firing aptitudes. A few of them, anyway. Enough to keep her from killing herself, the standard remedy in centuries past.

Margo's father Xing was a senior duster. Using laser scrubbers fired from a handgun, the beams destroyed all submicronous particulates off clothes, pavement, the well, domino track, casino tables, toilets, even the play-money chips. Cleanliness is next to NOliness, the motto had it. The policy of indiscriminate dusting had resulted in the total expulsion of all potential viruses and bacteria. No bugs in Mind. Nothing preserved save the colony, its council, and its NO.

Concomitent to this thinning out of critters, the language was made over. Words like *creep, slimy, buzz*—not to mention *frond, edible, spume, zebra, ancestral tree, ptarmigan*—no longer existed. The all efficient ease of enunciations better suited to the Chinese *Homo erectus* with their atrophied jaws, ughs, oohs, ichs, associated specifically with the blunt impressions of a giggle, a groan, a fart, were what endearingly marked the range of articulate candor in Mind.

Xing was bright, meticulous, even zealous in his job. He stood five feet three, had a cockatoo's tuft of yellow hair, wheezed from twenty-year's faithful service, spit 1000 times a day for a reason he would not, could not convey. His wife Odile had her own appointment – every citizen did, of course. Education conferred the rights of citizenship. At present, there were nearly 100 noncitizens. They comprised the young, mostly. In seven instances there were complications, the unassimilable, that is, the mentally insane. These were people immune to tapping.

Odile supervised the production of high heels. She also had her own secret, deeply, unreachably cached: her daughter, Margo, who was real. By the work shifts in practice during those years, Xing and Odile were separated. He was still unaware that Odile had delivered in a laundry room, eight-by-fourteen, jumbled with digits good for all exhortations. She'd gasped at the mystery, having sucked in her breath month after month, not knowing but sensing enough to keep quiet and out of sight, a discretion altogether to her credit during that period of the Last Turmoil. She rocked to older rhythms as her vaginal flood worked opposite the century, declared that it had waited for this, years to come forth, and was not about to be disenfranchised. She forced the child from her, bit off the dangly, concealed her for months thereafter, and then the fact was done. She never did question the incident. Between Odile and Margo, there was unspoken conspiracy, unmentionable.

"You must come!" Margo pleaded, urging Odile into the Tower, where Hardanger had assured Margo they could speak to one another at long last.

"No," Odile carped, with a welling of frustration. "Your father's comfortable. Go look after him as every night. Oil his back, will you! He's exhausted from so much good, good dusting all day."

"But Mother . . . "

"No!" And a tear came forth. For two decades she had felt the Friction, repressed duality, psychosis. The need to think against herself, against the monitors that listened in on everything one felt and said. Only from watching did they restrain, by vote of the Council, men who, it can be assumed, cherished their own privacy and so censored all such furtive voyeurisms.

She thought against thinking. But gestures of a kind were wanton, free for heavy show, which merely compounded the girl's fierce need to scream. And so she, too, was verging on the sickness of her age.

Those with Friction were erased, re-educated. But the young, prior to teaching, were allowed their excesses. Bliss spent whole nights eavesdropping on the thoughts of randy young nymphettes, secreted away in his furnished pew.

Bliss. Head councilman, spokesman for the NO Complex. He came from outside, knew more, had access, his very own past. He'd survived the Finality Era, risen to stardom in the Alsek Underground and then, when the lower Frontier was evacuated, he went tribal, maneuvering opposing forces with expert knowledge of the old disciplines. Bliss found himself isolated on the last spit with a gaggle of horny, malicious, vanquishing thugs, their brains deadened, lips cleft, eyes blank; they travelled by ice discs, wearing platinum chain mail; poured out of the mist and conquered the last military outpost, replete with its cloying medley of strung-up whores, generals, and poets. In undisputed control, Bliss brought the community into rigid and orderly subservience. Even by the time of his invasion, the gods had ascended to their pre-eminence. The Complex reigned supreme.

Bliss had no idea that Point Mind was it. Though when that dubious status did dawn on him, he set about to doubly insure a system invulnerable to caprice. He stood resolutely at the apex of known things, occupying the gilded throne of NO-mind. The Complex was tautologic, innerly assured, a perpetuum mobile. It shone chrome-molybdenum, titanium, was omnivorous, clicked, mightily, sounded sweet to his ears. It was a closed system of energy so compassed and powerful that Bliss, in fact, had little to do. The Complex maintained heat on the outside, to keep the spatulate, ablative terminus of the glacier in continual run-off. The voluminous cu/secs of icy water were naturally exploited, resulting in the generation of so many BTUs along the molten ridges of the spirulina pools; and nourished a sufficient ambience to foster the gourds from an artificial soil of varicolored metallic chips. There were, of course, inordinate megawatts to go around, given such confinement and diminutive population.

The confinement. Bliss liked to think of it as order. The outside was unmeasured, steaming, and aglow with all manner of wartime residue, the fake sublime, horizons bathed in acid.

But to backtrack a moment: A hallucinogen had infiltrated the language by 2400 A.D., had come back into vogue, that is, from a desert speech community comprising a mere dozen dry farmers of bean and mellon, to the esperanto of most physicists, replacing Basic and Mandarin. The outside was fully skewed, supplanted, had come inside. Various camps argued the implications for mankind, back when such concepts as humanity enjoyed some purchase. Phenodeterminists foresaw the Finality, image processed it in five-color graphics, resplendent to the eye, if bewitching. They acknowledged with some discomfort the fact of microbial exhaustion. All the rain forests, the tropics had been wiped out. Throughout the mountain ranges, zinc, uranium, copper, and oil exploitation, the total consumption of organic materials, and the final hourglass effect – erosion of soil into adjacent lowlands – secured an unprecedented depletion of certain gene pools known to be essential to the continuance of the human species. A hierarchy of proxemics (the extraspecial interactions from protozoa to Lepidoptera to Sherpa) once fundamental to all upland environments, was no more. The elements themselves were gone, replaced by desert. Cold, wind-burned, barren, in Brazil, Laos, Wisconsin. But no desert salamanders wriggling under shivering nebulai; no fine-smelling cacti.

Every exploitable resource had been disposed, not in accordance with human need (population estimates conflict but all were agreed on exponential negative growth) but to feed the Complex. All for NO.

Mind had become the environment, a scientific revolution prepared for in previous centuries by the empiricists, the meat consumers, the warmongers, the true believers. The steam engine, railroad tie, telephone, airplane, metal spiggot, internal combustion engine, the daily newspaper, shampoo – all these innocent forms of expression had led smoothly to the Finality.

Psycholinguists predicted the coming boredom amongst those survivors whose language would have suffered from lack of any image pool. Statistically, it was averred that the overriding majority of all words throughout history had derived from the outside – from shooting stars and swaying wheat fields and the scent of magnolia. From purring Siamese kittens, freshly cut grass, and the crashing surf.

But what with the internationalization of all lands, waterways, and sky, such propositions were not only untenable but foolish. It was during the early roundups and detribalizations that the Hopi's language was discovered and resurrected by scientists with a wave of near Biblical conviction. True enough, it shared similarities with Chaldean, forerunner of classical, that is to say, desert Hebrew. But regrettably, those scientists were exterminated during the Finality. And so it went.

Bliss knew of Hopi. As did Hardanger's father.

Bliss also knew Hardanger. Feared him. Whereas Hardanger knew, felt next to nothing about Bliss. Bliss had his own trove of manuscripts, and knew of the Tower's equal store. But such diversions represented no sure threat. Most men and women had stopped reading hundreds of years before. The books were left over abracadabras, vintage irrelevancy in an age gifted with the Complex. Hardanger was not so quick to compare the present with the past. He'd labored over antiquity with a loving volition, felt special affinities for Poussin, Ecclesiastes (Qoheleth), the limericks of Nash, the fables of Aesop, the renga of Sogi. He was not pleased with the era into which he was born, whereas Bliss was very contented. But when Bliss got out onto the concourses, mingled with his civilians, emerging from his abbatial den, naked, all six feet, eight inches of him, long-fingered, his fourth, right hand finger having been pulled off during a period of torture which Bliss stoically withstood, and became famous by, as one who could tolerate all pain; gleaming with a narrative brow, stunted balls, feline waist, a large, a bovine, a ballustrade of a man, but graceful; descending the seven-tiered pyramid of his abode weekly, to proclaim the State of the Colony, smile, smile evanescently, pat the heads of boys and girls, urging his subservients generously. He would pass out rectum gum, make light of life, Socratically wandering the geometrical haven, his hips instinct with a politician's carouse. Truly, his assays were warm ones, believed in. Whereas Hardanger was largely the loner, the fool, whom they left alone.

The council. Slibowitz, Dr. Fan 4, Wickiup, Ng, General Martinez, Opus 318. Each one of special past, superior quirk imparting just the extra edge of leadership. Theirs was a sovereign duty to NO. Mind was otherwise free of petty bureaucratic organs, though under prevailing hierophany there was no crime, no taxes, nothing to stagger, upset, or even remove.

Opus 318, endowed with her own terrible secret, had comely breasts, acquiline center, shaven skull. Twenty-eight and the youngest councilman, the only woman among them. Parents dead. Brother escaped, presumed dead. Opus had the gift of music and it was this one sphere to which the denizens of Mind payed unanimous attention. It was said that her capacities came to her from the spirit of her brother Noah, the only human being to have gotten out onto the ice. It was rumored that Noah was named after the Complex itself. He was the bastard son of Bliss, meant to rule, one day, the exiguous tribe. No one ever questioned his motives (you just didn't do that) save Opus – blue-eyed Opus – who found the egomania of Bliss just too tedious. Bliss, fearful of her daunting beauty, and of her scorn, had her retapped. She fought him down below, lost, lost the normal use of words. With a resulting (unexpected) uplift of glossolalia, chorales, screeches, blessings in disguise. Her voice quavered like the clarions of Strassbourg Cathedral some nights, arousing painful memories in sleep. Citizens bit their pillows, pissed all over themselves, vented private hurts and wants. Her power became recognized. She was to be silented, put under the auspices of NO. It worked. Whenever her taciturn nature rebelled, Bliss dragged her atop his pyramid, rammed her all the night, induced suffering. She'd be back in place the next day. Not a real amicable relationship. But in fact they were both arrogant as hell.

A week would go by. She'd lay off the soprano coyote sounds. Until all at once, say in a fortnight, the madness would get to her, she'd unleash like an animal beneath the infected aura of the bleak night; like some future muezzin in the mechanical dawn.

Though she did not know it, she had hopes, verbal conation, for a better world; simple desiderata, Pythagorean, visualized by means of line patterns she conjured with her pharynx – glottal stops away from the first utterance (Bliss suspected, programmed it, tested the howls of many Pleistocene species, knew her to be *touched*, that is enjoined) – egressions, counterpoints, parabolas of tone as philosophical as the mist that rose elliptic from the peripheries of town.

Sometimes Opus would stare out along the green glass, that fused the mote of land separating the Complex walls from pedestrian areas. She'd see colors, feel the incorrigible urgings of a song in her grunty gut, and with a sigh return to her duties of state, which were to entertain. How her brother Noah managed across the glass, through the barricades (assumed to be ionized) and into ice country, she could only fantasize. There was one way, of course. But to commit to it was unendurable.

Wickiup's name child, a girl of eight, had crossed the mote and slipped into the Complex (say two mothers who eyed the goings-on), and there was killed in a manner with little to recommend it. This same scenario had repeated itself many times before. Resulting in the general acceptance of censure. With time, the inviolate stature of the Complex carried the momentum of out-and-out reverence, the only mystery or impulse left this clan. In fact, there was no *real* mystery about the eliminations. For General Martinez was appointed that task.

Martinez. A ruddy Melado, first to be produced at Point Mind. Weight, 285; height, 5 feet, 9 inches. Vicious bruiser, temporal-lobe problems. Frequency of fits — daily. Unmarried. Might consume ten portions of intake a day. Ng detested him. Ng had to work with him.

Ng. His governess had been one of several Emotions developed by the early Floetech. She could not spank him but had the history of the world down to the period of the Macy Conferences and development of cybernetics, back in the period of the Frisbees, Solzhenitsyn, and cancer, all stored with doily delicacy and waiting for split-second conveyance to the Mongol darling, scion to moribund admirals in Ulan Bator who'd gotten their visas fixed through electronic computer break-ins, gotten out through the Buddhist cave route in northern China, ending up in Dayton, prior to the Finality Era. Ng had a past, all right. Everyone did, in fact. It was simply not pleasant to recall memories.

At four, Ng was fully equipped and entered Flotech's training program while other children his age still played with spitmen, understudied dung collectors, or slaved in the quarries of quartz. Imperious, slanting cheekbones, pyretic breath, fully developed facial hairs, he seemed like a kind man for all his learning. He was clear in his head as to his station in life. Ng was the only other technician, other than Bliss, qualified to oversee circuitry. So skillful were his monitorings that it was said Ng's blood mingled with that of NO. The circuitry did not seem difficult, but it was, and Ng at last grasped the arcana shrouding so many centuries of cumulative technology. He could not know precisely the period when the will of man faltered before the surer efficacy of machine, but falter it did. His Emotion had blipped harmonious lullabies to him when he was still in mercury suspended, receiving all the feedback his chakras could stand; and in those melodious songs — discphonics not onlike the mating calls of

Emu birds, of Bowheads – was stirring apotheosis of computers, fusion, pogo sticks.

Perhaps during the Great Folly, a decade that enjoyed successive waves of warfare, plague, instant annihilation, and the subsequent rise of the fusion men – human byproducts of the aftermath. They looked like acroliths, which is to say, they were rooted in the earth, but their minds steering clear. They guaranteed a stable future, communicated by way of intricate signals beamed off individual identity tags so as to instill personal trust throughout the collectives. The fusion men were represented by Alia, a freak Angora with human, above-average intelligence. But she had leukemia, died after five years of bizarre rule from a satiny cubby in Bel Air. The environmentalists (major camp up in Alberta) got to Alia, spawned serious dissent throughout the Alliance, hoping to wipe out the human imposition without hurting, at the same time, the earth. This was tricky business, full of unethics, double-cross, hypertension. They undermined Washington all right, but in so doing plunged the remaining twenty million residents of the planet into dismal melee; survivalist factions. No one machine had the power to gain over the warring machine clans. Machines fought and died to secure acreage that had not yet fully withered. By earlier standards, it all had. But standards changed as the need increased.

A commune of Chinese scientists and farmers (long-time troglodytes) emerged roseate countenaced, cherubic from Sung-like times, their machines and vocal chords sterling, unmarred. Ng was quatrarune trog. Dr. Fan 4, on the other hand, was of the old environmentalist lineage.

Dr. Fan 4. Fed up and nasty. Lawyer, last one. Four times called to invent proceedings that would otherwise have slipped by unacted upon. And given the widespread malaise and ennui, any event was worthy of investiture, if only to entertain in the brief twilight.

Endowed with a fantasy life exempt from eavesdropping, Fan was quite a fellow. One of those chaetabrains, like alloplasm, a regular cheesewart, that is, which in any century is liable – by accident, by oratorio, by connections, by squeamish, harried luck (like Kepler, Churchill, Eratosthenes, Don Quixote) – to land on the right histrionic or formula and thereupon make a go of it, reach bottom, give reason to believe, and pop pimples heretofore unpoppable. They got panache, these kinds of folks, and genuine smarts. But more than that, they don't mind getting their feet and legs all the way in tarnish, smut, radioactive sludge, you name it; and so they're usually the first to be hired. Not that Dr. Fan had much competition. No one was flocking to do what he was supposed to do.

Dr. Fan was a thin man, a stutterer, given to cynicism and nostalgia; a misanthrope suffering the limits of his era. Fascinated by Margo. Always on the lookout for worthy contestations, court possibilities, but subject to Bliss' sanction. A frustrated man, fifty years old, unwed, still craving companionship of a kind that didn't exist anymore.

Whatever love of nature had characterized his ancestors was now weaned from him, leaving but a feeling, an undecipherable smattering of sympathies. In his first celebrated case, he'd brought a wastes suit against the State. A family had been accused of hording its own excrement, of leaching ground nutrients for additional sustenance. The Doctor defended them on the grounds of evident malnutrition, careful not to implicate the Complex, which would be futile. The jury was selected by machine, Bliss at the helm. A well-fed, jolly crop of morons – not a one of whom had ever stopped to acknowledge his own waste – found the Rosenbergs guilty.

Later, Fan defended a young pixie-faced victim of slobber. The woman was attacked on her way back to the living unit in the half, sallow arc of evening. She was a wind therapist, wind that came at certain moments, infuriating Xing, came from off the ice, bringing with it curative properties. No one questioned the source of these winds. Cooling, mercurial, it was assumed they arose somewhere in the Complex.

He lunged out of metallic shadow, grabbed her, tried, tried to make something of his assault, a calculus of cunt in bright lights and beckoning buzzes from primeval times clanging about his berserker brain, umphing him into ardency. He clapperclawed and jabbed, with all too much cogitating; she stood quiet, obediant to the surprise and the hope for something. He fingered frantically, spun dreams in air, but the knowledge was lost, his tumescence a fluctuant one at best. All he could manage was a slobber, as the behavior has come to be known, spitmen.

Disappointed, she turned on him, pressed for conviction. But the Doctor lost again.

Then Fan tried to halt the ritual masturbations, arguing that such practice further estranged the body from its natural end, which was to propogate. No matter it was fully incapable of fostering progeny; no matter it wasn't in the scheme of things. For such views he could have been disposed – either by incineration or forced tissue decay. There were, admittedly, some archaebacteria that got in.

But Bliss found comfort in the Doctor's heat. His views were antiquarian, alive with charm of centuries before. The conflict was spelled out across his brow in the twitch of his left eye and flaring nostrils and shrivelled neck and clubfoot and palsied right wrist. In his dishevelled look, like a Mediterranean radical past his prime, wears oversized suits, badly matched tie, shuffles indignantly with high hopes of future resurrection. The fop was heckled some, then thrown out of court. In private discussion Bliss reeducated the good Doctor, reattuned his mind to the sixty-four Faustian precepts, the eight-fold darkness, the four predicaments. Fan repented, and within a year had his fourth case underway: a man had thought, thought suddenly and with some intelligence. Bliss brought him before the Pit.

"You are hereby charged with possession of ego!" railed Slibowitz, the special prosecutor for the case, an ugly polemicist who worked from Fan's brief. Word never got out what happened to the man, since he was never seen again, never came back up.

In Slibowitz and Wickiup the council had its alpha and omega of buffoon. Both men came from gutted pasts. Prone to uncontrolled night vision (most was manipulated) that welled up dense with images of the Great Folly, of 10,000-pound uranium shells, mutant humanoids roving the contested wasteland, of radio-resistant, disease-toting weevil hordes, mythologic uplifts along myriad Cordillera from Sweden to Melanesia; stabs of hovering crowds crowing for all the last spoils. The whole populous become a refugee in their night vision.

Despite his paraphernalia, Bliss could not reprogram, reduce the frequency, nor alter them in substance. Wickiup, psychologist for Point Mind's beleagured scantiness of conscience, was troubled by the added realization that humanity, as once understood, was no more. The roots of his night vision led straightaway to painful memories that weren't even his own but which had sure enough infected all thought; resounded like the ping of a bird that punctuated the glacial atmospheres after two centuries of silence; like a virus between words going back to atavistic grimaces and moans, discoloring the earliest meats, forcing viscera, megalithic grunts, the syllabic hysteria of fonder times, and the butchering of calves for veal.

Such night vision worsened, got to Bliss himself. And Margo. And with them came eczemas, forebodings. All was not so perfect on the ice floe, off which the winds came howling throughout the night, battering the inferno walls of the monolithic Complex.

ICE BIRD

Hardanger tossed all night for the hurtling winds that kept up into morning, rachetting the wormwood shutters against the brick. A firepit in his high lodge still sputtered. A clock ticked.

"Hardanger?" a woman's voice called distinctly from below.

"Ahh, Margo," he noted, shaking out his head and toppling well-practiced down the spiral stairwell of stone, into the entrance.

One-half kilometer away, in his sanctuary, Bliss sat between two thin, quivering tubes that projected on his own retina; in his ear, the sound image of Margo and her mother Odile starting up the Tower of Symbols, Hardanger in the lead. The mother was breathing fearfully, her mind forcing control. Bliss lost the image. Brick! The sensors could not penetrate earth-substance.

"Go ahead. You're free now," Hardanger advised.

"I don't understand?" She looked with perturbed surmise on the objects in his loft. "This? And these? What? I . . . "

"Butterflies," Hardanger mulled. "Wings of a flying animal lost 300 years ago."

She went to touch the filament of color. "No," he warned.

"And that? I don't understand. Please, I don't like this."

"Sit down, Mother."

"Books?" the woman repeated, the words coming from some past. "You've got books?"

"And look at this!"

The mother joined her daughter against the open shutters. "Huh?"

"Glacier," Hardanger explained. "Ancient ice, all around."

"How much of it? What distance am I seeing?"

"I'm not sure. Maybe several miles. Even more, perhaps."

"But how?" She stared like a mad woman, face into liberating wind.

"What are you thinking, tell me Mother!"

"I, I don't know," she turned, unwound by these revelations, sitting dizzied on the stone floor.

"Oh Mother!" the girl cried, on her knees, embracing the bewildered woman as Hardanger looked out, his hair flopping against the air currents.

"Don't cry, Mommy. No, no. It will be all right!"

"Dear one . . . " They smothered each other in smarmy outpour. Strange, Hardanger thought. "Something in the air!" With his spyglass he combed the ice, peering through the murky lense to each crevasse, following the curvatures of grey as he had done daily for two decades in the hope . . . .

"What's he got there?" Odile asked, reassembling her wits.

"Look." He handed her the instrument.

With strict seriousness she made survey, no gasps left, her curiosity for the other artifacts in the room now but concentrated on a distance never before granted her. She looked away as if to question him further.

"There would be no point," he volunteered abruptly. "Nothing lives beyond. The whole world has become the ice."

"The whole world?" she pondered. "The *whole* world?" She gaped at her daughter, then at Hardanger, despairing of such learning.

"Your thoughts are your own in here," Hardanger delivered. "Yours, you understand. These relics, these words – exoskeletons, quetzal feathers, petrified avocado, picochips, ionized neutrosesames, astrolabes – no one else has seen them. Bliss knows nothing!" he continued.

"Yes, nothing!" Margo furthered.

"It's not possible," the mother conjectured. "Bliss himself? But yes, true it is; I feel no pressure there!" Her hands to her temples.

"Go on then, explore for yourself!" Hardanger urged her, hoping to gain her trust, if only for his and Margo's sake.

The mother thought, just sat and thought all day, until her face bore witness to the radiant meditation, while Margo studied as she had done day after day, for two years. Hardanger had taught her eveything he had in his store – computations, folklore, dream. She had come to believe in certain verities long renounced – like walking, like mortality. Had come to *feel* this thing for Hardanger himself.

"I'm a plodder, just a plodder," he told her. "I have a stake in finding out the truth, in answering some basic questions. And I'll do whatever I have to do to answer those questions."

"Truth about what?" she came on. He paused, then readied: "We can't be all that's left. Of this much I am confident."

The concept didn't faze her at first, but then slowly she began to thrill to it. And why not? The thought had, it is true, never been allowed. The planet, they'd long been told, had been scanned by the *experts* for any living substance. No sir! What was left was what they had there in town. Opus thought differently. Margo was as won over. Hardanger stared out across the ice.

In the evening he repaired to his bed and lay awake staring at the stone ceiling. Margo and Odile had gone, the mother in a dangerously jubilant state.

"Absurd!" thought Hardanger. That they'd waited this long to speak openly with each other. But then it was certainly no easy thing to break through thraldom decades old, the law, the nature of the universe, and Bliss' thorough eye. The Complex had them mesmerized.

His light flickered around 3 a.m. and the thick plodder fell away to sleep. In his dreams he heard the cracking of thunder and the long, unearthly cry of a hapless and bound mastodon being surged into by hatchet-bearing pygmies in armor.

On the town side, nothing ever poured through Hardanger's window in the morning. Only the blank, characteristic hue of EMR (Environmental Modification Rite) one of NO's unusually exact provinces of techno-fix. Chewing gourd jerky, Hardanger closed the shutter, preferring the free sweep of ice on the other side. Behind him, a pile of books waited to be reread for the tenth time. When –

Something moved across the ice. His eye stopped, paralyzed. Yes indeed. Something moved there! It flitted a shiny blue, some distance out, in the greying glint of morn. He lunged for his best eye piece, a seventeenth-century Jesuit instrument made in China. Hardanger's eye fired to the crevassed region wherein he detected the commotion, then waited.

When again it did it. Damn! Zeroing in, Hardanger saw it all, followed it. More nearly the color of cleveite, of certain plants from his botany book. In fact he'd never seen such a color; such an animal. Any animal!

The bird disappeared in a crevasse, came out again, swinging wide easy, diagonal lopes across the overcast margins. It was frivolous, ascendent, and free, swiping every which way. Hardanger's fingers trembled, his mouth drew saliva, his chest fidgeted. The bird was no longer in view. It had come in towards the Complex. Suddenly it was there before his shutter, staring right at him, cross-eyed, with devilish, yes, with grinning, luminous, really silly attitude; holding at seventy vibrations per second. Hardanger reached out to grab it when the bird flashed downward, crashing into the bricks, tumbling toward earth, stunned, before reviving itself, fluttering back into gear, and away.

Hardanger raced down the Tower, out into the platform area adjoining the concourse; ran pell-mell into the danger zone. He didn't have his mind together. Bliss, from his claustralled niche, picked up thought coming hard from the quadrant in which Hardanger recklessly flung himself. Bliss alerted Martinez with the conventional alarm system, then waited.

Martinez travelled the circular belt around the Complex on an airfoil that shot his body comfortably within seconds to any point. Like a demon he emerged, flat-faced, malevolent, teeming with uncompromise. The two men faced one another without saying a word. Martinez went for his collar. Hardanger held his position.

"You resist me!" the General screamed.

Hardanger kicked him in the groin as Martinez threw a strangle hold around him and squeezed. They went down gyrating with the thick circulature of rutting behemoths fighting out desires. Both men wanted to kill the other. Martinez was practiced in this. But Hardanger in his entire lifetime had never exerted force upon another. Now four decades of idle nervousness blossomed with a malignancy the fat General was not altogether prepared for. No violator ever contested Martinez, the dutiful exterminator. But Hardanger was all heat. He tightened his choke hold, roared, and shoved the General into the metal retainer of the Complex. There Hardanger beat him, persisted, tore at his flesh, ripped into his bulk with carnassial premolars, until the jugular was chomped through, Martinez flailing spastically now, and Hardanger really getting into it, until it was done. Releasing him, the spouting body fell limp unto the pavement. Hardanger sped away, spitting out the taste of carotid.

Bliss viewed the struggle from his monitor chamber. Unpleasant evidence, he thought, continuing to behold the movements of Hardanger, who now broke through the final barricade, a simple fence, surmounted a chest-high barrier (no one did pullups, mantled, exerted in Point Mind) and leaped down to gravel bed spread alluvially into the yawning grey ends of the earth. Blood on his hands, his eyes searched for the bird, as he moved out.

The brilliance of glare stopped him at first. He cowered from the hot light and distillating sky. Ideas assaulted him. The glacier took on life of its own as he approached. Its digestion was augmented, it smelled fresh, gave out wild sentience. He went crazy with the freedom of being outside. Why hadn't he come years before, having been afforded so clear a prospect from his Tower? Such questions.

The bird was there, hovering up-current. It saw Hardanger, dove giddily toward him, lipped, then alighted on a stone of the moraine. Its voice was Circes sweet, of melodies unknown to human ear. Its blue head tufts shivered in the wind. It stretched its wings; gave out the aura of innocence. Twenty feet away, Hardanger stopped, staring intently now. The bird hopped from its rock toward the man, hopped twice; yearning, Hardanger perceived, for similar reunion. Hardanger approached; the bird hopped again. He lifted out his hands, slow, ever so slowly, when it flitted away, lifting off, catching the breeze and vanishing over the mishmash of rivulets coursing through the terrain.

Hardanger followed in his accustomed tennis shoes and woolen breaches – museum pieces. He touched ice, where the glacier was low and level, alive with run-off. He held on to it for goodly seconds, delighting in the cold, the texture.

Then he climbed, traversing through the jumbled amassment of nevé and floe. The winds were assuaging, all regelation, incessant. It seemed to Hardanger that he was breathing air for the first time, seeing aerial mists saturated with true sun, whose striations colored the clouds the way his Segantini book revealed them. He couldn't get enough air. A thirsty man, given water, can conquer the world. The water ran through ultramarine troughs, Roman aqueduct-like, whole polished networks. He drank on all fours, turned and looked back at the enclosure from whence he'd come. Sickening! Storm clouds seemed to hang over the dwarfed community. For the first time, Hardanger heard no whirring of the reactors. Instead, his senses were flushed with the wide, tumultuous cavalcade of water venting the earth's first and last force.

The bird again set down, a stone's hop from Hardanger. They stared. The bird shouted, Hardanger shouted, then both started up again.

He came to an impediment: a large darkness, gaping in a slant with a narrow arête ribboning through its center. *Au cheval* he straddled the elongated file like a child, discounting risk, banishing all thought of return, not at all cognizant of the exquisite danger, having never before risked anything. He attained the end, stood shakily, then slipped down the backside gully, ripping his pants, his palms and ass on the sharp granular ice, arriving thirty feet below in a depression of black ice, beneath churned-up gravel heaves. The entire glacier had this dull chalky gravel. But Hardanger revelled in the pain he'd caused himself, took stock, breathed in, and continued, like a man reborn.

The bird flew circles all afternoon, quaking with invisible wing snatches, hummed-up, perking to the lumbering giant that followed it across the multisplayed crevasse country. The ice was raw, no fresh snow atop it. The air was warm, gusty. Hardanger guessed that it must be summer. There'd been no way to calculate.

The terrain gathered inclines now. Symptoms of elevation change. A greater rumbling of water, submerged concussions, and seracs mounting in ever denser, loftier gradients. Fog now swept over the highest pinnacles of ice, coming from the sea of which Hardanger was as yet ignorant.

The chanticleer bird could be heard somewhere in the chaos of glistening towers, sweeping through one crevasse after another, topping off at high summits, then diving down again, obviously enjoying itself, sweet cajolings receeding further and further back into the labyrinth. Hardanger kept up, stumbling into pits, sliding down frozen mesas in the muddied debris of the Weischel. Surrounding the unthwarted man spread seas of splendid desolation that, to his unbelabored eyes, were busy with the spirit of new life. New color, air, sound. Such power had the man in thrall. He squatted, shat, watched it curl and lump upon new earth, under a new sky, the very sight somehow a palliative. Residents of Point Mind had mixed feelings about shit. On one hand, it was intriguing stuff, coming right out from inside you and such. Smelled bad, true enough. But somehow kept residents in touch with a separate reality. And it made for excellent fuel. Kept them on their toes.

All day Hardanger climbed up into hills, losing himself in a proverbial way, a neurological way, losing that regimen in his senses so long unadapted to the natural light.

"The bird," he thought. "Where has it gone?" But he went on, plying into uninhabitable stretches.

An outcrop. Standing far-eyed. Foehn cleansing the man's hank of curls. He looks far out over the glacial dunes he has crossed. To the edge, he notes with special anxiousness where the interglacial had cast up its black towers of ice, caked in sliding mud, in boulders fractured and shattering with the ongoing sulphurous tide of creation.

He surmounts the summital pinnacle, very precarious. No guest registers here. He gawks at tentative courses open to him. There is one apparent. So he pushes off against buttocks, careening down with the irration of ancesters, flops, plops, plunges. "Oh!" he screams, until bounding no more. He lies sprawled atop the trickling dark lateral moraine, where the grey, the white, the blue and black have merged.

And the green!

Hardanger's wrist is slashed. Blood congeals. To scrape it off would pinch his golden arm hairs. To leave it is unthinkable. Dismay stirs the man who passes around the decision, gets up, hastens to the color which no most exists in his memory, save by weathered inference from the books. The books that have told him everything he knows; how to hammer a nail, perform a brisk, eat an artichoke, were one to exist; how to manufacture kilowatts, tend claybank horses, collar a heretic, send an SOS, explain away the great famine that stifled the human world during Finality times with gaudy statistics full of sensible stuff, lifeboat ethics, and finally, reading into them, a refined, self-pitying derision. It was the same in earlier times. Such that the artisan networks – GROUPA 2500, intellects of Europe who debated the world's crises in on-line interactive cables, serviced by Radical League (illegal) transponders; RVA (Radioactive Victims for Absolutism); TDDS (The Tortured and the Damned for a Democratic Society) – all such groups eventually narrowed in impact, self-image, then fizzled out altogether. The dualism of bicameral mind – that schizophrenic collaboration embedded in the cerebral neurons – catered to the easy coersion of certain rarified comforts underneath political ones. The urban geography of the West was altered, and afterward all other regions. Hardanger knew the pattern. The government gets the individual feeling bad about himself; this negativism carries over into the movements, fouls parties, necessitates the same old totalitarian heart. But at the bottom of it, the individual is to blame; he's gone sickly, insecure, hates himself. The suicides were fostered by the State. Opium injectibles, depressants, the Sunday Morning Special, as it was called – these were supermarket items meant to give anybody who wanted the chance to wipe himself out. In India and the USSR such policies were celebrated.

In the meantime, ruminants lost their stomach microflora; the planet's tubers had no more nutritional value, no more taste; domesticated animals, beginning with the French poodle, the dingo, sheep, goat, all went nuts, fled their masters. Horses bit the plowman; chickens, it was learned, took to deliberately fouling accessible water supplies. The Muscovy duck died out. The red-eyed vireo sang its melodious 22,000 encores a day no more. Not even a one.

During the Finality, human response was all too calculable. Wildly oxidized amino acids incurred protein deficiency; increase of osmotic pressure in the stomach, of cholesterol, edema, excessive losses of water (10 percent is fatal enough); the end of hydrogen uptake, the last days of linoleic acid – a little thing in its own right, you may say. But no sir, for with it, the enhanced vulnerability to radiation, which of course was everywhere by then, most densely consolidated in high-altitude regions. Denver, Lima, Mexico City; they were glowing.

It was a time of exotica, Hardanger contemplated, blood on his finger, skin peeling from his torn knees. Of words like Kwashiorkor, Xerophthalmia, Marasmus, beriberi, pellagra – wonderfully sounding names. A time of disintegration of the human personality and the internal organs. Food also (grapes, legumes, soybeans, manioc, peanuts, sorghum, rice, corn, millet, cassava, barley, wheat) all exhausted in the general fracas to get more, to get ahead, well, in fact, to get *anything*, amid the general decline of soils and of energy. A leaf, cooling mud, in the end, to stick your mouth and nose into, when there was no more cloth to cover them with; no more shoes to eat.

By 2200 A.D., forty billion human denizens of the planet, clawing for the last tidbitties, like jackals, but unable to digest cellulose, or eat garbage effectively, or ward off the random outbreaks of – call it hate, vendetta, sadness – directed against each and every living thing. No one escaped, no one. By 2200 A.D.

Breathing slow, on his knees; arms limp, as in a pew; at the rakehell convergence of two mobile ice fields that glittered and sulked and heaved. A crackling overhead. Hardanger lunged to the side as a crashing serac tumbled. His feet were in run-off, the gelid cascades gurgling through smooth rock the color of smoke. Mud, gravel, reflection of blue firmament – "beautiful . . . you're *beautiful!*" he hollered, going outward steadily from his past. He followed the alluvium around the ice terminus, where towers hung high; marmoreal, jagged as a whale's dentistry. He was an escapee, at the end of the world, where small spiders deposited their eggy clusters in the crystaline interstices of snow and lichen; where the cave beetle burrowed deep into ice, and the sleeping eye of night awakened to the first sky-marooning chill. Infinitesimal clickings of rock being carried down-rivulet. The grace of whispering.

Hardanger scrambled atop a knoll several hundred feet over the last of the ice. Hundreds of insects were summiting here. Chunks of caulk lay across the umber surface in waning light. Soil! Pitted, as if to be eaten. He ran his fingers through the stuff, until the dark droplets spotted his hands, until, in his own mute fashion, there was the relief of thanks. "I've come home!" he thought.

He turned the earthy crest, descending into the freakish pasture of new forest that had grown up between two adjacent glaciers. It was a triangular affair, delicate and lush. *Abies lasiocarpa*, Tsunga trees, Larix, Betula, *Picea engelmanni*, Pinus, sweet grass, frailty everywhere; Alpine Sorrell, Diptera crawling through it along the dark creek that rushed through clumps of leaning dogwood. It was untouched, undiscovered, and in Hardanger's mind, impossible to assess. No dissertation had prepared him for such freedom. The very aeolian verdancy, the night, the sweetness – oh, and a palm tree, yes, here, after so many centuries of skewed surface area, and the ludicrous intervention of one Defense Department after another. The night air, the possibilities open to him. And the night itself – suffused with weird glow.

Cold, he sat down on the grass and hugged himself, laughing. He felt hunger, and the vague memory of fear. Memory dating, like the twilight.

He sat absolutely quiet, his senses denuded against the rippling night, so rich with scent. Imagine, for his first time. Even the taste of it right in his very mouth. A bevy of tiny buttercups (*Ranunculus glacialis*) littered with infant flea creatures, three of them standing out, with transparent wings, fumbling up and down the Aven carpet, along the *Dryas octopetela*. He was dizzy, concentrating so hard that heat and tears came out and he fell back as if he'd had two straight carafes of ouzo. And he heard a "coot-coot" sound, caught the flickering shade of animation in the shittim wood. A vole perhaps, or water pipit.

He lay perfectly at peace, his back against the earth that gingerly held on to him; and he examined the wide gallery of heaven. Unaccountable, numinous, poor Hardanger distraught with the angelica of the thing. In middle age. "That I've wasted so much time!" he rasped, scooping out water with his hand, chugging. While the unbridled borealis gave him the personal assurance that this one night was his, and full of the answers he'd been seeking. There'd been no mention of such glories in all those books, nor was he before aware of other galaxies. Bird migrations, for that matter. Or of breast feeding. Or paddle ball. So many things had come to an end after the Finality. The tabulation of baseball scores; dew on the grass. Medicine. Player pianos. Whippoorwills. Legal rights. Jumbo jets. The sanctity of carbon. Tampax. Fire engines. Spearmint gum. The Ghent Altarpiece. Wasps. The Cleveland Philharmonic. The Nile. Harvard's Widener Library. Saudi Arabia. Samoa. The Hilton Hotel in Athens. Lox & bagels. The *New York Times*. The evening news; the memory of rock & roll, of Sibelius, of Titian. Nothing was left. What Hardanger picked up were those discoursive bits from a few good reads. He was *lacking*, lacking so much. And here, in this place, he began to sense the moment of history, craved a vocabulary for it; craved to know the interface, the breakdown, between society and nature. Such hopes, useless. For the language had lost these things. New generations – born of uprootedness, born into gypsydom, diaspora, born of the leukemic period, after the Finality; surviving by no science, but by an unnatural selection given to unsubtle discriminations, bogus blood, meshuggener mentality, eyes dulled, dreams dead. The hecklers, pushover lunkheads, half-brain commune captains, the voice disks – these artifices were the magnets of learning, last vocables, to construe those dictates by which the leaden legions obeyed unquestioningly the tasks before them. Separatist chores, cheek scars – the zipper kind – marking friend or foe. For the eyes were the same on both sides, generally speaking, a mottle here, dark dents there. Eyes that had seen too much too fast. In a bright light, a lightning flume, exigency without a purpose or shade, as darkness whitened out, the last awkward instant, giving no senator, no movie star, no child, no nobel laureate, an afterthought, not even a second's worth of common prayer. In a civilization accustomed to summing up its verminous affairs, such abruptness must have been disarming. Or so the underground survivors later quipped.

Xing was the first to find Martinez. "This will not do!" he muttered, trying futilely to dust off the blood. No matter the mangled neck, the rigormortis, Xing — formidable duster — had his job to do. He'd never seen such vermillion stuff, nor could understand the seeming repose. How listless it all was, the flesh, that is.

"A strange thing," he considered, peering down from his mop upon which his chin rested, laser gun in the other hand. And given to such thought, he had, additionally, a jolt. So unexpected an encounter with Bliss himself who, accompanied by Ng and Dr. Fan 4, flew to the scene shortly after Bliss had viewed the combat on his monitor. Sirens wailing. the siren in Point Mind had an alarming effect on the unattuned ear for it was the sound of mothers screaming.

Surveying the mortality (with Xing arrested in an attitude of the dumbfounded, Dr. Fan rising with the zeal of litiginous prospects ahead, Ng all agog over the breach and its technical implications; maybe there was fallibility here, he asked?) Bliss cooly figured the sequence of events and plotted his recourse.

"Just as well," his inner voice pitched. "Didn't much care for him. Didn't *trust* him. And here's why!" But to dispose of it, the body that is — incinerate. No thought of Hardanger anymore, or very little. "Man can't survive the night!"

Little sense of the cartography beyond. For all of his proven cleverness, his next-to-NOliness, Bliss had little curiosity for the outback, having encamped once on the ice in the early times and survived it. He remembered only the uniformity, the dismal hills, the lack of game. Nothing to distinguish one hectare from another. "He'll be back," Bliss, of the Pharoah school, bargained, making ready for inevitable combat. For it was on the outside that lessons might yet be gleaned. Of this much, unhappily, Bliss was still in the know.

Opus 318 felt something, a twinge, as she performed her voice in the community amphora, a lapis vase the size of a bistro, giving denizens the one and only fully realized dumb-out, with blue to dote on, imparting a vibration from the Before. The arrangement was quite intimate. Soft spots on the curved surfaces upon which a dozen participants cozily reclined permitted a degree of schmooze — the sloppy exchange of desire abstracted to a halo of velleity — unrivalled in other spheres of Point Mind existence. Try, if you will, to visualize this scene: a man, wearing synthetic velvet rubberized booties, dark spectacles (moddish), and the requisite groin neutralizer (a plain net fitted with electrodes) squirms ever so blatantly to the pair of boobs crouched low beside him, unfurls a tongue color of maggots on the make, and primes himself for — he doesn't even know — while others abetting the orgiastic impulse, swaying like campgirls to the weird sacerdotal sounds of Opus, begin the old one-two with their fingers, legs propped up, heads straining to see in the hermit darkness, in the ghostly blue light, in the cushioned interaction. Fingers, legs, heads, tongues, cavorting without a clue all night long.

Opus engathered her audience, spread them like butter with perfected timbre so unlike the machine messianics, the socialist pieties, and Complex platitudes: "Eat nothing, breathe nothing, think nothing, say nothing, do nothing" – those utopian bromides that mankind had been always striving toward.

Opus dazzled them with her resonance, her exhibitionism, her crotch, nipples, tush, and solfège of old days, such that they were unaware when her jaw bones relaxed, eyes glazed over with the verity of an other world, nose hairs puckered, but something different about this moment. Her gut, heart, and soul trembled with the sensation of her brother, of something out *there*! Her singing stopped, she fell back, very Antigone. Her hand reached for the lung; Opus, lovely Opus, wordless, with all her sagacious access to revelation in the tyrade tradition of Cassandra, Madame Curie, and Betty Davis opened the bubble slot, plunged forth from her octave into the concourse. Oh she felt it all right. Something moved, *moved* out on the ice!

Ng went quickly to his operations room, scouted about with pliers, twisted a few fittings, stopped up a gap or two with sheer pins and dowels, analyzed the full network by visual scan only to learn technically what took other men 500 years to come by: There were shady areas, malfitted regions without proper wire that had escaped the overview. Hardanger had gotten out. NO had permitted it. Ng's logic swooning to new heights – that the gods were not infallible. Incredible! Or was there special purpose to the outcast's expulsion? He quelled his euphoria and set about to summarize the breakout.

Xing stammered, eyes ablaze and rheumy. Odile, who still had not recovered her composure, listened, was troubled, eyed from time to time young Margo whose Eleatic sense of inquiry prevailed upon the situation. She condemned the custodian, though he was her father (nothing new in that: In Point Mind, children and parents routinely turned each other over to the authorities for various wrongdoings. It was encouraged, mandated).

"You lifeless gabbering fool – wake up!" and spat on the floor, something she'd picked up from him, Xing responding with his usual mopping diligence.

"She's saying something," Odile gathered, arm extended to her man of thirty years.

"What?" he blurted. "What's that?"

"You saw something! Now what was it?" Margo went on, furious.

"A, a man, a man lies down, Martinez lies down." Pointing to the neck. "The same color as blood. Lots of it. His Eminence came. Some problem has arisen . . . " He stared cagily, then simmered out altogether, sliding softly away into duster dreams. In his fondest night vision, Xing saw himself as a window-wiper in Leningrad, Gulf of Finland summer galoshes, and bright new squeegee up on one of those skyscrapers of the olden days. And that's just where he put himself now, as Margo sat back, heavy with upset. The mother, too.

For the woman, existance was a given, unfestive, no special powers, no darkness to excite it. Whereas the man could call it madness ("I stare at madness and have to fight to keep from it each day," Hardanger had claimed) but woman knew the boring sanity of submission and had little sympathy with the artist in her. Odile had the artistry, no question about it. But she was trapped. And although Margo thrilled to new learning of the kind Hardanger imparted her way, she could not enter into pact with his deeper churnings. Her lack of multiple purpose, of mystique, of torment, her clarity Hardanger simply attributed to that cumulative deadening, less pronounced in the young, granted, but nascent like a cancer nonetheless and pervading all the women, all of their flesh and hearts, in the insular and terminal township of Point Mind.

Hardanger's special yearning was not shared. His unassimilable passion before the fuzzy mirror in his rickety tower: "What's wrong with me?" he'd ask himself. "Why can't I be like the rest."

There was very little to hold the man and woman together – no sex, no love, no mutual destination, no repertoire, no unison of reference. Fundamental impulse, before the stages of humanity, before the knowledge. Man and woman. Before the naming of genitals and the touch of lips and first sparring. Before extraction from his side. While they were still one, in fact, and all the attraction stemmed from something more sinnister, and blind: the awful straining to separate, separate for good. That was the primeval catalyst.

Such friction was, in itself, a glorious outcome, though few smiles accompanied the forbidding discovery of it in Point Mind. Margo liked to hear the thick man's lisp, his hoarse elocution and strange words. He had a malady, she reasoned, and malady alone – in so hygenic an environment – constituted fugitive charm, maverick form. Under the sick man's tutelage she'd come to pronounce all sorts of things (monad, estuary, apricot) and grown fond of the pure sound of learning, could feel the construction of words with her dainty white fingers in her mouth as they formed, words she never got at home, with a Mongol idiot father, and ever-deferring mother.

"You're a good man," she told Hardanger, after a few months of learning.

"Suck this," he said at once.

"Why?" she queried.

"Try!" he urged her, taking her head in his hands.

She did it. But all of her long-limbed, wet-tongued electric grappling could not induce the slightest. She wanted it as much as he. But nothing came. Just the hint, fervid poetry, of how it might have been. Like the rumor of Noah's deliverance. His carmine loins shrank back, her jaw throttled, buttocks went up, yet no sweat poured down the cleavage. No gleam. Nothing to suggest an ardor. But one more angle in the lifeless tower of brick. Still, both fumbled for a better future, Hardanger, his brow dry, repairing to his aeronautical charts from primitive times, or Boethius, or Dante; Margo, looking on, doing up her clothes, practicing her arithmatics. Night after night. Rote. "But why am I doing this?" she'd often asked. And once she memorized the answer she stopped asking him.

Though baptized in the ungoverned chant of voluptuousness, without an extra breath, a spittle, not even a spare sigh. No feeling. It had been stamped out of even them. No matter how hard they tried to be different.

Now she looked far and fast. It all came in one heap of exasperations. Motive for so doing. She ran from her family's confine, tore out of step from others, leaving behind the honeycomb of habitats, raced past the Pit, not watching (others never, never came so close), exiting the concourse and heading up into the Tower.

THE TRIAL

Five hundred years before, Tigaraqmiut hunters would have been out on the ice; a normal day, clubbing walrus, harpooning a whale or two (*Balaena mysticetus*) from their umiaks of stretched seal skin. The children – powdered white, rosy cheeked as port-wine – followed the elaborate tunnels of mice to certain amalgams of duff, edible tubers. Half the summer ground browse would have been left alone so as to keep in perpetuity the mice who were so critical to the overall schema. Whitefish and trout were netted, and the white whale and ringed seal. Seal oil would have nourished scant flames within the farthest sod-covered ebruliks, comfy havens festooned with whale bones, driftwood, and antlers. Families came together to stuff their sweetmeats, juicy fats, wild berries (*Rubus chamaemorus, Empetrum nigrum*). They'd toss each other on stretched skins, carve scrimshaw for the occasional tourist, go out after caribou in snow machines. By fall, the days were short, mists long, cold, stirred up with drench from across the frozen sea. All the harvest would be gathered, the village huts fortified, steaks smoked, oil and gas stored for three-wheeler use in winter. Everything else came in by Beaver plane, once, twice a week depending. The winter was not an unhappy time. In fact quite jake. Above the blue ceiling of ice mist, stars twinkled, lovers coupled down below in shantytowns protected by Samoyans, their fur white and noble and glittery in the cold, like the crystalline snowpack and the very stars.

The folks would wear tutuliks of caribou skin and caribou-fur socks along with dry grass innersoles. That or L. L. Bean footwear. Their parkas had wolverine ruff. Seal-gut rain jackets inserted with decorative labrets really did it. The winter was devoted to having fun, mostly. Fun that was focalized in the karigi. Anatquqs (shamans) restored lost souls in there throughout the festive carousings. Though shamans at that time were becoming melancholy.

In the spring, travelling inland to the lagoons, camping in domed kaloviks made of bull-caribou skins sewn together, spread over willow poles, they hunted ugruk (bearded seals; *Erignathus barbatus nauticus*), stalked along the ice after other sleeping creatures, hooked and netted for pike, tomcod, crab, and various sculpins; they went after migrating waterfowl with bolas and arrows. In the summer this store of fats and proteins was supplemented by young willow leaves and shorebird eggs. Sometime in July, they'd move back to the coast in their skin boats. With other Eskimo groups they'd trade seal line and oil, blubber and furs in exchange for iron lances, reindeer skins, and tobacco.

Then the dogs all died out of parvo. Gasoline was no longer a feasible purchase item. The caribou herds dwindled. They were unable to figure out how to cross various pipelines. Their migration patterns were ruined, young separated from parents. All but the most stalwart (a few dozen) of the Tigaraqmiut headed south for the first time in 20,000 years. To Pioneer Square in Seattle; and downtown Los Angeles.

Undistinguished, typical dawn enveloped Point Mind; and unflattering light, nonluminous, as flat and blank as the walls of the Complex. It was the first day of the trial.

Dressed in black jumper slacks of an organic accretion that regulated body temperature and hooded stylishly in rubberized slippers that merged nicely with the leglets and went with a synthetic fur cravat and dark glasses, Bliss struck a sleek, grim countenance before his mirror in the morning. Fingernails bitten down, cuticles sore from excess gnawing. His once firm musculature, built up on battlegrounds, now was largely shadowed, hairless, with green moles of sedentary life. An exemplary neoteric who did not relish the days ahead. He was on shaky ground despite Dr. Fan's superb array of allegations and Slibowitz's eager vehemence; despite Ng's reliable testimony and the monitor playback, there was no doubt about it: Two camps were gestating.

Wickiup had been infected – night visions of a new order. His sleep was intermittent, occasioned by fierce growls that came through him uncontrolled. He rocked his household with gasping at wee hours. Mud, his mate, lay paralyzed with fear, grasping both sides of her sand futon, and Wickiup's kids stared wide-eyed from their respective quarters into the chemfixed dark of the family's technocil. At vesper hours, shrieking Tibetan alephs, !Kung incants, muttering involuntary apocraphae in a dozen extinct tongues, eyes rolling, mouth full of drool, he careened the full madness, night after night, awaking with rings under his eyes, scalp wrinkled, discomfitted. No way to go to work in the morning. Especially this morning.

"But what do you see?" the lovely Mud asked him, her black olive eyes hesitant before the moving mystery of her man. The sixty-year old administrator deferred her look, afraid to dwell on his own ailment for fear of bringing night vision into day. In a society such as theirs, night vision was especially dangerous. If a tendency were picked up on the monitors, it could, under extreme circumstances, necessitate a retapping, both painful and awkward, for it meant the re-creation of a man's personality. He would no longer find communion with his family, his friends; he'd lose his reference pool of privacies. Councilmen were ordinarily exempt from these inveiglings, but the night visions were disturbing enough to upset protocol. Wickiup was forbidden from communicating their afterimage. Furthermore, he *knew* that Bliss knew.

"I see nothing, I tell you," Wickiup murmured, downcast, given to penitent demur.

Bliss sat before his access board, tubes to his brain, electric danglies in his nostrils. The Complex generated billions of signals, on-call data brought forth in seriatim that profiled successive predictions in proportion to relevancy. In terms of jurisprudence, the Complex had infallible precedents, zadig orations, Sanhedrin sagacities, to go on.

Bliss anticipated all possible arguments, programmed his judicial responsions so that the words would come like automata, as if spoken by NO itself; with the redoubtable and exalted stamp of the heavens, the certitude, spontaneous wisdom. Like the crack of a whip, or the amaranthine robe that Bliss wore to trial, the color of the signet imprinted on each reactor.

The problem was the *species* of knowledge he was making ready to deal with. Hardanger's suit was couched in black sheep prose, filed with unrecognizable dispatch, no doubt in the man's mind. That he'd come back at all!

Bliss steadied his hands to his ears as the tubes quavered in the still chamber where his algae got cold on the breakfast table and the mirrors creamed back at him the untoward possibility of conflict, the first in a generation. "He wants to dick me!" Bliss bittered, a certain relish flavoring the prospect, the hope of novelty; that something might happen in a world where nothing ever happened.

The tubes took nine seconds to warm his grey matter. Then the record plunged, history turning from mere gossip to specifics, like a barrelling locomotive with its headlights turned inward; or a butterfly alighting: Between these two poles of motion and experience, the massive commingling maw of knowledge, ravening, entered his head, refurbished his logic in the matter, gave verve to diatribe and just the right generosity. His people admired him, for all their lackluster; saw in him the true patriarch. Of greater importance, the tubes gave Bliss the wherewithall to combat his archrival, with whom he'd shared but one, standoffish legal exchange upon the latter's unceremonious re-entry into Mind, from off the ice fields. He was naked, moving like No-Man, skulking and on all fours, demonic, oblivious to harm, his muscles brazen and bruised. Especially dour, shifty, moving with calm purpose toward his aerie where Margo had kept up her vigil, crouched near the firepit. There the trove of excremental embers had made life toasty for her in his absence, over the interminable fortnight period of his disappearance. Bliss watched Hardanger re-enter, was enthralled by the man's slow gait, his brutush loins and charcoaled back; Bliss blew up the wounds, saw all the lacerations, and the dried blood; but more, witnessed the strange gleam in Hardanger's eye. And this is what Bliss feared most, recognized it from Finality times when some people still uttered the word *freedom*.

Wrapped in the serge blankets, pouring through each unread manuscript with a grip on her emerging vocabulary to learn the secret of his mission, in fits Margo's vision enlarged. Until, when Odile climbed stealthily the spiral tower stairs to visit her daughter, they could share the welling-up of desire neither had before elected to indulge. Until Odile was won over with her dangerous memory in which was stored the natural history of fertility. Margo learned of her true origin, the birth, that last night prior to Hardanger's return, and if knowledge ever were to invoke change in the physiology, this was it. She trembled, began to bloom right there. "Mommy! You mean –"

Yes, dear," Odile spelled out.

"I *want* something!"

"What do you want?" the mother prodded.

"What do you think, blockhead!" the twenty-one year old savored with gusto. "I don't know how to say it, I'm lost for it, help me, it's here, here!" with which she pointed to her Chi, encircled her own waist with her arms, spread her hips to reveal the capacious possibility for thrust, for concept.

They held on with equal terror, the mother's ague her daughter's ague, until morning, when febrile passion for that *man* – Hardanger – had both of them aroused, kneeled over with the first hope since ancient days.

"Xing knows nothing of these matters," Odile relayed. "His head has no meaning, poor man. He tries, though: everyone does."

"Tries *what*, Mommy?"

"To get through the day, I suppose," she stabled, no bones to pick, what with the flesh long ago flaked off.

"It's not enough," the young girl angered. "What's more, I plan to do something about it."

"What will you do?"

"I don't know yet." She sat in visible distress, planning for some siege tactic.

"What if he doesn't come back?"

"Mommy!"

They quieted, sullen. No point confusing matters with a speculation better suited to astronauts or escaped convicts. They had to get through the fact of their one, undifferentiated dilemma. For all of their microniche sufficiency, they were trapped, and the first – they assumed – to be aware of it.

"He's not there because there's something out there," Margo declared, scintillant with an idea that lay abroad.

"Something, maybe," Odile countered. "There was once, anyway," and the friction came back, singed her recall, made her cry.

"Be strong, fight it! If there was then, there is now. This is not a question of tenses, but of *being*!" the girl demanded, thinking high and constant.

"Maybe," Odile seconded, in some pain.

"Mommy what's out there? Oh tell me! What was it like?"

The memory was excruciating for Odile precisely because of the tapping, whose purpose was to do away with history, to conquer it. During the Great Folly, the Clean Future Wars, and the Ravages of the Reign of Alia, knowledge had pretty much exhausted itself. As the earth fell steadily to greedy rapacity of earthlings, information and unregenerative symbols were wiped out – destroyed by dint of their demonstrative link to topography. In the wake of felled forests, selva cut through, the tribesmen living there enslaved to the Alliance; Antarctic acreage dragged to dissolution by various sun-belt protentates, amid the final stones of Delphi, last bits of chert and arrowhead from Tennessee, first edition Joyce and Carlyle's from the British Museum; amid charred ruins, no Cycladic art, no Persian Safavid miniatures, no flying fish, no Lassie, moonshine, Gregorian chants, Three Stooges' footage – nothing! Nothing was left to inspire the oracles once memorized, the fairy tales, mythology, the glories that were Fresno. The land was destroyed and with it the wellsprings of language and art and ritual.

Those few hardened dialects that survived did so on the basis of an insulation from the world – a plastics Esperanto. And with the decline of language, of wit, pathos, there was, at last, a not so flattering union – nature and culture, both of them dead. Both indescribably eroded. It was a maelstrom no one expected – the inability to say what one needed to say, to see what one wanted to see in so massive an impairment across all cadres of the vanquished. DNAs reconvoluted, marine life licked; no cereals on the Altiplano; no heart-to-heart talks; no flirtation, no marriage proposals, no gab, no long-distance phone calls after eleven p.m. to the folks back home in Borneo. Instead, a time of frustrated silence. And with it Bastille Days, October Square nights.

The Cubists had their own revival. They were abstract revolutionaries this time, mathematicians from Boston and Prague. They gathered in the relative seclusion of bombed out student grills and hashed over principles from Suzuke Roshi to Enkidu – all in nanogrammatic, holographic tidbits of a computer animal. Not so much as a word was spoken between them – no time, no talent anymore for the cumbersome, the epistolary, the human touch. But the gleam of metal – hairiest koans in the annals of confusion. Final consensus: Forget it! The cleverest of them made straightaway for hideouts in Maui. The rest, well, history has lost them.

Let's face it: Greece was gone, emblem of all civilization. Matter of fact, she'd been gone for some decades (gutted) despite the Alliance's good 'ol boy efforts to blast through the Turko-Libyan lines and get some of those homosexual isles afloat once more.

After a decade of such worldly antics on the part of short-sighted Man, there emerged a gentle breeze to lull survivors to sleep and in the meantime tell a story, just as it was. Not really, in candor. It just sort of seemed that way to those ravaged few who dreamed from out their worn hovels with not a notion as to where next and what on earth.

The camps, as previously described, arose unchallenged. Started speaking all over again but this time had no chance. No gram of uranium, no ounce of guayule oil, no Motorcraft, no assembly line, no wing, no vestige from before.

Over and over again. Until the final blast was finally done, the seriousness of the onslaught at last appreciated in its subtler sense. No more babies. No more making out. No more pet stores, wonder bread, bicycles for two.

"You see," Odile went on. "We had so little choice in the matter. The individual, I mean."

Margo paused, susceptible to Mommy's loathing and logic. "Is there such a one, an individual?" she questioned.

"Oh my dear . . . "

The two consoled all that long night, picking up the fragments from a world made once of untainted stained glass – frescoes radiant – now shattered, the shards strewn in her hell-bent head.

"It was our fault," she finally conceded. "The individual's."

At dawn Odile got up, shat in the embers, and gave Margo to understand that what she was feeling had a word, and that word was not ever to be uttered or thought outside of the Tower.

"But Mommy!" the young girl dizzied.

"But nothing, you understand!" Odile continued. "What you've got you'll keep. They can't take it from you, can't even eavesdrop. You're free with it. You *feel*; and it shimmers (*shimmering!* she remembered the word), "and shimmering, you'll just get on with it."

Margo waited, watching from the window for her man, as a warm phosphorescence sponged the patchwork of glacier. And then the figure emerged, at first minute, hardly moving, then definitely. My god, there he was – alive, her man, coming home just as she'd believed in, with the humeral swagger of strength.

"What are you waiting for, girl! Go to him!" Odile incited.

Margo studied herself in a marble mirror, adjusted little details she'd ne'er before considered, and *feeling*, in one fell swoop of air, feeling incredulity, feeling one aboriginal hurray, she threw herself down after him.

The Pit had been readied in preceding days. An inverted pyramid hewn of one great glacier quartz the size of a hashish den. The San Diego aviary, a Saskatchewan silo, it was an unmarred surface, the white of a queen termite's young, smooth as pus, with three rectilinear veins, perfectly equilateral, through which the energy of the Complex percolated in a piceous flow of juices straight from the proscenium into the heart of the machinery, where an orifice admitted the otherworldly light of diffused sun, gave evidence of the gods, of NO. The proscenium, where judge, defendant and prosecutor stood, on points. Three points. Five-hundred feet down; points representing the forces that congealed on the lonely epicenter. The juices taunted, vexed, clean-cut with precision, were capable of destroying a man's mind in a microsecond or of bestowing a reprieve tantamount to divine intervention. But the defendant had very little chance. The judge and prosecutor, much greater one. The truth, after initial speeches, was to be won in the calligraphy, the dance of truth (foresight). A cunning liar had no chance – cunning was single minded. Nor was there the intelligence enough – double intelligence – to think outside the Friction. This close to the core reactor, with quartzite bringing down the caving mass of the one great piedmont well, abase the greatest mountain chain on earth. In all that energy, there was little chance to outfox, outthink, the Pit. But Bliss was fearful nonetheless, expectorating the morning's accumulation of phlegm, walking from his technocile into the plaza, passing through the guard's post.

"Mornin' yer Eminency," the sentinel slouches saluted. This day the guards were nervous. Xing had gotten the word out.

Into the elevator shaft whose slow fizzle noise Bliss kind of favored, its dignity of approach and careful descent rate. There was, to be sure, persuasiveness in the man's pomp. He knew the Pit intimately, knew its quirks, and feared this day like never before. As if approaching destiny, knowing that such an occasion was bound to have occurred, he went forth with it.

Of course he had him beaten. Punitive transgressions. Warranting expeditious incineration. But then Hardanger, the alien with untapped mind, perhaps was worthy of something more interesting.

Bliss wanted to understand Hardanger, had studied the murder on the monitor screen to seize on elements of style and repressed personality. Bliss revered the fellow's blunt body type, recognized pre-Finality gestures in the atavistic use of hands and the Byzantine capacity for burgeoning of a sudden, as when the mouth exploded across Martinez' throat, and the blood gushered and Hardanger still did not falter.

Bliss applauded the fact of such initiative, but sensed encroachment about the suit. Hardanger was filing civil suit against the Complex. Unthinkable, but there it was. He intended to prove that the Complex – its heat, its method – was in fact hastening, not curtailing, the inevitable destruction of Point Mind by the advancing ice. In his report, drawn up the very night of his return, Margo by his side, he stated that an imbalance had been perpetrated, that NO was in the wrong.

Whatever merits the case against Hardanger enjoyed, Hardanger himself was in terrible possession of data, and Bliss suspected the worse. He shuddered, thrilled to consider it.

The Pit accommodated all the councilmen in an arc of pews, of comfortable foam recliners outfitted with direct access keys, tubes, and hearing aids to online combat. Behind them, arranged in thick row, twenty selected citizens by invitation only sat bystand. Bliss, basically friendless, invited no one. The other councilmen and woman each solicited their favorites. Hardanger made his own bids for support, as was the defendant's right, and naturally brought Margo and her mother. But where was Margo?

Hardanger arrived early on, unaccompanied, clad in faded corduroy held up by a Trotsky belt from the museum cases. He wore a violet lambswool sweater, said to have been worn by Samuel Johnson. It was riddled with holes 800 years old. And in his tennis shoes he was an anachronism like never had the gallery of observers seen in all their merry lives.

"It will put him on the defensive," Hardanger thought, stepping through the crystal porter into the main pyramidal chamber wherein he was awaited.

A dozen or so citizens were there in their fanciest rubber booties and plebeian happy-day looks – eyes dazed, tongues hanging out, or tart and loquacious as pidgeons in Times Square. All the councilmen were seated in their safe positions sixty feet back from the defendant's curious circle: a circle within a triangle within a Pit.

Hardanger stood at the side, lingering, very Studs Lonigan, strapping, all cynicism, surefire, until the councilmen all motioned at his annointed stance.

Opus 318 stared with unequivocal draw at the handsome, bullish strategist as he assumed his position with the easy steps of a Touareg. Hardanger found himself on the outer point of the circle, the one least likely to win. "Slobberjobbers, pinheads, ghouls," he thought. To which a subtle emission of colorful steam spurted from nowhere in the primitive, mechanical way of a calliope. "Farce!" he thought again. This was his first time in the Pit. You only went once. He studied the place in quick jerks of hydrocephalic airing out. He'd of course *heard* things about it, all his life, enough anyway to inspire no joy in the matter. Now he gained in strength; "I mean what the hell," he reckoned. "If you've got to go . . . " and all that. But beneath such hauteur and boldness came the sense that all of Point Mind, the entire legacy of workings, were a hoax, oblivion oriented.

Bliss stood there before him now, peremptory, scanning the arena with resolve, averting contact with his rival.

The pyramid filled up. Odile arrived, looking fretful.

"Where's Margo?" Hardanger called out, innocent enough. But before she could reply, the imperial voice of Bliss broke with sinnister command: "Order in the Pit!" and all were silent as petunias under water.

Again, squinting at Odile, Hardanger tried to learn what had happened, for Margo was not going to get in. The shaft had closed. Odile, in her turn, was stymied, could not speak, and was not *about* to think, especially not in there; nor *could* she express her emotions, after so many years inhibiting them. She lowered her head in the attitude of mournful obeisance which was the general rule before Bliss.

Hardanger turned around, faced the pews with a smirk, then stared directly at the hard Stalinesque visage of his contestor. Opposite was Dr. Fan, trembling, having forgotten all about Hardanger, what he looked like, what he was. As Hardanger, recall, seldom ventured from his privileged nook into the trunicated bedlam of town, and few even remembered that he existed. To some of the young, Hardanger was a legend, a ghost in a haunted tower. Not a one of the kids, though, ever considered attacking the idea head on. Never such initiative.

Like waging electrons on display, cloaked in distinct bubbles, the three men – judge, prosecutor, defendant – faced one another now before the tiers of humble, would-be jeerers. Fan resembled Haman; Bliss, well, he likened himself to the unspeaking, anonymous hero of "Reverie," that last motion picture ever to be made and released; the antistory of all stories, no humor, no intelligence, no ethic propounded, but the sordid, epic look at a brat, supreme narcissist, Buddhistic, reared in a Himalayan serai surrounded by adoring maidservants, and the ubiquitous trade in nuclear contraband that made the Gobi Desert–Paris–Dublin circuit. His mentor, a Dr. Pangloss type, imparted to the child all the ways of worldly survival; he grew up, enmeshed in one cosmic squabble after another, until it was his sole knowledge would make or break. Film was produced on location during the regime under Alia. Discs of it were preserved, it being the last film. Bliss loved the character of Radix, whose supinate muscles were wont to extend very Pope-like, lending rancid religion to the masses, giving hope and concealing great enterprise. That was Bliss, a regular Hector, poetaster, no-nonsense killer. Hardanger, looking very Oedipus, or Attilish, judged the strange room's contents, size, and relative invulnerability, as the pyramid's vaulted depth was silented and Bliss assumed his unbeatable air, looked to the councilmen, spread his arms in a benevolent gesture that encompassed the shining geometry of quartzite all around, then offered peroration.

"Citizens," he started up. "We are survivors in a world that has not made it easy, until now. We have survived as a result of NO, and because of my unique relationship with NO. The Complex has given us our destiny. It feeds us, you will agree, clothes us, lends us a share in its own perfection. We are beyond the inclement vicissitudes of this small planet; we have, in our modest and harmonious way, transcended the vagaries and now stand free and noble, lacking nothing, in possession of total energy, pure water, pure NO. This happy circumstance has come about through no accident. We were destined to breathe, to act, at the pinnacle of order, rationality, and perpetual life. We are the only life form on the planet. It will always be so. We will survive. And because it is NO's wish, it is incumbent upon us to uphold the order long ago established that fosters our privileged dependency on NO. For this reason, disorder can not be tolerated, must be expunged!"

Hardanger's jaws were open. He had his bite. Thought rapidly how best to strike back. The tactic was all important. One wrong bit of timing and he could lose the mettle, the gut of his comeback. He had no precedent for the embroilment, the cross fire he was stepping into. Knew nothing of court days, manner of defense, points of prosecution. He remembered that he'd killed a man, though in a manner less clear to him.

"Therefore," Bliss coldly ventured, big guy, full of his manhood, "we enter the Pit to re-establish the order, which it is our livelihood to sustain."

"Long-winded son of a bitch," Hardanger glared.

There were no guards in the Pit, nothing to suggest the possibilities of restraint. There were no jails, no weaponry, no thorizine, nor overt provision for subjection. The tapping – that was the control, and the diet, the very environment, instituted to absolve the individual from his paltry decision-making lack of talent, from his pettiness, his feelings, his possibility for robustness, his animal.

Hardanger was at tremendous disadvantage. The councilmen supposedly knew the outcome from the beginning. Bliss would have earlier communicated the case against Hardanger, circulating the state's brief. But Hardanger seriously doubted that Bliss would have been so bland as to risk revealing Hardanger's own suit, for it was tantamount to a personal admission of guilt on Bliss' part. It would not look good for him, even in eyes glazed over. So Hardanger decided he would wait, see what the good, the ludicrous, the pitiful Dr. Fan had up his dacrillium wristlet.

From within the Pit there was no more the incessant whirr of the Complex, but rather a true grumble, like that of an ocean liner parting anchor, its propellers churning deep mire.

"There is only one rule here," Bliss' prolegomenon furthered. "And that is the rule of Truth, which propells its own mandates by its own momentum, resolves all conflict in fairness, and thus insures our holy prospects for the future."

"Grandiose phraseology," Hardanger mulled, upright in the nudity and dislocation of his stance. "For such a schmuck!"

There in the one-on-one configuration of sepulchral rock were the *perfect proportions* of Leonardo's crucified body of a man. There was unmistakably in Bliss the pose of such rendition, Hardanger thought. "The man is mad, aspiring."

"If there are comments at any time from the councilmen, for whom, I hope, the loss of a compatriot – General Martinez – will aggrevate no pre-existing prejudice, then by all means utter your indictments. He [pointing at the accused] can't hurt you!"

"Bastard!" Hardanger let loose.

"Spare no relevant facts," Dr. Fan motioned.

Hardanger stretched his fingers, slipped one hand into his large trouser pocket and stroked its contents.

"If the Doctor, chief prosecutor for the Complex, would read the complaint."

Fan was banking on a change of name. This was his fifth case and he wanted to win it, for once. He was, it is true, backing the Complex. Before, he was always defending the accused, while Slibowitz prosecuted. But no more sentiment this time, no more wisdom. He was out for blood, prepared to grovel, gouge, dig out the blatant fact of this Hardanger's – this alien's – guilt. He had the evidence, the film, the record of thoughts, the very transcript (for every word was recorded, conserved, renewed, in Point Mind, for use at a later time, this practice apparently dating from the Jonathan Swift theory that every spoken word diminished just so much the time allotted the lungs.)

Coughing, unable to look the defendant in the eye, "Yes . . . quite so, well then!" he stammered.

"Please, please get on," Bliss inveighed, egging the incompetant.

"The accused, Mr. uh, Hardanger. Accused this day, of murder by biting, chewing, and, as the data shall suggest, swallowing. There was, it is true, no drooling, or at least none the monitors could catch. But drool or no drool, the accused has taken the decisive bite, incurred death, and then, more importantly, breached the Complex walls, gone outside. Where we cannot say. No one can say!" Fan collapsed then, exhausted by his miraculous spate of well-spoken denunciations. Energized, like one of those flesh-eating creodonts with a brain the size of a Tunisian rat's scat, he lowered his head, a victor.

"If Lazarus could, he didn't, tactfully," interjected the guying Bliss. He spoke directly to the tensed and figuring man in the limelight, the animal upon whom everyone's attention was now directed, instinct with enough malevolence, in any case, to make them hiss. Save for Opus 318, whose langourous look was all fascination. She couldn't be sure but she thought she heard something coming from the man, something she had never heard.

In the Tower a woman groaned, stared at herself. Her chest ached, her belly gasped, her head reeled. She'd stripped off her synthetic carapace, examined the pink, enflamed nipples before the old mirror and then, in horror 8 (the horrors were affixed in Point Mind, 8 being fairly dismal), she'd seen the dark red fibrous mass flow down her leg, the viscous particulates, the whole stringy glob, all at once.

"Oh my NO!" On the floor, her fingers dumbfounded. "What, what is it?" she clawed at herself. No anger yet, as it was the first time. "Mommy!" In her agony for hours, unable to stand, to run for help. Ashamed.

Hardanger waited, waited all through the day for the opportune moment. Fan was dithering about to finalize the case quickly. Every bit of data was evidentiary. Then Xing was brought in, but Xing played dumb, subject to shock, would not speak, was dismissed, thereby setting back the proceedings. Good old Xing.

Fan was of no school. His scrupulosities derived from pure invention. He was, simply, assigned his duties as a result of his higher-than-average sense of mission and a vituperative if often incoherent, diction.

"The Complex is unbreachable. You have breached the Complex. Therefore," he gloated, "therefore, *you* are unbreachable!"

Bliss winced at the end of his patience. "No, shithead!"

"Then who?" the befuddled lawyer ignited. And mustering his ingenuity, "The Complex did not breach Mr. Hardanger. Mr. Hardanger breached the Complex, isn't it so? Ask yourself!"

"The prosecution has concluded," Bliss forced, done with stamina. "The evidence has been presented."

"May I?" Hardanger began.

"Are there any objections?" Bliss queried the councilmen. "There are none, then. So be it. Begin."

"To respond," Hardanger stepped up, "to the Doctor's positively brilliant syllogism. In fact, the Complex *has* breached me!"

"What's that?" Fan alerted.

Bliss was prepared for this. A single thought sufficed to ensure his championship.

Suddenly, the triangle in which the three men stood separated from the tiers, sunk deeper into the Pit. The veins of rock were now totally exposed, like deadly vipers, and the power trickling through them. The power issued from the center, slanted near to each face – within nose reach, a single grimace away – and sinewed revoltingly up in animate urgings controlled, it seemed to Hardanger, by Bliss. The power, alive like a toxin, the stuff of genes, color of urine, of ichor, suffused into the ungiving rock, and then into the fusion reactors themselves.

The top of the pyramid was a screen, convex, looking down to the spectators, enlarged optically to provide intimidating dialogue with the threesome, who stood on an island of crystal in the sea of depths, yawning, mineralized space on all sides. It was eccentric space, hooked up with the main computers, the very air conspiring, heavy like an acid.

Hardanger was not able to move without touching the flow. And this was the point: trial by combat. The spectators gathered to the white hanging edges, peered down, as into the well.

"No emotions, no volatile, sudden movements; clean objectivity, only," Bliss asseverated with deft matter of fact. "I would suggest you weigh your thoughts."

"Old bones gather no fuss!" the foppish Fan postulated, rigid to his spine, eyes bright.

"We seek the truth, here," Bliss volunteered. "There is no question you've killed a man. Not much of a man, sure enough. But a man nevertheless." He looked with some fraternity at Hardanger. "You seem to think there is rectification affording your innocence. Your thinking is biased. The screen shows it to be so."

Six eyes stared into Hardanger's brain as it was captured on-line, sliced into three-dimensional co-ordinates and photographed through and through; magnetized, dyed, looking like a dendrite flow, a satellite photograph of arable lands, like hunter-gatherer phagocyte cells riding the white water of lymph to the site of an injury and cannibalizing the invader cells. Anatomized so that the neurons fired thoughts as thoughts fired neurons, a profile emerging of a system organized to promote self interest.

"There's the extent of your defense, Mr. Hardanger!"

It was a trick which Hardanger had not been prepared for. He was stuck to his position. A single movement would kill him, tap him instantly. Recovering composure, he spoke out. "Pretty, ain't it!" smiling weakly, forcing the disparate from his mind; dispelling with his hatred of all those around him.

"Let's see what else you've got in there," Bliss forwarded.

The screen focused entirely on a small portion of Hardanger's right hemisphere, scanned according to a program that translated otherwise incomprehensible pictures into English encapsulation: motives, fantasies, calculations. The mind of most *Homo sapiens sapiens* could register one, sometimes two images at once. Rarely two. Certain symphonic conductors had a slightly larger facility. The computer, on the other hand, had no stake in any one thought, with resulting clarity and compass.

Hardanger read off the screen, mutely resisting reaction, when finally he could resist no more. He watched for the purple explosion of clouds and fusion of color, of wriggling tissue, and hyperspeculai, a throbbing piece of organ, sheered through so fast that the thought alone remained pumping its arterial design in digits of quintessence, flesh of quintessence.

"You're full of shit!" Hardanger bellowed, his head shaking, trying to repress steam.

"Happily, you can't think that as fully as you wish," Bliss stated. "Watch the screen. You're thinking defense, your next action. Nerves to your hands. Fingers twitching."

"I bring suit against you and your goddamned contraptions, you and that cabal of maniacs, you and your false gods. In the name of NO, I say fakery!"

An astounded reflex permeated all those present, save for Bliss who could not be so easily thwarted. "Machine, eh?"

"Contraption, simple nuts and screws. You'll kill every last member of the species!"

Commotion from above.

"And how do you suppose I'll manage that, Mr. Hardanger?"

Opus 318 suddenly broke out with a rash of melismas, in response to a sound, a sound that only she and Hardanger could hear.

Bliss, not expecting her same old voice, shifted position, stupidly: The vein caught him, stole his mind, projected it onto the screen, and in that precise moment, the bird – a steely blue hummingbird – squirmed forth from Hardanger's pooped-upon trouser pocket, and flew up into the open, admitting its high pure whistle of a cry.

"Shit!" Bliss cawed, clamoring for air, his body hanging there, like Hardanger's, in a spastic effigy, without mind.

"There's your outside!" Hardanger boldened.

The bird shot up into the Pit's apogee, stopped in mid-flight directly before its friend, old Hardanger, who had carried it miles across the ice, fell back, then shot away once again, looking frantically for an exit.

Bliss glowered at himself on the projected screen, his curses flooding the arena. There was blackness on the screen, blackness irrupting into little burps of gaseous color, like stains and undigested stems, vegetal matter, peas, water cress, in his feces.

"Lies!" Bliss roared.

"What is it?" the terrified Fan sang out shrilly.

"An ice bird!" Hardanger said, calm now. "There is animal life. I saw large mammal prints, other birds, insects, plants. A new sky, new stars. I saw enough to judge the error of your assumptions." Forgetful, Hardanger had turned, as had Fan, to catch the miraculous flight of the bird. And in so doing, he'd caught the vein of energy, disrupted an inertia which had glided carefree into the computer systems for centuries. Now both he and Fan joined Bliss in the torment of dissociation. Hardanger went from image multiplication, to image severance. The three minds went side by side, no tango this, but painful permutation, on-line, all focused on a screen.

Trial by mental combat. Not infallible.

Opus clarabelled her tongue out, trying to replicate that which was the purest sound she'd ever heard. The bird ignored her. She pouted, sat back down, while Ng, fearful for the unconventionality, the unpredictability of the bird's parabolic and sporadic flight patterns, repaired to the elevator shaft and fizzled up, past the sentinels; and from top, he hastened back to the comfort of his technocile. Slibowitz, on all fours, looked over the edge, then to the screen high in the nave, along the arc-boutants of rock, then again over the edge. He sensed, dared to think of Bliss' defeat; saw the possibility for seizing control, what with Ng having copped out, and the other councilmen in something of an incapacity; he waited, thinking freely now that the screen was accounted for. Meanwhile, Wickiup rasped, grew tight around the throat, gaped at the bird – at the first animal in his life – then at Hardanger, until there merged in his mind's previous night, and all the other nights, the figure and bulk of Hardanger, a figure looming in his night visions, perceived through the impossible white-out of a blizzard; a figure moving slowly, with the stride of a T'ai Chi master, over an unfamiliar landscape. Stooping to ward off something, roaming freely in the blackout of time.

Squirming unanaesthetized, as in open surgery, the three men stood incorporeal as their minds wrestled, were dissected and projected.

"Bird, there are no birds. The PF factor was too high at the end. Did away with all warm blooded. Then the lizards. Sea Horses. Lilacs. Nada, nada, nada. We were desperados, all of us," the mind of Bliss burst, its cerebral perturbations sparring well on their own now, full of spectres, as the husk of Bliss hung dumb and withered as by a puppeteer's string.

"The bird is there. Shows no, no ill-boding signs. A plenum of other species out there. There is life evolving, life we could be a part of; you and your gods have kept us from our origins!"

"One bird proves nothing; check its genes, genes!" Fan said.

"It's a healthy bird, bird," Hardanger's brain poured.

"Bird!" the observers all shouted, forming in their cobwebbed mouths the first organic consonation in a lifetime. "Bird!' they called again, rhapsodic, arms outstretched. "Genes!" they all shouted.

What the bird sees: Its eyes flit through glint after glint of tilted surface matter, picking up speed, free and lifted on invisible eddies. Glare against rafters. Eyes sticking out, legs descending with no end to roots, must be below the very bottom, eyes like prune pits, wobbling through head after tufted head. Swings precipitously, searching for a way out, alights here, zooms there, poops, poops again, hiccups in its little way, sees everything for the first time in there. Frantic to get back to the wood and herb and wind of its sky nest. Field of scented grass, cushions of scented air. Moist. Warm. Mint and colorful career, sky-born, tree tops, one glen and fifty years is all, a dozen generations. Fairies, fairies, sees hands the biggest goddamned talons; waving, flocking with no predictable awkwardness. "These beasts!" its minutiae brain cells conjecture in mid-flight. "What malodorous, what dangerous, what gracelous wombs? We hate you all!" The bird smells a paté of ca-ca that it made in Hardanger's trouser, dives down for a look smell see, jerks away from certain collision when the man's fingers – riding the paroxysm of dissociation – tremble. The bird overshoots, beak and all, smacks into a vein of quartzite. Momentarily stunned, it staggers in the air, falls onto the Pit's platform, where its mind is magnificently enjoined to the others, up on the screen.

"Movement of beasts, bred to the earth, eating all the color, eating its own movement; pop-out, pop-out, one-a-second babies popping out, eating everything," the bird's brain bellowed unexpectedly. To which Bliss' own irrasible head went choleric. "A trick! In the name of NO a trick!"

Fan: "Is it true? An outside after all?"

"We are the wind," says the bird.

"You are without the steel," says Bliss.

"We are flesh of the great organ."

"Stupidity. You blurt nonsense. There is only NO!"

"Listen to it," Hardanger's brain echoes. "Bliss, hear what it says!"

"We've been forever in this place. The sphere is a garden. Little progeny. Little life. We breathe the song that moves us. Our eyes are food. We hover because we're close to the beginning. Closer than the rest of you who stumble backward, thinking forward. Thinking, thinking, always paving the air with your brain. Big dummies! Leave us be!"

"You're up to this, Hardanger!" Bliss veered.

"I think it knows what's what," Fan thinks.

Hardanger: "The Complex is not an open system. It thins out, grows top-heavy. The Sun is not infinite. What are the statistics? I demand the statistics! What of felicity? The institution of granola, salad, mammary glands, chocolate cake! Culled from the cartographies, pharmacopeia, from deep within the sheltered canyon lands, reed baskets, tamarisk thickets, Egyptian bakeries, Disneyland, Charlemagne, bordellos, Tyeho Brahe's little girl . . ." Hardanger enumerated for fourteen or so minutes. Everything he'd read; all the breviaries, the goodies, dreams, archanthropines, artifacts, mesotechnobilia.

While Bliss gritted, stuck to his position, the transposed heads carrying on like mews or angels in an updraft.

"Selfish. NO is prepossessed, feels not for man. We made it, *made* it after all!" the alien, of the north, blasted. (Hardanger's lineage descended from an assistant to the VP of a PR firm in Hemsedahl; another ancestor was a ticket agent, a schizophrenic, an impotent priest, a bad poet-adventurer who ran out on his wife and kids, worked odd jobs to the East, remarried in Vladivastok, vanished; a lonely dentist, acrobat in Berlin, another acrobat, this one with the Omaha circus; six children died young—plague years—fisherman, woodsman, Oslo aristocrat dealing in radioactive scrap, all the scions of his clan angry, mordant, caught in the upheavals, dispersed far and wide). "But they didn't *mean* to manufacture so heinous a complex," Hardanger went on. "They were victims, caught between two rocks, two clashing seas, two allurements: comfort, or drudgery. What could they do? When presented with two choices, chose neither. Invent a third—that's freedom, says I!"

"Yes, yes—" the bird repeated.

"I begin to see," Fan parsed.

"In so glorious an end, an end, an—" Bliss reviled. "Freedom has no room. No algae, no extremity enough to satisfy itself. Freedom from what? Chaos, famine? If it's statistics you want, well they're right here. Forty-three billion deaths!"

"That's a whole lot of stench!" Fan's neocortex readily acceded.

"You're a typical romancer," Bliss went on. "A waltzer."

"Lousy dreamer," Fan furthered.

"Dreams, damnit! They're what's good about man!"

"And simpleton," Fan accelerated.

"Simplicity is good! Your imagination overconsumes, wants too much," Hardanger screamed.

Bliss configured: "He's tongue-tied, frothing. Brain is all ruck; time to bring it home."

"You don't know what you're saying—"

"I know exactly what I'm saying."

"He knows," the bird poked in.

"I've studied it," Hardanger went on.

"You've lived alone in a tower. You're removed, antisocial."

"Antisocial?"

"They don't fear you, they don't want you; you're no myth: they hate you! You're finished, Hardanger; prepare for exile!"

"I am the one who hates, do you understand! I hate you!" his mind raged, spilling pain in an echoing of drum, the opprobrium ricocheting from mind to mind, no salutary conventions, this foursome.

"I'm the only one who feels here, me and the bird," Hardanger tried to say. The actual man stood cold, clenching fists in frost.

"I feel sorry for you. Go back where you came from. Go to your lifeless bricks," Bliss impugned. "Stay there and all will be forgotten. The choice is yours!"

"Do you admit to wrong doing?" Hardanger was seen to impugn the blister of a man.

"Wrong?" Now there was address, Bliss open-souled, brows extended.

"You've tampered with evolution. I have little experience with the past, having been born in my tower. But I've seen a plot of true earth, true blue, and this one plot over the duration of days and nights is enough to know that the fittest aren't necessarily the best."

"Oh really? Now listen here, Hardanger. I've done what anyone granted the power would have to do. Evolution is finished, come to its grand denouement. We're *it*, thanks to our technology. I've taken the survivors and programmed them to go on. They had nothing back then, nothing! A few probably remember. You speak of evolution. *You* are lodged back then; *you* haven't evolved; you're out of tune, misplaced. I feel sorry for you, miserable wreck, miserable mind; go back to your tower, take the plaudits and run!"

"Plaudits?"

"You love it, Hardanger. You relish being the romantic, the throwback, the last man. False modesty, false genes. The pre-Finality was leading to the Finality; in essence, there was no difference. Yours are the genes that chaos and destruction come from. You're living in the horse latitudes, those calms before distemper. You'd thrill to the simplest expressions of logic; you're limited, Hardanger. I tell you this because you think you've learned something from books. But no, my friend; you haven't. Books went out hundreds of years ago, along with learning; along with art, philosophy; ethics; love. You're a dreamer, and I pity you for it!"

There was silence between them. Then, "Hardanger. It's a nice name," Bliss quieted. "I wonder where it comes from?"

"I come from nowhere. Midwest, or something like that," Bliss said. "I come from the amalgam. Textile electronics. Silicon futures. Refractions, war, disease, promycelia. I came from the need for order."

"I know all that," Hardanger replied. "But your order is wrong. Wrong logic, wrong paradigm. Funny thing is, a man of your clairvoyance, thinking you're right and all. Where has all your science gotten you?"

"To hell and back. Always back, where I now stand."

I'm sorry for you. Guess that makes two of us."

"Repentent, he's repentent!" Fan's half-wits dribbled mumbo-jumbo in effusive throbs of the punch-out. "Bliss is wrong. NO is flimsy. Flimsy, that's what they said. And Hardanger's mettle, that occipital that marrow, it's prettier guk; wrong! wrong! woebegone. Could it be? Yes it could!" he fluttered on and on. "Science, the Alliance, the Mayans, the constellation Orion – it's all gobbledygook!"

"Science, the Alliance, the Mayans and Orion – it's gobbledygook!" rang the corus of beleaguered bystanders upstairs.

"You're a misfit, Hardanger. I knew this all along. That's why we avoided you, envied you, allowed you your romance."

"But have you no curiosity? You, who came from the before? My Tower, the museum? You're interested in none of it?"

"Saw it all before. No big deal. Bad memories, sad memories. Sever them, that's it!"

"That's it, that's it!" Fan pledged.

The doom, with its hour after hour of legal antiphone, persuaded no one. Least of all the bird, which revived, went whipping towards the light at the pyramid's zenith, one hope pumping its soft heart. Only to disappear in the Complex. But, when at last it did, its mind still worked upon the screen and suddenly the tireless defendant, prosecutor, judge – excited by their fateful encounter – stared in awe upon the transformation that the little bird effected.

"NO!" Bliss convulsed, the bird brain pecking at his own, puncturing it, spilling it, correcting it. "For god's sake stop!"

"That's better," Hardanger enlisted.

Bliss fell over, seeing his first and last revelation: the whole natural history of the bird – of the outside, there before him, flooding into his mind through the fontanelles the bird had opened. The outside: It was Real! In the flesh. And reality worked!

Bliss grabbed Hardanger. But physically, it was no contest.

Twitching tail feathers, beak of brittle bone, lungs easily crushed as chestnuts, eyes like caviar, the ice bird was sucked mercilessly up, up into unfathomable brightness, where the steady drone of cyclotrons and rods and plutonium pistons spit out those vague gigawatts day after day; was sucked up with one, emotional peep farewell.

When all at once the awesome process by which such fusion operates rejected it, went afoul with the foreign substance, so tuned and antibiologic a machine that even a few feathers were enough to wipe it out.

There was a grim and rippling whine, like the groan of a manatee come into collision with an outboard motor, like the first fracture line of a slab avalanche on Huascaran; and then a computerized wail, a warning siren, high and mighty, like the chorus of howling Eumenides that bit the bullet right in tune now; the Pit's dank atmospheres rent the stymied collusion hall, sent balls of fire up the elevator shaft, across the concourses, into the clustered technociles, throughout each glittery gallery way. The fires were put out almost as soon as they arose, because of fancy extinguisher prototypes built into Point Mind's system. But what they could not squelch was the break-down within the reactors.

As the bubbling hydrogen sound grew in intensity, all went fleeing into the night's inferno. Aeneus' darkness; darkness of the Dordogne caves; of mines near Lake Baikal. Darkness, hard and circular.

THE BEAR

Hardanger was full of instinct. He needed Margo and needed to get out. He ran.

Glow, like Sudan's summer solstice, infiltrated town, stripping it of the tar, the bedrock. Gassy miasmas hurtled down terraces singeing metal cobblestones, igniting vomitously into a violate mushroom cloud. Such mushrooms will stay the same. The effusion poured over each coordinate, melting computer outlets, buckling the very mote of glass on the outer concourse where fumbling denizens socked futilely about.

Hardanger ripped off his shirt in the incendiary night, saw the Tower sway, plunged up past the crumbling bricks toward the museum. There was Margo, down, eyes closed, body limp, hair beginning to smell. She was barely breathing.

Without a word he picked up her 120 pounds in his hairy arms and sprinted through the bluster of increasing firebrake. Little time. Like an unbleating lamb slung abreast both shoulders, the honey-haired woman lay still as her man fled toward the ice. As he went over the small retaining wall, others plowed into him, faces wrenched.

Hardanger kicked his way over. Others tried but with muscles so deteriorated and minds so against escape, even the mantle proved to be excessive. They all fell back, hair burning, eyes afloat on the skillet of the outer concourse, as Point Mind was engulfed.

Bliss. He had his own devices: a quicksilver shuttle that plummetted further down the well from the Pit, cutaneous layers unto a track, there from the beginning when Solidarus, First Architect, construed just such an escape hatch. But Bliss hit the track too hard — it was his first time — and the apparatus ricocheted uncomfortably into the mineral black rock. With the earth quaking, there was no time to dally; he got out, slicked hair tied with a beaded orange scarf his mother, long ago, had given him, and he moved by foot precariously toward the subterranean headwall of ice. It was here, in the foggara, a technology from ancient times, that the water poured out into polyproplyene ducts that fed the single tap in Mind. "Go, go!" he moaned, as the rock began to cave in, seams bursting, strident polyphony in his head, rock baring the burning face of ice beneath, Camaroon black ice; and alone, 3000 feet down, he thrashed toward unknowns, still strong enough.

The terminus was vibrating, the heat coming in waves through the tunnel from above, and gas stinkier than mustard. Bliss pushed away rubble that had long accumulated at the pendacled entrance, entered an unused duct, its diameter three times his own size, and raced through to the trough beneath glacier. Looking back, he saw his utopia begin to slurch: The ducts melted, and the ice – rearing into blanket heights of cave and alloy – began to come apart. He fled, deeper as the flush of molten mineral sheered through the terminus, fizzling like a steamy espresso, and the guts of the Complex were ruptured. It was a meltdown. But by then Bliss was out of danger. The escape tunnel was well provisioned. There had even been speculation, once, as to whether Noah may have found his way in there, surviving off the rations in the underworld.

Hardanger carried Margo into daylight without once stopping. His limbs were dogged. He kept on as sun cleansed his temples and stippling eddies licked his chest hairs. His back muscles shone, calves were pumped, hard as birch knots. The woman in his arms remained unconscious. In spastic jerks from time to time her joints quivered.

In the early afternoon Hardanger rested, far into the crevassed maze. The gelid surface was granular, brought blood from the man's bare feet. He breathed heavily. She was curled about him, exposed to the blinding light, which bred the same undeviating horizon of beauty. No matter where the eye turned. Save backward. And there Hardanger saw what he had not predicted possible: Point Mind was gone! In its place, a faint, ever so pinkish mist, level of the radiation belt that encircled the Complex, veiling a hole the size of Vatican City, that very minute fast-filling back up with water, sludge, mud of underneath, and raw ice; a diaphonous striation that gave no clue as to the 400-year-old civilization that once prevailed there.

He pondered her face; put a finger to her cheek, stroked there ever so quietly, an urging all about him. Kneeling down – the coarse gravelled ice digging into his palms and knees, uncomfortable as hell, after a night like that – he looked at her in a moment of swell clarity, as though the soul were made of gonads and the eyes burned Cocteau bright. There is gold dust churning in his arteries, a topsy-turvy figment of fellatio brimming from his fingertips unto her sparkling cheek where tears have left a saline solution that he licks off. No tear before, no words for what he needed now. She was not dead, he knew. Nor was she very lively, however. He stroked her neck. He felt something, felt for her. And as he stroked, his lips drew closer, until he was no more the observer.

Her blood-stained fingers came alive as Hardanger nuzzled gently. Roused, they timorously absorbed the pulse, grew restless. Her lips quivered, explored, came to rest moist and firm against his neck. For the first time, Margo's chest heaved, chokes broke the spell, her eyes opened, stared up at him and they embraced, hot and bestially, stubborn about this long-won cavort, this nifty snatch touching that no one prized in centuries preceding.

"Hardanger, oh Hardanger!"

"You bring me to the earth. I want you, want all of you – how can I have you?" he pleaded, furious, inept, his teeth clenched with the anger of lost technique.

They both stop: there is an animal sound, long and far-away. No meaning.

The two virgins stood up, their ardor abandoned. He with "The Thinker's" perplexed brawn, she with the thirty-six inches of Praxiteles leg, servine ankles, callipygian bottom, face like lambent angels, the kind Renoir would have ravished, then painted, when ravished all over again, day after day.

An animal sound again. Strident. Like no sound the two had ever heard. Its repetition irks them, bringing some foreboding. They move on.

Later, "Stuff came out of me!" Margo confessed, as they hastened along their traverse. "Red stuff."

"Blood," he told her. "I've read of it. See," showing her his arm, where some was dried.

"No, from here," she harped.

"From inside there?" he speculated. The far-off moan of an animal; the hot noon drawing beads of sweat from his overgrown crop of curls. The distance. And a fear of the all-embracing totality – the sheer completion of their situation.

"I feel," she gloried. "Feel, feel, feel!" her voice trailing off as the two – through the eyes of an eagle – were seen to meander in fits, fillups, plops and tiny spurts across the shambled tapestry of glacier.

An eagle (*Aquila chrysaetos*) follows them all through the day, from high in the cooled hollows of space. Toward the glimmering green and maroon stream and quiet peace – peace – that Hardanger recalled, had been upon him imprinted.

They labored across thin islands protruding, stemmed perpetually until their antics, hoists, grunts and myriad glissade were learned, became a game to them that they played surprisingly well. At one point Margo noticed little moving things on the ice – Collembola, mites, glacier fleas – and picked them up, blown-away, it should be added, this being her first encounter with another, on the outside; but alas, all of the creatures died by the very warmth of her palm.

So they set on off again, over ventifact and lag deposited on the icy knolls, and by haloed dark they trudged into the aromatic garth. "It is the same," he motioned.

She experienced everything for the first time, in no modest exclamations. This was her personality – to go all out; emotional exhibitionism. She wept with gladness, though her body was peppered with complaints. She touched the arenacous ferns, the dogwood, the delicate varieties of deciduous tree, dazed to the irridescent sheen of blue lupin, the stream, the wind – both cold and warm. And the odor of it all. She put her tongue to everything, lapped up the dirt, the seeds, insects; she painted her face with mud and grass, stuffed things in her vagina, rolled in it, splashed, ate, went lovingly into contact.

"It's so *good* here!"

Hardanger, learned man that he was from all the books, applied his skills, broke wood, started fire that, atop dew-studded sweet grass, they cuddled for the night.

"Put more wood on," she demanded minutes after. "And more! I want more, yet more!" she revelled. "There's so *much* of it. As if it were put there just for our usage. Give me, let me touch it; oh to feel it. Wood, you say! Wood! I love the name. Love! Mommy told me about Love!"

They kept the flame large and with the blaze there came in each of them the desire for the other, unabated now.

"Give everything to me. Spread yourself, open yourself," they sung, their touch combustible.

"But the blood, beware my blood. It's coming again, some. Not much. But oh yes, do do that more," she undulated, clammoring for it, as he stirred up her womb with ungainly fingers as if it were egg batter. He mounted her, feeling first time pain in his side; she on her back bone, on the cold wet sonorous earth, breathing, on her sterling haunches, and free and resignedly to that which toppled out of the heavens, squirted in the unabashed perforation of fast lights which her eye, turning to take in the upward slaking of her ravisher, just scarcely caught, fleet through the downsloping darkness.

"Something there," she groaned, squirming so that her excited, contracting anus was cleaved against the grime and poked repeatedly by slippery stands of green – green things undescribable – like little blades. "Something there," she again cried out.

Drenched in the two of them, unhearing, no mind now, he turned ever so slightly, saw a flickering too.

"Shooting star," he soughed, before replunging in the grab and grip and white skin that shook, and in the powerful rape which each clung to she saw many stars bore into his bones the night he bore the same into hers.

"Star!" she repeated. "Star. How nice the word is –"

"Shh," he pillowed, his mouth, his love muscle divine, from her pubes to inner ear. Until the fire was gone and the embers too.

A high pinched caw. Hardanger opens his eyes. "Margo?" He stands abruptly, runs naked into the copse of high alder. "Margo?"

"Ha!" her voice comes back at him. She has been up for hours, eating, stooped, voraciously devouring flowers, blue bells, rhododendron . . . the prettiest ones, clicking seeds between her teeth. She squats, perhaps excited by him, pees on the flowers.

Hardanger notes details, points out a spider's gold abdomen, tiny gnat-like things rappelling down the grass; a track, size of a large pepperoni pizza; mullein, sourdock, thistle, pine. All of their combined years of study weren't worth a horse oat compared to such wild things, *real* things.

Nature had gambled on the human consciousness, a flagrant experiment and, given it ample time to prove itself. What it got were bodies so thick one could promenade upon a carpet of human carcass from Quebec to Lake Titicaca. Nature had lost out on man. Only the sun was worth mentioning, after all was said and undone. The sun, which produced the energy that drove the water cylinders, that in turn nourished Point Mind's algae reservoir, sustaining life, the echo entertainment, the free-for-all paintings, sand sleeps, speed balloons, masked balls, water therapies.

If she'd memorized *Phaedra, Dharma Bums*, the *Ion*, it was only because she relished the *sounds* of learning, the accent. How could she *get* the *Dharma Bums*? It expounded various logical types, dating from a period of *beatnik, pot, Big Sur*: What the hell did those words mean in the twenty-sixth century?

"Hardanger?" she called out. "What's that?" examining a flower.

"I don't know, if it's the name you're after."

"All right then!" She flopped about ever so sprightly, unaware as yet of her parents' desire, of everything she'd ever known. Now there was only this contact with a stretch of verdant acreage, between two glaciers. She had enough, everything she'd ever need; enough to satisfy her heart. She had her man now, and adored the taste of his sperm and of the sweet herbs all around. In the happy morning sunshine, with a big fluttering bird overhead, warning them, she commenced: "This one, let's call Margo!" she frollicked, yanking out a bright yellow buttercup and chewing it up without hesitation. "And the next one we can name after you."

Unpleasantly large, with holes. Hardanger studying the markings in the mud, beside the creek. His brow sullen with an indecisiveness, a memory from all his reading. He'd seen this track before, had suggested what it bode to Bliss, in the Pit, and now, again confronted with it, felt vulnerable, unalone. He heard his heart beat.

In the afternoon, after barfing, they sought shelter. Clouds had come and with them wind, cold, the heightened pace of resettlement, unaccustomed effort, digging with fingers that were gnarled and smashed from the ice crossing; breaking limbs too green to snap, gathering the earth in greedy armfuls.

Until it was completed and as the rain poured down on them they lay tangentially beneath the dripping canopy of shrubs which they'd fashioned round the base of a drooping juniper.

"I'm cold. Hold me. Hold me!" she screamed.

"I'm holding you."

"But I'm cold!"

It rained for three days and three nights. They didn't like it. They got sick. Margo had a bad time with her belly, which threw up everything she put down in it. Hardanger grew weak. They had a fight.

"This place is unreasonable. I want to go home!" she demanded.

"It's not there anymore."

"Where's Mommy?"

He paused, then: "Odile died. The town blew up. Disappeared."

She glared at him and all her loveliness dissipated. "I don't believe you," she cried.

Later, "Please."

"I can't do anything more," he said, hardened to the misgivings she'd aroused in him. "What do you expect, anyway?"

"Expect? Why everything, of course. I want to sleep. I want a soft ripe gourd to rip into. I want to suck seeds by a warm fire of solid shit. Not this smelly runny stuff we keep making. We're not supposed to shit this way. It's bad here."

"I thought you liked it here?"

"Water's never come from upstairs like this. I hate it. I'm cold."

"You're acting stupid. I told you it was rain. We didn't have rain before because the Complex modified the weather. It wasn't natural before. I taught you."

"What's natural?" she griped.

"Everything, now."

"*Well I don't like it!* Do you understand?"

"You're tired. Try to sleep."

In the morning there was again a mellow sun to bake the earth. Birds sang once more, the kind Hardanger had caught with his own two hands. Margo meandered up the creek repentant, the sound of waterfalls beyond rekindling the goodness in her. Then, in a shallow pool where she was bathing, adoring her spent body, she noticed a sliver, then another, flash by. She jumped, came back for closer inspection, threw her hands into the tiny splashes, put her face to the water, and saw them: pink, red dots, dozens of them, coursing the sandy bottom in smooth comminglings.

"Hardanger!"

They gutted the sockeyes, bones and all, until their bellies were bursting.

Then they fortified the shelter, thinking ahead. "This is all there is, all you must expect," he warned her. "This is a good day. Others are likely to be worse. Be prepared."

He warned her again that night, to get it deep in her recalcitrant head, while she gasped beneath his 240 pounds of entering entrails. Hardanger's face looked different, she thought, beneath the pregnant moon that warmed his softening features. She'd never seen a total moon, thought it embarrassing to do what they were doing directly under it.

Hardanger was no more worried about Margo, as the days followed one another like sheep on scissored cliffs. With as many fish as she could devour, and plenty of flowers to be consumed, and her bathing hole that caught enough sun to stay slightly warmer than the rest of the creek, and a patched shelter that didn't leak during the periodic drizzles, Margo was in good mental shape. She had her protein supplement, got stronger, complained no more. Her twenty one years no longer existed behind her; had faded away like a bad sting of onion in the eyes.

Hardanger, however, was troubled. He missed his Tower, found the duration of his excitement over newly discovered plants, melolonthids, insects, archegonia, strewn tracks to be quickly diminishing. He was no scientist, not really; what for, out there? To start a new world? He wasn't stupid. He knew the situation they were in, appreciated it with respect to their relative place. Earth had been demoted. For 250 years. All the other planets, save Mercury, sustained greater fanfare, madding conurbations; had been populated from the time of the very first emergency shuttles. Until, in post-Finality times, when the last teletype machines were silented, it was an established fact that earth had not only lost its pre-eminence in terms of what the human species was able to obtain for itself, but that the rest of the solar system offered more — more room, more mineral, greater thermal energy, wind power, even compressed oil; across subterranean Europa, a society of musicians had seduced and conquered the resident bacteria, transforming that tranquil moon of Jupiter into a paradise of raucous jazz. On Neptune, and along Mar's jagged Cordillera, other successful adaptations had occurred, forever dulling the ancient code of Gaia back on Earth, where Hardanger and Margo — sole inhabitants of a long-forgotten Eden — stared into the unknowable future. "This will be interesting," he mused. No geometrical precursors on the Rhodian shore; no excuses, alibies, or deferrals to detract from the elemental fatigue of their situation. Adam and Eve at the beginning of the world. But with an infuriating twist: memory. Burdened by the hybrid confusion of their past, genetic whispers, the caudal tails of sin and aspiration.

She came to him with things in her hands and a request, which he fulfilled; delicately doing her under-eyes with charcoal from the fire, drawing berry juice on her lips.

"Tell me I'm now beautiful," she insisted.

"Now you're less so," he said.

"You're humorless," she deplored.

He sensed a change in the air. Colder each day. He became more troubled. But Margo couldn't help him now that she was into her First Project. No longer interested in many of the flowers (she actually preferred the taste of some of the uglier ones). She was busy make tiaras, braiding grass, weaving, doing those infintessimal artistries that would take up whole days, otherwise idle. While Hardanger caught their fish in the early mornings when he could do so easily with his bare hands. Granted, there were no other kinds of hands, not on earth, anyway.

One day he ventured to the upper edge of the new forest, not so inquisitive now, just to be mobile, the itch to move, where the black glacier laterals cast up crystals such as he remembered from the Pit, crystals, stones – stones?

Of a color and ebony smoothness unknown to him. Rind-weathered, case-hardened patinas of oxidation, iron and manganese silicate; thin stripes of lavender, orange, grey and purple around the orotund white rocks. Dark chunks of impediment, thick and substantial as Missouri gravy; cipolin; fang-sharp serpentine in dusty sheets; dalmation granite. The universe in microcosm.

He'd build a Tower!

Margo liked the idea a whole lot too. Together they fashioned a formula that might work for any couple, in any era, given the minimum of naps, the right set of predicaments, the incentive that comes from sensing that you're the last couple on earth, blessed building blocks, colorful ones, and a man with hands sizeable enough to lift seventy-five pounds in a palm.

Their routine went something as follows:

Margo dried grass and twigs near the fire they stoked continually. She then amassed great mounds of fine loess from beside the creek that tumbled out of the moraines; she rewet it, dried it, wet it again with the addition of the grass and pieces of shittim wood found there. With a long bole she stirred up the mixture. Hardanger, meanwhile, had cleared the place best suited: a flat, open piece of land exposed to four sides, like a Norman or Carmelite outpost, affording the eye a grand distance across the trammelling ice floes, in view of the walkable sea, frosted mountains, and taupe overhead. A superb spot, with vegetal matter, water, and the new abundance of animals. Once the grass was pulled, a hole sunk with the work of wooden stakes, Hardanger began the travail of rock moving, fitting, erecting, and at the end of each day, racing against the palpable transition to autumn, cemented the structure with Margo's coagulant.

They worked hard, ate lots of fish, foraged. The nuts and berries and fried salmon brought taste back to their mouths, and with so sudden an orgy of enzymes on their tongues, a pleasure part pain, as with the steely sweet nip when a tongue tests a battery. But the raw gut pleasures exceeded all else as they slept on wicker, their stomachs rumbling with the fish bones, between full swills of bacteria water, water of the highlands, bringing with it algal filaments – *Anabaenopsis* – orange pyroxene, gray Plagioclase, purple olivine minerals.

They said less and less to one another now, made love with greater infrequency and fused together into the glow-worm firmament each night, suddenly proud of their ordeal. The glaciers creaked, spoke in languages of the body. The whole earth moved, came alive. The ice crashed into the sea all day long, less at night when the stillness embraced inertia. A yellow warbler shot hither; dragonflies came curiously; jays, bees, a host of buzzes. The two beings labored like ants, transporting the stones, shaping each fit, and raising their structure.

Knowing that this was all there was, they each worked like lunatics on the fringe (Bliss had been called that), and within a month, two things had happened: The Tower, with spiral stairs, and one circular room atop it, with two openings as before, was done; some fifty rocks high, twice that many around. A civilization unto itself, glory-gathered. And the eagle. The eagle that had first warned them, now seemed to take to their company, made its nest in the high branches of a fir tree at the edge of the clearing, and seemed always to be watching them.

They moved into the Tower, Margo first. She'd caught all the fish that were left – three dozen. They smoked them, left two for the eagle. The eagle, in turn, left Margo and Hardanger the heads of various rodents, right at their doorstep.

One morning, they awoke on their mat bed that Margo had made (woven grasses and pine needles atop sand) and saw that it was snowing. The Tower was cold, damp. The fire did nothing to allay this dilemma. They were still largely naked. So they covered the gap near their bed with fir branches.

Suddenly, Margo sat up with a start: "Why not go back, just to see!" For her part, this was an inspired notion.

"I've told you countless times; the town disappeared."

"But I see it!" she bid him, looking out the lower glacier porthole.

"You see memories."

"But they are real. The people, the ease, mostly; how warm it always was."

"Mucilaginous, foul, metal color, everything a bore. Deadly. Workings of a rebel monster. You forget."

"I suppose," she sank back, resting on her bunched shoulders that had grown and fetched strength for themselves.

"Are you happy?" she asked him later, after they'd done it repeatedly. Such a question she'd never posed.

"I'm relieved to be moving."

"Moving? Moving where?" she perplexed.

"I mean as an animal in the universe, vulnerable to adaptation, to evolving."

"Evolving where? Where do you want to go? What do you mean by all this? Tell me!"

Hardanger could get maudlin this way when given a chance. "Wherever it's in our path to go."

"You're so vague, Hardanger. You always were like that. Say what you mean," she said, feeling her stomach which had grown large.

"We're free to take part in some plan out there. Before – now think back – before, there was no chance of change, of being reborn. Now with the stream, and the flowers . . .

A shrew (Sorex) zipped across the chamber, stood up, looked intelligently at them, shrieked, baring its razor teeth through which its powerful toxins came, then leaped sideways fifty times the length of itself. The shrews lived in the chinks of the Tower. Hardanger usually managed to find some foodscraps for them. But today there was nothing.

The snow accumulated. The stream began to freeze. Margo was too cold to watch the process. There were no more fish to eat. No clothing to be gotten. No more plants. Tracks of Lagopus, snowshoe rabbit, of Microtus making ready for the long darkness.

Margo was feeling something. She wanted to vomit but could not. Her breasts hurt. She had some bleeding. She fainted. She could not work. Her throat was sore. She coughed up sputum green and white. She farted a lot.

"I'm hungry. Help me."

Hardanger knew. He took her by the cheek, ribboned his tongue on her neck. She held his arm, moaned, kissed the hairs on his wrist, looked up at him beseechingly: "Help me!"

He felt her stomach. She looked at him, then fell back, clutching at her gut.

"I'll be back. It may take several days. But I will find food," he avowed.

He went out into the snow. The eagle was up in its tree, feeding its young. Seeing Hardanger, it went into a panic, cawing in the snow winds.

"Give me food!" Hardanger shouted up.

To which the eagle circled around, made do-do, then repaired to its young, one of which was raising itself, wings attempting to spread and lift heavenward. The mother eagle knocked it over with her beak, and settled back down beside the nest, talons nobly fixed to the cold limb of fir.

"They have it together," Hardanger thought, as he trumped up into the forest, not a clue as to the next step, the fulfillment of the most basic chore. He was no hunter, no fugleman. The idea of killing animals, even the fish, was not agreeable to his mind. But he *was* hungry, and was she. He studied the fish place. Only tracks on either side. The kind he'd seen frequently. Tracks that replaced one another, until there were deep grooves. Tracks that his mind was not fully unconditioned enough to revere or logic out.

He dug into a tree where woodpeckers had hammered. He searched for piles, molewarts, fungi, worms, anything; gathered lichen, remaining alder leaves, sought out the softened roots of grass in the soil against the creek. But no sarcologist, he. Nothing in his past behavior to recommend such toil. And anyway, snow covered the ground. He was desperate.

The eagle found him, circled several times, volplaning in for low sweeps.

Hardanger set off toward the West, from whence seemed to come the odor of salt, uncommonly fresh breezes. He had an instinct, one acquired no-doubt from sea-faring ancestors in centuries past. But as Hardanger disappeared over the far-distant knoll of new forest, a commotion invaded the trees, not 500 meters from the Tower.

A man stood on the edge of a clearing staring at the Tower.

"It's them!" he stunted, and set about his preparations.

The stranger moved back out of sight, scanning the present prospect, when he heard a woman cry out. The sound was hypnotic. It attained crescendo timbre.

He sharpened his spear on stones underfoot.

In the Tower Margot heaved in steam, her forehead burning with fever, her bowels breaking. Tossed back in convulsion, her hands clutching at the rock wall behind her, grasping for support, she tore her nails in to her very palms, grovelled and despaired.

"No, No," she pleaded.

When it started, after the third prolonged contraction.

And when she was done, there were now two of them, the human pitch at once recognizable to the emboldened stranger who started up from his secret crouch off in the forest.

And as he approached he made out the oddest collaboration he'd ever deciphered, in the form of a bright blue bird riding merrily on the back of a big, furry animal the likes of which the stranger had never seen.

"It must be tame," he reckoned. "And the bird," he thought. "It's him!"

Overhead, an eagle was shrieking. "Shut up!" the man hurled.

In the Tower, Margot quietly wept, having held the blond, fussy creature at arm's length, out-of-her mind with awe and the deepest feelings she had ever experienced.

"Are you me?" she said to it, bringing it to her breast, where she could feel its tiny warmth and beat of its own heart.

"Sun flower. That will be your name."

She ran her finger along the bizarre cord that stretched from her to Sun Flower, tasted the outpouring of slime and blood and water that surrounded her on the wicker bed, and the nuzzling gurgles of her new friend.

"Hardanger?" she querried. "But it looks like Hardanger?"

The stranger moved slowly towards the Tower. The bird and its host creature were no longer in sight. When something happened, a pulse or shock in the very atmosphere around him . . . .

A strange motionlessness of feeling came over him, as if everything were a dream, that he'd wake up as before in the warm summer with hanging fish waiting to be eaten. And he missed the leaves and the gourds. And the absolute confidence in the power of the Complex. And he remembered life as a child and his father's anger – he could never understand such continual wrath. His father with the long white beard and wild crop of hair and sophistication. But mostly his upset. Those were hard times. And the picture of his mother. But so vagued out in the primitive recall. He thought, "How simple times were. To sit with the warm hum of NO always by my bedside. No need of prayers. No lack of sustenance. How lucky I was, the Power!" when, from 500 feet away, near the clearing, in the direction he had to go to reach the Tower, his blind eyes met the furry mass of a brown sow and her cub. *Ursos arctos.* The mother was probably 1300 pounds, pronounced snout like a conic section, hips as large as a washing machine. And riding her shoulder, the bird. Clicking, the bear aimed her gunsight, her enraged face. Her stare was unnatural, the stranger thought, unclear in his fascination, starting to walk backward, then forward, then to the side. There was stench in the air. Her hair was caked in snow. Both mother and cub had been through it, too. No niceties. The creatures gauged him, a giant sense of salvation, that is, of altercation, coming up through the sow's inners, like lava, ready to connect with the idea of this guy.

She reared up, snorted, champing her teeth harder in a readiness to charge. To attack with that locomotive, strange unnegotiability that propels such a bear past all other animals and had for millenia. Only large wolf packs dare to approach them.

The cub was crying like a vexed infant. Everyone was hungry that year. The bird rose like a gyrocopter and flew to the nearest branch.

Tears came from the strangers eyes, involuntarily. He felt supremely happy about this, nervous but happy. In a cloud of unknowing, walking forward now with steady vision, spear raised.

The bear started toward him, its eyes consumptive with the outrage of a woman who has made up her mind, cannot be appeased, reasoned, or spoken to.

Mammoth, a hulk with the meaning of folklore which all children once accepted from the Moscow circus and halcyon days; that the bear actually thought like a child, could be cuddled, prevented forest fires, would stand on his rear legs for an animal cracker; furry, so as to instill the desire for extra innocence, in fact, its odor, the size of its excrement – larded with half-digested roots, hamsters, tin cans – the huge marks of its pasturage, uprooted trees, all suggested the unthinkable.

The bear charged from ancient into modern times.

"No!" he groaned, flailing, swollen shoulders, shuddering, rubbery ankles, hands worn raw; a crippling agony in his ribcage preventing him from running, leaning in, his hands to his side, "No!" he invoked again, burning; mayhem in his uphill legs, weakness. With sudden boost, but not miracle enough. Rolls over an abutment; marathon heat, lightning tumbles, punctures, grass slipping, she's there, entangled, enough time to lodge the spear, a slow wound, intensifying its rage, CHOMP, and good-by.

She first heard the ice bird screaming, then dozens of other birds. She stood weakly, opened the fir coverage, peered down into the snow. "What? Hardanger?"

Then realized; collapsed.

All day. They devoured slow, body part by gross body part, bones and all, happy as the perennial adjectives would have us know. It was better this way. It *was* the way. As the sow ate, so did she die, hemorrhaging.

The stranger had dragged himself to within twenty feet of the Tower, directly beneath the shutter opening.

The first day the sow and minuettish cub, propped quaintly up on rumps – Peking zoo style – ate his belly, gorged on the neck poking around with six-inch claws as if the meal were crab ciappino, lacking only bibs.

At night Margo couldn't tell whether they were still out there. She thought she heard munchings in the wind. This kept her from going down.

She lay in a stupor. Sick, sick to death, nursing her child.

Next day the bears were there as usual, feasting on the chine, crunching each vertebrae, foraging all day off the wet corpse. The cub licked its claws like it were chocolate roe. The sow hovered over the stranger, weaker by the day, fighting off its own end with voracious wrath.

Even the raptor eagle got a few pieces. And the ice bird, just in for the occasion, got some licks.

All Margo had left were some dried berries and a wooden container with water. Hardanger had carved the crude implement for her as a gift on the occasion of their first month outside.

The bears stayed a week, every so often eyeing the Tower. They got every last bit of the man, save for his right hand.

Throughout the week, Margo had to deal with the screeches and hollerings of birds, birds from everywhere, fighting over remaining crumbs of her man. The shrews too. And vole, foxes, marmots, creatures of a menagerie; creatures of the past, creatures of the future.

She never looked out after the third day. Never moved, expecting the bears to come in after her.

One late afternoon, on the seventh day of their attack, Margo was alerted by the wimpering of the cub, as it tried to waken its mother, but the sow had lost too much blood and lay dead, next to the stranger's skeletal pile. The sun had managed a scant illumination in a manner no doubt befitting. Margo, barely able to move now, somehow got herself down the stairwell, crawling, entered the clearing, stooped down.

Strangely, the skin cleaned off, they'd left a hand, all bones, clasping tight to something. The bears might have finished him off, another chew, that's all. But they left it alone. An offering? Such musings were alien to her.

"Dear man, what do you have there?" she spoke softly to him, to her hope abandoned. She slumped down, hallucinating. No strength in her to pry open the bones, when, upon one meager exertion, they crumbled apart. A clump of frozen soil in them, earth, which Hardanger had grasped at the last. But as she turned to scoop up the orphan cub in her arms, she recognized destiny in a flash: the crumbled hand comprised but four fingers. "Bliss!" she exclaimed.

Later, she is still sprawled atop the frozen turf, atop the seeds, and spores, and pollen; atop the dead Hymenoptera, the vascular plant fragments. Margo has opened up the sow's gut where the spear had entered, her hunger toying with its penetralia. She will suckle both her greedy infants now, waiting for him, lost in prayer.

It is beginning to snow. Cold morning. Not a sound. From the looks of it, the snowstorm will set in for good.

When she sees, sees something, eyes enlarging.

There is a lone figure approaching from across the crevassed distance, moving slowly, like a tired man.

Selected Bibliography

Abbey, Edward. *Desert Solitaire: A Season in the Wilderness*. Ballantine Books (New York: 1977)

Armen, Jean-Claude. *Gazelle-Boy*. Universe Books (New York: 1974).

Barney, Gerald, (Study Director). *The Global 2000 Report to the President*. Council on Environmental Quality, with the Department of State (Washington D.C. Government Printing Office: 1980).

Barrett, S.A., and E.W. Gifford. *Miwok Material Culture – Indian Life of the Yosemite Region*. Yosemite Natural History Association, Inc. (Yosemite: 1933).

Brandt, Willy. *North-South; A Programme for Survival*. Pan Books (London: 1980).

Brown, Lester with William Chandler, Christopher Flavin, Sandra Postel, Linda Starke and Edward Wolf. *State of the World 1984*. W.W. Norton Publishers (New York: 1984).

Brown, Lester. *Building a Sustainable Society*. W.W. Norton Publishers (New York: 1982).

Brown, Tom, Jr. *The Tracker*. Prentice Hall Publishers (Englewood Cliffs, N.J.: 1979)

Cahill, James. *Chinese Painting*. Rizzoli International (New York: 1970).

Cioran, Émile. *A Short History of Decay*, trans. by Richard Howard. Viking Publishers (New York: 1975).

Coburn, Broughton. *Nepali Ama – Portrait of a Nepalese Hill Woman*. Ross-Erikson Ltd. (Santa Barbara, CA: 1981).

Clark, Kenneth. *Landscape into Art*. Harper & Row (New York: 1976).

Cousteau, Jacques-Yves. *The Cousteau Almanac – An Inventory of Life on our Water Planet*. Doubleday Publishers (New York: 1981).

Daumal, Réné. *Mount Analogue*, 4th ed., trans. by Roger Shattuck. Penguin Metaphysical Library (Baltimore: 1968).

Eibesfeldt, Irenaus Eibl. *The Biology of Peace and War: Men, Animals, and Aggression*, trans. by E. Mosbacher. Viking (New York: 1979).

Ehrlich, Paul, Anne Ehrlich and John Holdren. *Ecoscience: Population, Resources, Environment*. W.H Freeman & Company (San Francisco: 1977).

Ehrlich, Paul and Anne Ehrlich. *Extinction: The Causes and Consequences of the Disappearance of Species*. Random House (New York: 1981).

Fukuoka, Masanobu. *The One-Straw Revolution: An Introduction to Natural Farming*. Rodale Press (Emmaus, PA: 1978).

Glacken, Clarence J. *Traces on the Rhodian Shore: Nature and Culture in Western Thought From Ancient Times to the End of the Eighteenth Century*, 2nd ed. University of California Press (Berkeley: 1976).

Gourhan-Leroi, André. *Treasures of Prehistoric Art*. Harry N. Abrams Publishers (New York: 1967).

Hsi, Kuo. *An Essay on Landscape Painting*, trans. by S. Sakanishi. Wisdom of the East Series (London: 1936).

Humboldt, Alexander Von. *Cosmos: A Sketch of a Physical Description of the Universe*, 4 vols., trans. by E. Otte. Harper & Row Brothers (New York: 1844).

Kazantzakis, Nikos. *The Odyssey – A Modern Sequel*, trans. by Kimon Friar. Simon & Schuster (New York: 1958).

Kazantzakis, Nikos. *The Saviors of God*, trans. by Kimon Friar. Simon & Schuster (New York: 1960).

Kazantzakis, Nikos. *Buddha*, trans. by Kimon Friar and Athena Dallas-Damis. Avant Books (San Diego: 1983).

Kennan, George. *The Nuclear Delusion: Soviet-American Relations in the Atomic Age*. Pantheon Books (New York: 1982).

Knipe, David. *In the Image of Fire: Vedic Experiences of Heat*. Motilal Banarsidass (New Delhi: 1975).

Kuck, Loraine. *The World of the Japanese Garden*. Weatherhill Publishers (Tokyo: 1980).

Lach, Donald. *Asia in the Making of Europe*, 2 vols. University of Chicago Press (Chicago: 1965).

Lovelock, J.E. *Gaia, A New Look at Life on Earth*. Oxford University Press (New York & London: 1979).

Mansfield, Susan. *The Gestalts of War: An Inquiry into its Origins and Meanings as a Social Institution*. Dial Press (New York: 1982).

Margulis, Lynn. *Symbiosis in Cell Evolution*. W.H. Freeman & Company (San Francisco: 1981).

Montagu, Ashley, ed. *Learning Non-Aggression: The Experience of Non-Literate Societies*. Oxford University Press (New York & London: 1978).

Moran, Emilio F. *Developing the Amazon*. University of Indiana Press (Bloomington: 1981).

Muir, John. *Travels in Alaska* (Boston: 1915).

Myers, Norman. *The Sinking Ark: A New Look at the Problem of Disappearing Species*. Pergamon Books (New York: 1979).

Naeff, Weston J., in collaboration with James Wood and an essay by Therese Thau Heyman. *Era of Exploration: The Rise of Landscape Photography in the American West 1860-1885*. Albright-Knox Art Gallery, The Metropolitan Museum of Art (New York: 1975).

Nance, John. *Discovery of the Tasaday, A Photo Novel: The Stone Age Meets the Space Age in the Philippine Rain Forest*. Vera-Reyes, Inc. (Manilla: 1981).

Nash, Roderick. *Wilderness and the American Mind*, 3rd ed. Yale University Press (New Haven, CT: 1982).

Novak, Barbara. *Nature & Culture: American Landscape Painting 1825-1875*. Oxford University Press (New York: 1980).

Okakura, Kakuzo. *The Book of Tea*. Dover Publishers (New York: 1964).

Olschak, Blanche, C. *Ancient Bhutan – A Study on Early Buddhism in the Himalayas*. Swiss Foundation for Alpine Research (Zurich: 1979).

Patch, Howard. *The Otherworld According to Descriptions in Medieval Literature*. Harvard University Press (Cambridge, MA: 1950).

Perrin, Noel. *Giving up the Gun: Japan's Reversion to the Sword, 1543-1879*. David Godine Publishers (Boston: 1979).

Poirier, Frank E. *Fossil Evidence – The Human Evolutionary Journey*, 3rd ed. C.V. Mosby Company (St. Louis: 1981).

Reischauer, Edwin O., and Albert M. Craig. *Japan, Tradition & Transformation*. Harvard University Press (Cambridge, MA: 1978).

Salkeld, John. *A Treatise of Paradise and the Principle Contents Thereof*. N. Butter Printer (London: 1617).

E.E. Klimburg-Slater, ed. *The Silk Route and the Diamond Path – Esoteric Buddhist Art on the Trans-Himalayan Trade Routes*. UCLA Arts Council (Los Angeles: 1982).

Sartre, Jean-Paul. *Nausea*, trans, by Lloyd Alexander. New Directions (New York: 1964).

Schell, Jonathan. *The Fate of the Earth*. Avon Paperbacks (New York: 1982).

Sheahan, Richard T. *Alternative Energy Sources: A Strategy Planning Guide*. Aspen Systems (Rockville, Maryland: 1981).

Shelley, Percy Bysshe. *The Poetical Works of Percy Bysshe Shelley*, 1st ed., ed. by Mary Shelley. Porter & Coates Publishers (Philadelphia: 1839).

Shepard, Paul. *Nature and Madness*. Sierra Club Publishers (San Francisco: 1982).

Smith, A.G. Cairns. *Genetic Takeover*. Cambridge University Press (London: 1982).

Stafford, Barbara Maria. *Voyage into Substance: Art, Science, Nature, and the Illustrated Travel Account, 1760-1840*. M.I.T. Press (Cambridge, MA: 1984).

Thompson, William Irwin. *The Time Falling Bodies Take to Light: Mythology, Sexuality, and the Origins of Culture*. St. Martin's Press (New York: 1981).

Thoreau, Henry David. *The Journal of Henry D. Thoreau*, ed. by Bradford Torrey and Francis H. Allen, 2 vol. Dover Publishers (New York: 1962).

Tobias, Michael Charles. *Voice of the Planet*. Adrian Malone Productions (Los Angeles: 1984).

Tobias, Michael Charles, ed. *Deep Ecology*. Avant Books (San Diego: 1985).

Tobias, Michael Charles. *Deva*. Avant Books (San Diego: 1981).

Tobias, Michael Charles, ed. *The Mountain Spirit*. with Harold Drasdo. Viking-Overlook-Victor Gollancz Publishers (New York and London: 1979, 1980, 1983).

Tobias, Michael Charles, ed. *Mountain People: Profiles of Twentieth Century Adaptation*. University of Oklahoma Press and Interprint (Norman, Oklahoma, and New Dehli, India: 1985).

Toynbee, Arnold. *Mankind and Mother Earth: A Narrative History of the World.* Oxford University Press (New York: 1976).

Tuan, Yi-Fu. *Topophilia: A Study of Environmental Perception, Attitudes, and Values.* Prentice-Hall Publishers (Englewood Cliffs, NJ: 1974).

Tunnard, Christopher. *A World with a View: An Inquiry into the Nature of Scenic Values.* Yale University Press (New Haven, CT: 1978).

Twain, Mark. *The Diaries of Adam & Eve.* Coronado Press (Lawrence, Kansas: 1971).

Ward, Captain F. Kingdon. *The Riddle of the Tsangpo Gorges.* Edward Arnold & Company (London: 1926).

Watson, Burton, trans. *Cold Mountain: 100 Poems by the T'ang Poet Han Shan,* 2nd ed. Columbia University Press (New York: 1970).

White, J.C. *"Journey Through Bhutan,"* National Geographic Magazine (Washington D.C.: 1909).

Wilson, Edward O., *Biophilia – The Human Bond with Other Species,* Harvard University Press (Cambridge, Mass: 1984)

World Conservation Strategy: Living Resource Conservation for Sustainable Development. International Union for the Conservation of Nature and Natural Resources (Gland, Switzerland: 1980).

Zampaglione, Gerardo. *The Idea of Peace in Antiquity,* trans. by Richard Dunn. University of Notre Dame Press (South Bend, Indiana: 1973).

Index

B

C

G

H

Y

Z

BOOKS ON THE ECOLOGICAL CUTTING EDGE

ARCOSANTI: An Urban Laboratory? by Paolo Soleri

Paolo Soleri created a great deal of excitement in the early 70s with the publication of *Arcology: The City in the Image of Man* (MIT Press), which proposed a radical alternative to modern city planning.

At the same time Soleri began work on a model arcology in central Arizona: Arcosanti.

Arcosanti represents Soleri's first comprehensive statement on the project. It explores 63 topics – on ecology, city planning, social life, evolution and other subjects – in clear brief commentaries. Each topic is discussed in general, and as it relates to Arcosanti.

80 pp., 5½" ×8½", illustrations, photos, $5.95
Trade paperback 0-932238-27-0

DEEP ECOLOGY edited by Michael Tobias

This landmark collection of new essays by a group of the world's leading ecological thinkers heralds the arrival in the U.S. of the "deep ecology" movement . Arne Naess, the eminent Norwegian naturalist philosopher, coined the term as the new school of ecological thought. Other contributors include George Sessions, Norman Myers, Emilio F. Moran, William R. Catton, Jr., Garrett Hardin, Murray Bookchin, Paolo Soleri, and Alan Grapard.

"This is a moving, disturbing, enlightening, apocalyptic and beautifully written collection of essays, memoirs, and poems."
— *Publishers Weekly*

304 pp., 6" x 9", illustrations, photos, $12.95
Trade paperback 0-932238-13-0

THE LIFE AND ADVENTURES OF JOHN MUIR by James M. Clarke

Written in a lucid narrative style, this book presents both the heroic and very human sides of the legendary outdoorsman. It has been praised by the Sierra Club and environmental sympathizers, mountaineers and those who appreciate all that this man did to preserve and promote enjoyment of the natural earth.

"Perhaps the best introduction to the natural world. . ."
— *Library Journal*

336 pp., 7½" x 10¼", illustrations, footnotes, appendices, index, $14.95
Hardbound 9-932238-01-7

TO ORDER: Total cost plus $1.50 shipping/handling (CA res. add 6% tax) to:
Avant Books 3719 Sixth Avenue, San Diego, CA 92103